A
TEXTBOOK
OF
MATERIALS
TECHNOLOGY

A
TEXTBOOK
OF
MATERIALS
TECHNOLOGY

LAWRENCE H. VAN VLACK   The University of Michigan, Ann Arbor

 ADDISON-WESLEY PUBLISHING COMPANY
Reading, Massachusetts · Menlo Park, California
London · Amsterdam · Don Mills, Ontario · Sydney

This book is in the
ADDISON-WESLEY SERIES IN METALLURGY AND MATERIALS

Consulting Editor
Morris Cohen

ISBN 0-201-08066-4
  JKLMNO-VB-898765432

Dedicated to Frances, Laura,
and Bruce — partners throughout

PREFACE

This text has been prepared for those initial courses in materials which need the problem-solving approach of the technologist and the engineer, but which must fit into curricula designed for those who have a minimal background in the sciences. The problem-solving emphasis is supported by more than 100 example problems within the chapters for the students' guidance, and almost 250 study problems for assignments. An Instructor's Manual contains an equal number of supplemental problems, which the instructor may use for additional assignments as he deems necessary.

The necessary background in science is available from high school courses in chemistry and physics. If a student has not had those two courses, or feels that he is weak in those areas, he will be helped by a college general chemistry course. Calculus is not required.

The first two chapters are devoted to the engineering requirements of materials. Since the book is written on the premise that the student has only an abbreviated background in general chemistry and physics, early attention is focused on metals (Chapters 3–6), because the structural prototypes of metals may be simple, with only one kind of atom. Polymers and ceramics follow in Chapters 7–9 and 10–11, respectively, since each of those categories contain compounds that are at least biatomic. The last two chapters concern semiconducting materials and composites.

Because, for many students using this text, this may be the only materials-related course they will take, attention is given to introductory processing principles, as well as to structure and properties. The sequential arrangement of metals, polymers, and ceramics allows processing to be integrated with the nature and behavior of materials.

A textbook must be adaptable to a variety of audiences, since seldom, if ever, do classes in any two schools contain students with comparable backgrounds, curricular goals, and available time. This text has been written with the assumption that the courses it serves may be categorized into one of four general patterns: (1) a course with a broad overall emphasis on materials, (2) a course with an emphasis on metals, (3) a course with an emphasis on mechanical properties and structural design, or (4) a course with emphasis on electrical properties and applications. The notations given in the table of contents are a guide to the instructor who wishes to establish a course syllabus. The meaning of these symbols is as follows.

T   Parts or all of these sections are recommended if the course is to have a broad materials emphasis.

e   Parts or all of these sections are recommended if the course emphasizes electrical materials and their use.

ⓔ Optional in the *e*-syllabus; omit if time is limited.

m   Parts or all of these sections are recommended if the course emphasizes mechanical behavior and structural applications.

ⓜ Optional in the *m*-syllabus; omit if time is limited.

M   Parts or all of these sections are recommended if the course emphasizes metallic materials.

Ⓜ Optional in the *M*-syllabus; omit if time is limited.

Ⓞ Optional if the focus is broad, but time is limited.

Of course, the individual instructor must further adjust the details of the syllabus to accommodate his local situation.

The appendixes include reference data which, I have found by experience, to be most useful for preparation of problems by the instructor, solving of problems by the student, and subsequent reference by the student-turned-engineer. At the request of my students, I have included a glossary of commonly used terms in the field of materials science.

A 1971 study revealed that many instructors in first courses in materials across the country are not themselves specialists in materials science and/or engineering. Rather, they come from other engineering and science fields, with only peripheral interest in materials. This is particularly true of instructors of those courses which demand a less-rigorous science background. To this end, I have prepared an Instructor's Manual which gives more detailed teaching information, visual aids, examples of applications, etc. References for further study may be found in my other books, such as *Elements of Materials Science* and *Materials Science for Engineers*, also published by Addison-Wesley.

During the time that a manuscript progresses toward publication, an author becomes indebted to many other people. As with my other books, this has been true for *A Textbook of Materials Technology*. It is impossible to mention everyone individually, because they include innumerable students, each of my colleagues at The University of Michigan, the contributors of figures and technical data, and the ever-helpful employees of Addison-Wesley. I would be remiss, however, not to give specific recognition to the following people: to Professor Morris Cohen (M.I.T.) for his long-time counsel concerning my writing activities; to Dr. J. P. Bosscher (Calvin College) and Dr. John Ritter (University of Massachusetts) for their critiques of the manuscript and their contributions to the accompanying Instructor's Manual; to Mrs. Allene Crowe for her excellent secretarial help; and finally (but not least) to Fran for her patience and understanding.

Ann Arbor, Michigan   L.H.V V.
August 1972

CONTENTS

* See preface for explanation of notation.

T e m Ⓜ
T e m Ⓜ Ⓞ

T e m M

MATERIALS
OF
ENGINEERING

1–1 MATERIALS IN INDUSTRIAL PRODUCTS

All products that come out of industry consist of at least one—and often many—types of materials. The most obvious example is the automobile; however, the reader will quickly think of many other products, such as radios, textiles, light bulbs, motors, airplanes, and home appliances (Fig. 1–1.1).

A car contains a wide variety of materials, ranging from glass to steel to rubber, plus numerous other metals and plastics. Furthermore, a variety of steels is used in a car. For example, the thin sheet steel of a fender must have different *properties* from the steel of a transmission gear. The steel in the fender must be soft and ductile so that it may be rolled to a thickness of only 1/20 of an inch without cracking. The steel in the gear must be strong and tough, with a hard surface, so that its teeth will not snap off or wear. Note, however, that when the gear was made, it had to be shaped into a gear blank, and then machined into its final form. These operations could not have been performed if the steel had had its final desired properties at the time when it was being shaped and machined. To meet the final specifications, the steel of the gear had to have its properties *changed* after processing from a softer, weaker material to a harder, stronger one.

The number of materials which are available to the engineer in industry is almost infinite. The various compositions of steel alone run into the thousands. It has been said that there are more than 10,000 varieties of glass, and the numbers of plastics are equally great. When to these numbers we add the materials' capability of having their properties modified, as in the case of a gear (Fig. 1–1.2), the task of knowing the properties and behavior of all types of available materials becomes enormous. In addition, several hundred new varieties of materials appear on the market each month. This means that individual engineers and technicians cannot hope to be familiar with all the properties of all types of materials in their numerous forms. All he can do is try to learn some principles to guide him in the selections and processing of materials.

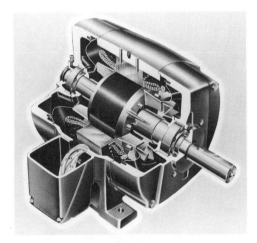

Fig. 1–1.1 Cutaway view of an electric motor, which constitutes a materials system. This requires a wide variety of materials, each of which contributes to the total performance of the motor. There is also a structure within each material that involves atoms electrons, crystals, and phases. The performance of the material depends on the internal structure, just as the performance of this motor depends on its components. (Courtesy of General Electric)

Fig. 1–1.2 Steel drive gear, whose properties must include strength for transmission of loads, hardness for resistance to wear, and toughness for resistance to fracture. During production, however, the steel had to be weak and soft so that it could be forged and machined. In order to change the properties, one must change the internal structure of the steel. (Courtesy of Climax Molybdenum Co.)

The purpose of this text is to familiarize the technical student with those factors which affect properties and to give him an idea of the service potential of various materials. When he is equipped with this knowledge, the technical person should be able (1) to select more readily and intelligently those materials which are best for his purpose, and (2) to avoid service conditions which might be damaging to materials.

The *properties* of a material originate from the internal structure of that material. This is analogous to saying that the operation of a TV set depends on the components and circuits within that set. The internal structures of materials involve atoms, and the way atoms are associated with their neighbors into crystals, molecules, and microstructures. In the following chapters we shall devote much attention to these structures, because a technical person must understand them if he is going to produce and use materials, just as an electronic technician must understand radio circuits if he is going to design, assemble, or repair a radio. Before we talk about structures, however, let us look more closely at the variety of materials available; we shall also, in Chapter 2, consider the more important properties of materials and how they are measured. This will enable us to discuss the relationship of properties to structures more intelligently.

Example 1–1.1*. Take a cord from some appliance, such as a toaster or coffee maker. List the chief materials used and the probable reason for their selection.

---

* Example problems will be given throughout the text, usually followed by both a solution and comments. The purpose is to give the reader additional points and information. Thus they should be studied along with the text itself. Additional problems appear at the end of each chapter for purposes of individual study.

Solution

a) Copper wire: Copper offers low resistance, so that wire with a small cross section can be used. The small diameter of the wire strands lends flexibility.
b) Brass prongs: These prongs must have greater dimensions so that they will be rigid. With these larger dimensions, brass may be used, even though it is more resistive than copper. Brass is also stronger and cheaper than copper.
c) Rubber insulation: Rubber is a nonconductor which has the added advantage of flexibility. Also it may be readily applied as a coating to the wire during the initial stages of production.
d) Bakelite receptacle housing: Bakelite is a good insulator, offers stability of temperature, and is rigid.
e) Steel spring: Steel is light, flexible, and not subject to permanent bending.

Comment. At this stage we can only note some of the characteristics of materials. Later we shall examine *why* copper has a lower resistivity, and *why* a bakelite-type plastic does not soften, but other types of plastics may soften when they get hot. ◀

1–2 TYPES OF MATERIALS

It is convenient to divide materials into three main types: (1) *metals* (Chapters 3 through 6), (2) *plastics* or *polymers* (Chapters 7 through 9), and (3) *ceramics* (Chapters 10 and 11). In addition to these, we shall give separate attention to *semiconductors* (Chapter 12) because of their nontypical characteristics, and because they have gained major importance in our engineering technology. The last chapter (Chapter 13) will consider *composites* which utilize more than one type of material. Before we focus on these materials separately in subsequent chapters, let us state some generalities about metals, polymers, and ceramics, and the bases for their differences.

Metals. A person recognizes a metal as having high thermal conductivity because it may quickly conduct heat to (or away from) his hand. He also knows that it has high electrical conductivity, and is opaque. Often, but not always, it is heavy and deformable. Usually it may be polished to a high luster.

What accounts for the above characteristics? The simplest answer is that metals owe their behavior to the fact that some of the electrons can leave the "parent" atoms. Conversely, in polymers and ceramics, electrons are not free to roam to the extent that some electrons can travel through metals. Since some electrons are free to move in a metal, they can quickly transfer an electric charge (Fig. 1–2.1). Each electron carries 1.6×10^{-19} coulombs (or 1.6×10^{-19} amp · sec, since a coulomb is the charge transferred by one ampere of current

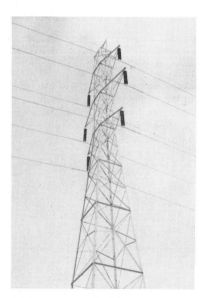

Fig. 1–2.1 Electrical conduction. The current is carried by individual electrons, each having a charge of 1.6×10^{-19} coul. In a power line such as this, only a fraction of the electrons are able to respond to the voltage gradient and leave their parent atoms to transmit charge.

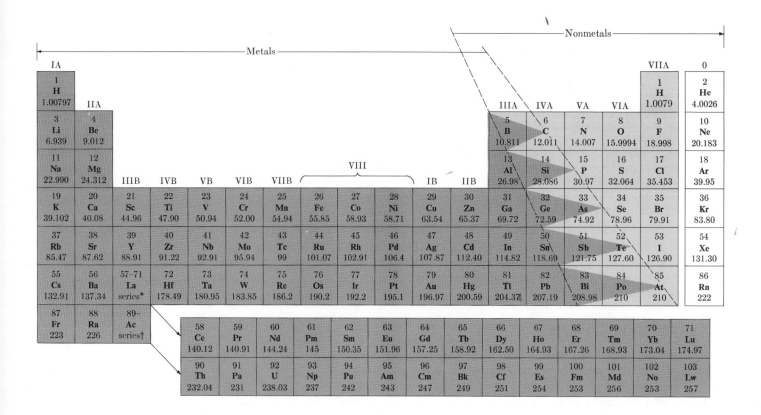

IA																	VIIA	0
1 **H** 1.00797	IIA												IIIA	IVA	VA	VIA	1 **H** 1.0079	2 **He** 4.0026
3 **Li** 6.939	4 **Be** 9.012												5 **B** 10.811	6 **C** 12.011	7 **N** 14.007	8 **O** 15.9994	9 **F** 18.998	10 **Ne** 20.183
11 **Na** 22.990	12 **Mg** 24.312	IIIB	IVB	VB	VIB	VIIB		VIII		IB	IIB		13 **Al** 26.98	14 **Si** 28.086	15 **P** 30.97	16 **S** 32.064	17 **Cl** 35.453	18 **Ar** 39.95
19 **K** 39.102	20 **Ca** 40.08	21 **Sc** 44.96	22 **Ti** 47.90	23 **V** 50.94	24 **Cr** 52.00	25 **Mn** 54.94	26 **Fe** 55.85	27 **Co** 58.93	28 **Ni** 58.71	29 **Cu** 63.54	30 **Zn** 65.37	31 **Ga** 69.72	32 **Ge** 72.59	33 **As** 74.92	34 **Se** 78.96	35 **Br** 79.91	36 **Kr** 83.80
37 **Rb** 85.47	38 **Sr** 87.62	39 **Y** 88.91	40 **Zr** 91.22	41 **Nb** 92.91	42 **Mo** 95.94	43 **Tc** 99	44 **Ru** 101.07	45 **Rh** 102.91	46 **Pd** 106.4	47 **Ag** 107.87	48 **Cd** 112.40	49 **In** 114.82	50 **Sn** 118.69	51 **Sb** 121.75	52 **Te** 127.60	53 **I** 126.90	54 **Xe** 131.30
55 **Cs** 132.91	56 **Ba** 137.34	57–71 **La** series*	72 **Hf** 178.49	73 **Ta** 180.95	74 **W** 183.85	75 **Re** 186.2	76 **Os** 190.2	77 **Ir** 192.2	78 **Pt** 195.1	79 **Au** 196.97	80 **Hg** 200.59	81 **Tl** 204.37	82 **Pb** 207.19	83 **Bi** 208.98	84 **Po** 210	85 **At** 210	86 **Rn** 222
87 **Fr** 223	88 **Ra** 226	89– **Ac** series†																

58 **Ce** 140.12	59 **Pr** 140.91	60 **Nd** 144.24	61 **Pm** 145	62 **Sm** 150.35	63 **Eu** 151.96	64 **Gd** 157.25	65 **Tb** 158.92	66 **Dy** 162.50	67 **Ho** 164.93	68 **Er** 167.26	69 **Tm** 168.93	70 **Yb** 173.04	71 **Lu** 174.97
90 **Th** 232.04	91 **Pa** 231	92 **U** 238.03	93 **Np** 237	94 **Pu** 242	95 **Am** 243	96 **Cm** 247	97 **Bk** 249	98 **Cf** 251	99 **Es** 254	100 **Fm** 253	101 **Md** 256	102 **No** 253	103 **Lw** 257

Fig. 1–2.2 Periodic table of the elements, showing the atomic number and atomic weight (in amu). There are 6×10^{23} amu per gram; therefore the atomic weights are the grams per 6×10^{23} atoms. Metals readily release their outermost electrons. Nonmetals readily accept or share additional electrons.

during one second). Also, as electrons move rapidly from a hot region of a metal to a cold region, they transfer some of the thermal energy they picked up at the higher temperature; i.e., they provide thermal conduction.

Typically, most metals are denser than nonmetallic materials. However, high density is not an inherent requirement of metals because there are light metals ($\rho_{Mg} = 1.74$ g/cm^3; $\rho_{Na} = 0.97$ g/cm^3). Rather, the basic feature of metals—that they free electrons—is especially characteristic of the heavier elements. Figure 1–2.2, the periodic table of the elements, reveals that approximately three-fourths of the elements are classified as metals. However, more than 90% of those with atomic weight of 40 or more are metals. Thus high density, *per se*, is only a coincidental aspect of metallic behavior.

The reflectivity of a metal is related to the electron's response to vibrations at light frequencies. This is another result of the partial independence from parent atoms of some of the electrons. In short, *metals are characterized by an ability to give up electrons.*

Polymers (often called plastics). Plastics are noted for their low density and their use as insulators, both thermal and electrical. They are poor reflectors of light, tending to be transparent or translucent (at least in thin sections). Finally, some of them are flexible and subject to deformation.

Unlike metals, which have some migrant electrons, the *nonmetallic elements* of the upper right corner of the periodic table (Fig. 1–2.2) have a great affinity or attraction for additional electrons. Each electron becomes associated with a specific atom (or pair of atoms). Thus, in plastics, we find only limited electrical and thermal conductivity because all the thermal energy must be transferred from hot to cold regions by atomic vibrations, a much slower process than the electronic transport of energy which takes place in metals. Furthermore the less mobile electrons in plastics are more able to adjust their vibrations to those of light, and therefore do not absorb light rays.

Materials which contain *only* nonmetallic elements share electrons to build up large molecules. These are often called *macromolecules*. We shall see in Chapters 7 through 9 that these large molecules contain many repeating units, or *mers*, from which we get the word *polymers*. Further, these molecules usually have structures which can be deformed; hence the term *plastic*. By learning more of the architecture of these molecules, the scientist and the engineer have been able to produce plastics which have the desired properties for an ever-expanding set of technical applications.

Ceramics. Simply stated, ceramics are *compounds* which contain metallic *and* nonmetallic elements. There are many examples of ceramic materials. They range from the cement of concrete (and even the rocks themselves), to glass, to spark-plug insulators (Fig. 1–2.3), and to oxide nuclear fuel elements of UO_2, to name but a few.

Each of these materials is relatively hard and brittle. Indeed, hardness and brittleness are general attributes of ceramics, along with the fact that they tend to be more resistant than either metals or polymers to high temperatures and to severe environments. The basis for these characteristics is again the electronic behavior of the constituent atoms. Consistent with their natural tendencies, the metallic elements release their outermost electrons and give them to the nonmetallic atoms, which retain them. The result is that these electrons are immobilized, so that the typical ceramic material is a good insulator, both electrically and thermally.

Equally important, the positive metallic ions (atoms that have lost electrons) and the negative nonmetallic ions (atoms that have gained electrons) develop strong attractions for each other. Each *cation*

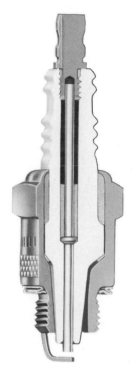

Fig. 1–2.3 Ceramic insulator (spark plug). The main material in this electrical insulator is Al_2O_3. Each Al^{3+} has lost three electrons and each O^{2-} contains two extra electrons, which it holds tightly, so that they are not available for electrical or thermal conduction. The positive cations and negative anions are strongly attracted to one another, as evidenced by the high melting temperature ($\sim 2020°C$ or $3670°F$), and great hardness of Al_2O_3. (One name for Al_2O_3 is emery.) (Courtesy Champion Sparkplug Co.)

(positive) surrounds itself with *anions* (negative). Considerable energy (and therefore considerable force) is usually required to separate them. It is not surprising that ceramic materials tend to be hard (mechanically resistant), refractory (thermally resistant), and inert (chemically resistant).

Example 1–2.1. Magnesium atoms weigh about 50% more than oxygen atoms (24.3 versus 16.0 *atomic mass units*, *amu*; see Fig. 1–2.2). What weight percent (w/o) of MgO is oxygen?

Solution

Basis: 100 Mg^{2+} + 100 O^{2-}

Mass: Mg^{2+} = 100 × 24.3 amu = 2430 amu;   w/o Mg = 60

Mass: O^{2-} = 100 × 16.0 amu = 1600 amu;   w/o O = 40

Total = 4030 amu.

Comments. MgO contains 50 atomic percent (a/o) Mg and 50 a/o O. Since the oxygen atom is lighter than the magnesium atom, its weight percent (w/o) is lower than its atomic percent. Composition percentages of liquids and solids refer to w/o, unless specifically indicated otherwise. Other abbreviations include *l*/o as linear percent, v/o as volume percent, etc.

MgO forms a very stable ceramic material (its melting temperature T_m = 5000°F or 2800°C) because of the strong attraction between the Mg^{2+} cations and the O^{2-} anions. ◀

Example 1–2.2†. A 100-watt light bulb is on a 110-volt dc circuit so that it draws 0.91 amp. How many electrons move through the bulb per minute? Each electron has a charge 1.6×10^{-19} coul (or 1.6×10^{-19} amp·sec).

Solution. The total required charge is 0.91 amp × 60 sec, or 0.91 × 60 coul.

$$\text{Electrons/min} = \frac{(60 \text{ sec/min})(0.91 \text{ amp})}{1.6 \times 10^{-19} \text{ amp·sec/electron}}$$

$$= 3.4 \times 10^{20} \text{ electrons/min.}$$

Comments. This problem, like most in this text, emphasizes setting up the problem without reference to routine formulas. Careful attention to units is not only helpful, but essential for obtaining dimensional consistency and a clear solution.

You will save time if you can handle the basic multiplication and division procedures on a slide rule. ◀

---

† ⓜ See preface for explanation of this and other symbols.

REVIEW AND STUDY

1–3 SUMMARY

The properties and behavior of materials arise from their internal structures. If the engineer wants a specific set of properties, he must choose his materials appropriately so that they have suitable structures. Should the internal structure of a material be changed during processing or service, there are corresponding changes in properties.

It is convenient to categorize many of our engineering materials as metals, polymers, or ceramics. Their properties and behavior may be related in part to the freedom of the outermost electrons of the atoms. Characteristically, metals are opaque, ductile, and good conductors of heat and electricity. Plastics (or polymers) usually contain light elements, and therefore have relatively low density, are generally insulators, and are flexible and formable at relatively low temperatures. Ceramics, which contain compounds of both metallic and nonmetallic elements, are usually relatively resistant to severe mechanical, thermal, and chemical conditions.

These introductory evaluations of materials are not sufficient for engineers, or technicians, inasmuch as they must know much more about the behavior of materials. Therefore, in the chapters that follow, we shall give detailed attention to each type of material. We shall also note that some materials have characteristics of more than one category. For example, is tungsten carbide a metal or a ceramic? (Fig. 1–3.1). It is a conductor like a metal, but it is also a hard refractory compound like a ceramic. Likewise semiconductors (Chapter 12) have intermediate conductivities. The categories of materials are thus not rigidly distinct, but blend from one to another.

Fig. 1–3.1 Internal structure of a cemented-carbide cutting tool, ×1500. The white matrix is metal (cobalt); the gray particles are tungsten carbide (WC) with characteristics intermediate between metals and ceramics. (Metallurgical Products Department, General Electric Co.)

Terms for Study

Atomic mass unit, amu	Nonmetals
Ceramics	Periodic table
Compounds	Polymer
Coulomb, Q	Plastics
Electron charge, q	Property
Internal structure	Semiconductor
Metals	

STUDY PROBLEMS

1–1.1 Cite the characteristics of the materials used in the following parts of a ball-point pen: (a) ball, (b) pocket clip, (c) ink tube, (d) ink, (e) metal supporting ball, (f) pen casing.

1–1.2 For the electric motor of Fig. 1–1.1, list the various materials and their prime characteristics.

1–1.3 Steel is used in a variety of applications. What characteristics are demanded for the following steel items: (a) automobile wheel rim, (b) bridge cable, (c) soft-drink can, (d) electric razor blade, (e) twist drill, (f) steel wool, (g) kitchen skillet?

1–2.1 Examine a two-cell flashlight. How many distinct materials can you list? Which are metals? Plastics? Ceramics?

1–2.2 How long will it take for 10^{20} electrons to pass through a 100-watt light bulb which operates at 32 volts dc? [*Hint:* Recall that $E = IR$, and $W = EI$.

Answer. 5 sec.

1–2.3 A metal alloy of 92.5 w/o Ag and 7.5 w/o Cu is called sterling silver. What is the atomic percent copper? From Fig. 1–2.2, note that silver and copper have 107.87 and 63.54 amu, respectively. [*Hint:* Base your calculation on 10,000 amu of the alloy.]

Answer. 12.1 a/o.

1–2.4 Examine a car. (a) List all the polymer materials (not parts) you can cite. (b) Repeat for ceramic materials.

1–2.5 Polyethylene, a common plastic, contains two hydrogen atoms for each carbon atom (and no other type of atom). Carbon and hydrogen have 12.01 and 1.008 amu, respectively. What is the weight percent carbon?

Answer. 85.6 w/o.

1–2.6 From data in the appendixes, compare the electrical resistivities of magnesium, Mg, and magnesium oxide, MgO. Explain the difference.

PROPERTIES AND THEIR MEASUREMENT

2–1 INTRODUCTION

This chapter presents the more common properties so that later we may make comparisons between different materials, or between different internal structures of the same material. It is convenient to consider three types of properties: mechanical, thermal, and electrical.

Mechanical properties are those responses a material has to the application of mechanical forces. Of specific interest are: (1) *deformation* (*strain*) which occurs when a *stress* is applied, (2) *hardness*, and (3) *toughness*. The latter refers to the amount of energy which a material can absorb before it fails by fracture. We shall not only discuss these three properties in the next few sections, but also give them considerable attention later in the text, because they readily illustrate how the internal structure of a material affects its properties.

A material's *thermal properties* become apparent when energy, which is introduced by the inflow of heat, causes the atoms to vibrate more vigorously and raises the temperature of the material. The results are (1) *thermal expansion*, and (2) *thermal conductivity*, both of which we shall discuss in this book.

Electrical properties are exhibited by materials which are subjected to applied electric fields. An *electric field* is usually measured in terms of volts per centimeter (or volts/meter). The more common electrical properties include: (1) *electrical resistivity* (and *conductivity*), and (2) the *relative dielectric constant*, which relates the accumulated charge across an insulator to the electric field. These properties receive additional significance each year, since engineers must continually develop new and more automated products which use electronic equipment.

Example 2–1.1. List the dimensional units for the two thermal properties cited above.

Solution

a) Thermal expansion, (in./in.)/°F, or with unit cancellation, $°F^{-1}$; or in metric units, (m/m)/°C, or with unit cancellation, $°C^{-1}$.

b) Thermal conductivity, $\dfrac{\text{Btu}}{(\text{hr})(\text{ft}^2)(°\text{F}/\text{in.})}$;

or in metric units, $\dfrac{(\text{joule})(\text{m})}{(\text{sec})(\text{m}^2)(°\text{C})}$.

Comments. The United States is the only major country which is neither on the metric system nor has formalized its plans to change to it. (See Lord Ritchie-Calder, "Conversion to the Metric System," *Scientific American*, July 1970.) Britain will change over in 1975. It is generally expected that the U.S. will introduce the initial legislation before that date, so the average student today will spend much of his technical life with metric units. The examples in this book will:

a) Use the units which are currently most common in the U.S.
b) When these units are not metric, the latter will be added parenthetically *if* such an addition aids the student in becoming "bilingual" with respect to dimensions.
c) The SI (Système Internationale) pattern will be followed unless other metric units are more convenient (g/cm^3, rather than kg/m^3; or angstroms rather than m^{-10}); or unless other metric units are already widely used, for example, cm, or ohm · cm. These exceptions usually involve changes only in decimal points.
d) Electron volts, eV, will be used for electron energies, and in Chapter 6 calories will be used for thermal energy.

Pertinent conversion factors are given in Appendix A. ◀

Example 2–1.2. [m] *Electrical conductivity*, σ (which is the reciprocal of electrical *resistivity*, ρ) is a property of a material. The resistance R of a wire depends on the resistivity, the length l, and the cross-sectional area A:

$$R = \frac{\rho l}{A}.$$
$$(2–1.1)$$

What are the units of electrical conductivity?

Solution

$$\sigma = \frac{1}{\rho} = \frac{l}{RA} = \frac{\text{in.}}{(\text{ohm})(\text{in.})^2} = \text{ohm}^{-1} \cdot \text{in.}^{-1}.$$

Since

1/ohm = mho, σ = mho/in.;

or in metric units,

σ = mho/cm.

Comments. In correctly worked problems, dimensional units must check out. Therefore it is useful to insert units into the problem solution to provide a check on one's work. This should always be done; for example, see Example 2–2.1 at the end of the next section. ◀

MECHANICAL PROPERTIES

2–2 STRESS AND STRAIN (ELASTIC)

Stress and strain have distinct meanings to the engineer. Normal *stress*, σ, is a perpendicular force per unit area while linear *strain*, ε, is the resulting deformation. The respective units are:

$$\sigma = \frac{\text{lb}}{(\text{in.})^2}, \qquad \text{or psi\dag;} \tag{2–2.1}$$

$$\varepsilon = \frac{\Delta l}{l} = \frac{\text{in.}}{\text{in.}}, \tag{2–2.2}$$

or dimensionless after canceling units.

In many solids, the amount of initial strain is proportional to the amount of stress (Fig. 2–2.1). Furthermore this initial strain disappears when the stress is removed. The strain which can be recovered by removing the stress is called *elastic* strain. The ratio between the stress and strain in the region in which they are proportional is called *Young's modulus*, Y, or simply the *modulus of elasticity*:

$$Y = \sigma/\varepsilon. \tag{2–2.3}$$

$$\text{Units:} \quad = \frac{\text{lb}/(\text{in.})^2}{\text{in.}/\text{in.}} = \frac{\text{lb}}{(\text{in.})^2}, \qquad \text{or psi.} \tag{2–2.4}$$

Table 2–2.1 lists the moduli of elasticity for several materials. This property, together with the shape and dimensions, accounts for the deformation of a spring. When Y is high, it takes considerable stress to deform the spring; conversely stated, an aluminum ($Y = 10,000,000$ psi) spring is only one-third as rigid as a steel ($Y = 30,000,000$ psi) spring of the same dimensions.

Example 2–2.1. What is the stress in a wire whose diameter is 0.05 in. and which supports a load of 30 lb by tension?

Solution

$$\sigma = \frac{\text{force}}{\text{area}} = \frac{30\ \text{lb}}{(\pi/4)(0.05\ \text{in.})^2} = 15,300\ \text{psi.}$$

Comment. Note that the dimensional units on each side of the equation must cancel each other, unless derived dimensions are used, e.g., volts = amp · ohm. ◀

---

† The comparable metric units are newtons/m^2. As shown in Appendix A, nearly 7000 newtons/m^2 equal 1 psi.

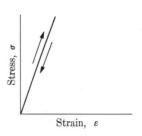

Fig. 2–2.1 Elastic strain. The initial stages of strain are generally proportional to the applied stress. Stress σ is a force per unit area. Strain ε is the fractional amount of deformation. Elastic strain is recoverable.

Table 2–2.1
MODULI OF ELASTICITY (YOUNG'S)*

Material	Elastic modulus	
	English units, psi	Metric units, newtons/m^2**
Aluminum	10,000,000	7×10^{10}
Copper	16,000,000	11×10^{10}
Steel	30,000,000	21×10^{10}
Window glass	10,000,000	7×10^{10}
Polyethylene	14,000–200,000	$1–14 \times 10^8$
Rubber	600–11,000	$4–80 \times 10^6$

 * See Appendix B for additional data.
** 1 psi = 6900 newtons/m^2. See comments accompanying Example 2–1.1 regarding metric units.

Example 2–2.2. Suppose that the wire in Example 2–2.1 is copper and is 100 feet long. How much will it stretch? Repeat for a steel wire.

Solution

From the modulus data in Table 2–2.1, the answer of Example 2–2.1, and Eq. (2–2.3):

a) $\varepsilon = \dfrac{\sigma}{Y} = \dfrac{15{,}300 \text{ psi}}{16{,}000{,}000 \text{ psi}} = 0.00095 \text{ in./in.}$

$$(\text{Total deformation})_{Cu} = (0.00095 \text{ in./in.})(100 \text{ ft})(12 \text{ in./ft})$$

$$= 1.1 \text{ in.}$$

b) $(\text{Total deformation})_{Steel} = \left[\dfrac{15{,}300 \text{ psi}}{30{,}000{,}000 \text{ psi}} \right] (1200 \text{ in.})$

$$= 0.61 \text{ in. } (= 15.5 \text{ mm}).$$

Comments. Steel deforms less than copper for a given stress because of its higher modulus of elasticity.

The answer was carried to two significant figures only because the original data in Example 2–2.1 were only that accurate. The usual slide rule can calculate to three significant figures when the data warrant that precision. ◀

2–3 STRESS AND STRAIN (PLASTIC) ⓔ

As the stress is increased to higher levels, the material may become permanently deformed, i.e., all the strain is not recovered when the stress is removed. This permanent strain is called *plastic* strain. Although during elastic deformation atoms still keep their original neighbors, plastic strain causes some of the atoms to move and to obtain new neighbors.

We use the term *yield strength*, YS, to describe the stress level which initiates the yield for plastic deformation. Beyond this stress, the strain proceeds rapidly with added stress (Fig. 2–3.1). The start of yielding may be abrupt, as shown in Fig. 2–3.1(a), giving a pronounced *yield point*, YP, or gradual, as shown in Fig. 2–3.1(b). When it is gradual, it is helpful to use an *offset* to consistently specify a yield strength (Fig. 2–3.1(c)). Thus, if a small strain such as 0.2% is tolerable—that is, $\varepsilon = 0.002$—it is much easier to measure a reproducible stress for engineering specifications than if it were necessary to detect the very first plastic strain. The offset procedure simply shifts the strain line of Fig. 2–3.1(c) 0.002 in./in. (or 0.002 mm/mm) to the right of the straight elastic portion of the stress–strain curve, producing the leftmost dashed line. The offset yield strength is the stress at the intersection of the two curves.

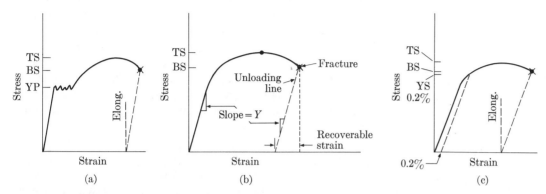

(a)   (b)   (c)

Fig. 2–3.1 Plastic strain. (a) In many steels, the first plastic deformation occurs with a marked yield, or extension. (b) Gradual initial yield. (c) Offset method (see text).

The additional strength (and hardness) which develops with added strain beyond the yield stress of ductile materials is called *strain hardening*; we shall discuss this in Chapter 4. The maximum stress of the stress–strain data as presented in Fig. 2–3.1 is called the ultimate strength, or *tensile strength*, TS. Ductile materials may continue to deform beyond this maximum because of local necking (Fig. 2–3.2), with the nominal stress (calculated on the basis of the original area) decreasing to the *breaking strength*, BS. The breaking strength has little significance unless the reduced area of the necked-down zone is taken into account.

Ductility. This is a measure of plastic deformation prior to final failure and may be expressed as either elongation or reduction in area. *Elongation*, El., is calculated as follows:

$$El = \frac{(in.)_f - (in.)_o}{(in.)_o}, \qquad (2\text{–}3.1a)$$

$$or\ El = \frac{(mm)_f - (mm)_o}{(mm)_o}, \qquad (2\text{–}3.1b)$$

where o and f are the original and final dimensions.

From Fig. 2–3.2, however, note that since the elongation is commonly localized in the necked-down area, the amount of elongation depends on the *gage length*. Whenever reporting elongation, one must be specific about the gage length.

Ductility may also be measured by the amount of *reduction of area* at the point of fracture:

$$Red.\ of\ area = \frac{(in.)_o^2 - (in.)_f^2}{(in.)_o^2}. \qquad (2\text{–}3.2)$$

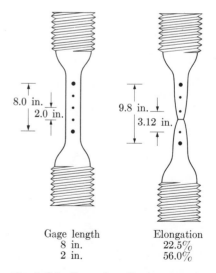

Gage length   Elongation
8 in.   22.5%
2 in.   56.0%

Fig. 2–3.2 Elongation. The gage length must be specifi since the deformation is localized in the necked area. For routine testing, a 2-in. length is common.

[The metric data, using $(mm_o^2 - mm_f^2)/mm_o^2$, give the same value.] The reduction in area is the measure of ductility preferred by some engineers because it does not require a gage length, and because it can be used to determine the true stress at the time of fracture.

One cannot establish an exact correlation between elongation and reduction in area, since ductility may be highly localized. Elongation is a measure of plastic "stretching," whereas reduction of area is a measure of plastic "contraction." Of course, a highly ductile material has high values of each, and a nonductile material has near-zero values of each.

Example 2–3.1. A test bar of aluminum has a diameter of 0.505 in. Yielding causes a 0.2% offset when the load is 1700 lb; the maximum load during testing is 3120 lb; and the load is 2100 lb when the bar breaks. (a) What is the yield strength? (b) The tensile strength? (c) The breaking strength?

Solution. Initial cross-sectional area $= \pi(0.505 \text{ in.})^2/4 = 0.200 \text{ in.}^2$

YS $= 1700 \text{ lb}/0.200 \text{ in.}^2 = 8,500 \text{ psi.}$

TS $= 3120 \text{ lb}/0.200 \text{ in.}^2 = 15,600 \text{ psi.}$

BS $= 2100 \text{ lb}/0.200 \text{ in.}^2 = 10,500 \text{ psi.}$

See Fig. 2–3.3.

Comments. After necking starts, we do not know the true stress, i.e., the force per actual area, because we do not know (from this problem statement) what the true area was. This situation is typical of engineering design situations in which a bridge, for example, is designed on the basis of the original cross-sectional area of the beams, and not on the basis of their dimensions just before a conceivable failure. ◀

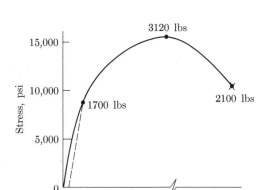

Fig. 2–3.3 See Example 2–3.1.

Example 2–3.2. Before testing the aluminum test bar of the previous problem, an engineering technician placed several gage marks at 2-inch centers along the bar. (Cf. Fig. 2–3.2a.) After the bar was broken, these spacings were 2.11 in., 2.93 in., 2.52 in., and 2.07 in., respectively. (Fracture occurred in the second spacing.) Also the diameter of the bar at the fracture was 0.168 in., as compared to the initial diameter of 0.505 in.

a) What is the 2-inch elongation? 8-inch elongation?

b) What is the reduction of area? The actual stress at fracture?

Solution

a) $\text{El}_{2 \text{ in.}} = 0.93 \text{ in.}/2.00 \text{ in.} = 46.5\%$

   $\text{El}_{8 \text{ in.}} = (9.63 \text{ in.} - 8.00 \text{ in.})/8.00 \text{ in.} = 20.4\%$

   Red. of area $= \dfrac{[\pi(0.505 \text{ in.})^2 - \pi(0.168 \text{ in.})^2]/4}{(\pi/4)(0.505 \text{ in.})^2} = 89\%$

   Actual breaking stress $= \dfrac{2100 \text{ lb}}{(\pi/4)(0.168 \text{ in.})^2} = 95,000 \text{ psi}$

Comment. The figure for 2-inch elongation is always greater than the figure for 8-inch elongation. The actual stress in the neck is always higher than the nominal stress calculated as the breaking strength in Example 2–3.1. ◀

2–4 HARDNESS ⓔ

The hardness of a material is important to the engineer in several obvious ways; in addition, it is easily measured and may often be related to the mechanical strength of a material.

The index of hardness may be reported in various ways. Most commonly, an engineer uses either a *Brinell hardness number, BHN,* or a *Rockwell hardness number, R.* Both these are determined by forcing a hard indenter into the surface of the material to be measured (Fig. 2–4.1). The Brinell indenter is a steel ball 0.39 in. (10 mm) in diameter, usually pressed under a load of 3000 kg. (A 500-kg load is used for soft materials.) The diameter of the indentation is related to the hardness;* a hard material naturally exhibits a smaller indentation than a soft one. With the Rockwell system, the indenter is much smaller, and the index of hardness is related to the depth of penetration. A number of scales are available, depending on the load applied and the exact shape of the indenter. The most common scale for reporting the hardnesses of steels is the Rockwell-*C* scale, R_C. It will be widely used in this book. Table 2–4.1 gives approximate relationships between various hardness scales.

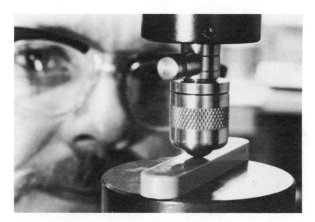

Table 2–4.1

HARDNESS CONVERSION SCALES†

BHN	Rockwell hardness			KHN[4]
	R_C[1]	R_B[2]	R_F[3]	
	80			
	70			972
614	60			732
484	50			542
372	40			402
283	30			311
230	20	98		251
185	10	90		201
150		80		164
125		70	97	139
107		60	91	120
100		55	88	112
83‡		50	85	107
75‡		41	80	98
63‡		23	70	82
59‡		14	65	75
50‡			55	65

† These conversions vary from material to material. For more accurate comparisons for specific materials, consult the *Metals Handbook* (published by ASM).
‡ Load = 500 kg; higher BHN values use a 3000-kg load.
[1] Indenter: diamond "brale"; load, 150 kg.
[2] Indenter: ball, $\frac{1}{16}$ in. diameter; load, 100 kg.
[3] Indenter: ball, $\frac{1}{16}$ in. diameter; load, 60 kg.
[4] Knoop hardness number. This hardness measurement is made by indenting small areas of the material under a microscope.

Fig. 2–4.1 Hardness testing. The hardness is determined by the size of the indentation. (Courtesy of Wilson Instrument Division)

---

*   The BHN is the load P in kg divided by the *surface area* of the indentation, where d is the indentation diameter (mm) and D is the ball diameter:

$$\text{BHN} = \frac{P}{(\pi D/2)(D - \sqrt{D^2 - d^2})}. \tag{2–4.1}$$

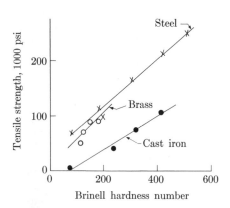

Fig. 2–4.2 Tensile strength versus hardness (BHN). A rough correlation exists between the tensile strength of *metals* and their hardness. This correlation is less specific for nonmetallic materials.

Figure 2–4.2 shows several correlations between hardness and strength. It is apparent that the same relationship does not hold for all materials; however, for a given type of ductile material, the strength can be estimated from the hardness data within 5 to 10%. Since the hardness test may be conducted without machining a test bar, the hardness test is widely used as an indicator of strength. In fact, certain ductile products such as steel beams can be tested for strength without preparing a sample, as would be necessary for tensile testing.

Brittle materials, however, provide erratic data with respect to hardness and strength because cracks may form which serve as stress-raisers for propagation of fractures. Therefore, in brittle materials, good correlations do not exist between hardness and strength.

Example 2–4.1. Basing your solution on the data for steel in Fig. 2–4.2, establish an empirical formula for the tensile strength of *steel* based on its Brinell hardness.

Solution. The slope of the curve for steel approximates 500 psi per 1 BHN:

$$TS \cong 500 \ (BHN). \tag{2–4.2}$$

Comment. This formula does not necessarily hold for other materials. (See Fig. 2–4.2.) ◀

2–5 TOUGHNESS (e) ◎

The product of force f times distance d is energy E:

$$E = fd. \tag{2–5.1}$$

Units may be expressed as in.-lb. The product of stress σ times strain ε is also energy, but energy for a volume:

$$\sigma\varepsilon = \left(\frac{lb}{in^2}\right)\left(\frac{in.}{in.}\right) = \frac{in.\text{-}lb^*}{in^3}. \tag{2–5.2}$$

The areas under the stress-strain curves of Fig. 2–3.1 define the energy product just described. A material which is strong (i.e., which can withstand high stress) and has considerable ductility (i.e., can withstand great strain before fracture) consumes a significant amount of energy during fracture. We call such a material *tough*. If fracture in a given material occurs with negligible plastic strain, only minor amounts of energy are consumed, and we speak of such materials as being *brittle*. Note the contrast between strength, as defined in Section

---

* Metric units are joules/cm³.

2–3, which is the *stress* required to produce failure, and toughness, which is the *energy* required to produce failure. This contrast is especially important in view of the fact that toughness is often inaccurately called "impact strength."

The consumption of energy necessary to cause fracture is not uniformly distributed and depends considerably on the size and shape of the test specimen and on the rate of impact loading. Therefore standardized test specimens and procedures are prescribed for toughness or impact testing. The most commonly used test specimen is the Charpy test specimen, in which (a) a V-notch, or (b) a keyhole notch (Fig. 2–5.1) is cut to locate the fracture. These specimens are then broken by a weighted pendulum. Any energy consumed by the test specimen when it is struck is not available for the follow-through swing of the pendulum (Fig. 2–5.2). This difference in energy can easily be measured by measuring the height of the upswing (Example 2–5.1).

Materials such as glass, gray cast iron, and certain plastics are called *brittle* because they fail with insignificant plastic elongation; ductile metals and rubbers (and even glass at high temperatures) are called *tough* because they undergo considerable strain—and therefore consume much energy—prior to fracture.

Example 2–5.1. An impact pendulum on a testing machine weighs 20 lb and has a center of mass 30 in. from the fulcrum. It is raised 120° and released. After the test specimen is broken, the follow-through swing is 90° on the opposite side. How much energy did the test material absorb?

Solution. Refer to Fig. 2–5.2.

$\Delta E = 20 \text{ lb } (30 \text{ in.})(\cos(-120°)) - 20 \text{ lb } (30 \text{ in.})(\cos 90°)$

$= 20 \text{ lb } (30 \text{ in.})(-0.5 - 0) = -300 \text{ in.-lb}$

$= 25 \text{ ft} \cdot \text{lb } \textit{lost}$ by the pendulum ($= 33.9$ joules)

$= 25 \text{ ft} \cdot \text{lb } \textit{absorbed}$ by the sample.

Comments. In general, materials have more toughness at high than at low temperatures. In fact, there is an abrupt decrease in the toughness of many steels as they are cooled below ambient temperatures. The temperature at which this discontinuous decrease occurs is called the *transition temperature.* ◀

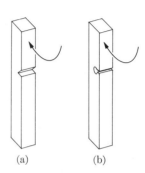

Fig. 2–5.1 Charpy test for toughness under impact. The V-notch (a) and keyhole notch (b) are means of standardizing the fracture conditions.

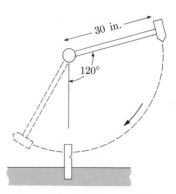

Fig. 2–5.2 Impact testers. The energy absorbed by the standardized sample can be calculated from the height of the follow-through swing of the pendulum. (See Example 2–5.1.)

THERMAL PROPERTIES

2–6 THERMAL EXPANSION

The expansion which occurs during heating arises from the greater amplitude of thermal vibrations of the atoms within the materials. To a first approximation, the increase in length, ($\Delta l/l$), is proportional to the change in temperature:

$$\Delta l/l = \alpha_l \Delta T. \tag{2–6.1}$$

Closer examination shows that the *linear expansion coefficient*, α_l, generally increases slightly with temperature (Fig. 2–6.1). For our purposes, however, we may assume that the value of α_l is essentially constant over small temperature ranges. Table 2–6.1 gives typical values.

The *volume expansion coefficient*, α_V, can be determined to be approximately $3\alpha_l$ if the expansion is uniform in all dimensions.*

Higher temperatures, in addition to producing a greater amplitude of vibration, introduce more disorder into a solid so that the atoms have more open or "free" space among them. This increase in volume is particularly important for those properties which depend on the movements of atoms. We shall discuss these in later chapters.

Example 2–6.1. A 62-inch copper wire is stressed in tension with 24,000 psi, and is also heated from 68°F to 122°F. What is the total change in length Δl? (The yield strength is 27,000 psi.)

Solution

$$(\Delta l/l) = \varepsilon + (\Delta l/l)_{\text{th}} = \frac{\sigma}{Y_{\text{Cu}}} + \alpha_{\text{Cu}} \Delta T.$$

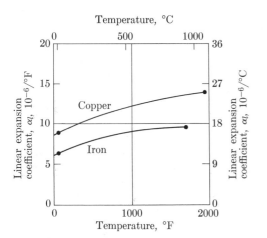

Fig. 2–6.1 Thermal expansion coefficient versus temperature. The volume expansion coefficient α_v is approximately three times the linear expansion coefficient α_l. The values increase slightly with temperature, and are greater for materials which melt at a lower temperature.

---

* Since the thermal volume expansion is $\Delta V/V$, as a unit cube expands,

$$1 + \frac{\Delta V}{V} = \left(1 + \frac{\Delta l}{l}\right)^3,$$

or

$$1 + \frac{\Delta V}{V} = 1 + 3\left(\frac{\Delta l}{l}\right) + 3\left(\frac{\Delta l}{l}\right)^2 + \left(\frac{\Delta l}{l}\right)^3.$$

Discarding the minor second- and third-order terms, we have

$$\frac{\Delta V}{V} \cong 3\frac{\Delta l}{l}; \tag{2–6.2}$$

thus

$$\alpha_V \cong 3\alpha_l. \tag{2–6.3}$$

$$(\Delta l/l)_{\text{total}} = \frac{+24{,}000 \text{ psi}}{16 \times 10^6 \text{ psi}} + (9 \times 10^{-6}/°F)(+54°F)$$

$$= 0.0015 + 0.00049 = 0.002 \text{ in./in.}$$

$$\Delta l = (0.002 \text{ in./in.})(62 \text{ in.}) = 0.12 \text{ in.}$$

Comment. We shall use $(+)$ for tension stresses and strains, and $(-)$ for compressive stresses and strains. In metric units,

$$\Delta l/l = \frac{17 \times 10^7 \text{ newtons/m}^2}{11 \times 10^{10} \text{ newtons/m}^2} + (16 \times 10^{-6}/°C)(+30°C)$$

$$= 0.0015 + 0.00048 = 0.002 \text{ m/m.}$$

$$\Delta l = (0.002 \text{ m/m}) \left(\frac{62 \text{ in.}}{39.37 \text{ in./m.}} \right) = 0.0031 \text{ m.} \blacktriangleleft$$

2–7   THERMAL CONDUCTIVITY

Thermal energy, *heat*, is measured in *Btu*'s (or *joules* in the metric system). The rate of thermal energy transfer, J_{th}, is proportional to the temperature gradient, $(T_2 - T_1)/(x_2 - x_1)$:

$$J_{\text{th}} = -k \frac{(T_2 - T_1)}{(x_2 - x_1)}, \tag{2–7.1}$$

where k is the thermal conductivity. The minus sign is necessary because heat flows "downhill" only; that is, $T_2 < T_1$. (See Fig. 2–7.1.) The most common dimensions of English units are:

$$\frac{\text{Btu}}{\text{ft}^2 \cdot \text{sec}} = -\left[\frac{\text{Btu} \cdot \text{in.}}{\text{ft}^2 \cdot \text{sec} \cdot °F} \right] \left[\frac{-°F}{\text{in.}} \right]. \tag{2–7.2}$$

The metric units are:

$$\frac{\text{joule}}{\text{cm}^2 \cdot \text{sec}} = -\left[\frac{\text{joule} \cdot \text{cm}}{\text{cm}^2 \cdot \text{sec} \cdot °C} \right] \left[\frac{-°C}{\text{cm}} \right]. \tag{2–7.3}$$

Table 2–6.1

LINEAR THERMAL EXPANSION COEFFICIENTS (AT 68°F)*

Material	α_l	α_l
	English units, per °F	Metric units,** per °C
Aluminum	12.5×10^{-6}	22×10^{-6}
Brass (70 Cu–30 Zn)	11×10^{-6}	20×10^{-6}
Copper	9×10^{-6}	16×10^{-6}
Iron	6.5×10^{-6}	12×10^{-6}
Building brick	5×10^{-6}	9×10^{-6}
Glass (window)	5×10^{-6}	9×10^{-6}
Concrete	7×10^{-6}	13×10^{-6}
Polyethylene	$60{-}100 \times 10^{-6}$	$110{-}180 \times 10^{-6}$
Polystyrene	40×10^{-6}	72×10^{-6}
Rubber (vulcanized)	45×10^{-6}	81×10^{-6}

* See Appendix B for additional data.

** $1 \dfrac{\text{in.}}{\text{in.} \cdot °F} = 1.8 \dfrac{\text{cm}}{\text{cm} \cdot °C}$ (or $1/°F = 1.8/°C$).

See comments with Example 2–1.1 regarding metric units.

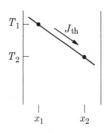

Fig. 2–7.1 Thermal flux. The heat flow is "downhill" and proportional to the steepness of the thermal gradient, $(T_2 - T_1)/(x_2 - x_1)$.

The value of k, the *coefficient of thermal conductivity*, is significantly higher in metals than in ceramics or plastics (Table 2–7.1). We noted in Chapter 1 that insulators such as glass and polythene can conduct heat only by vibrations of atoms. A metal, on the other hand, also has free electrons, which carry heat from hotter to colder parts of the material. Thus we shall see in Chapter 4 that in metals there is a direct correlation between thermal and electrical conductivity.

Example 2–7.1. One Btu equals 1055 joules. What is the conversion between the English and metric units of thermal conductivity shown in Eqs. (2–7.2) and (2–7.3)?

Solution

$$1\left(\frac{Btu \cdot in.}{ft^2 \cdot sec \cdot {}^\circ F}\right)\left(\frac{1055\ joules}{1\ Btu}\right)\left(\frac{2.54\ cm/in.}{2.54^2 cm^2/in.^2}\right)\left(\frac{1.8^\circ F}{^\circ C}\right)\left(\frac{1\ ft^2}{144\ in.^2}\right)$$

$$= 5.192\ \frac{joule \cdot cm}{cm^2 \cdot sec \cdot {}^\circ C},$$

or

$$1\ \frac{joule \cdot cm}{cm^2 \cdot sec \cdot {}^\circ C} = 0.1926\ \frac{Btu \cdot in.}{ft^2 \cdot {}^\circ F \cdot sec}.$$

Comment. There is a common error made in the units for thermal conductivity. The units, Btu/ft²/sec/°F/in., are *not* correct unless the thermal gradient, (°F/in.), is indicated as a unit. To be correct, one should use Btu/ft²/sec/(°F/in.), or (Btu · in.)/(ft²)(sec)(°F). ◀

Example 2–7.2. A stainless steel plate 0.11 in. thick has circulating hot water on one side and a rapid flow of air on the other side, so that the two metal surfaces are 205°F and 65°F, respectively. How many Btu's are conducted through the plate per minute?

Solution. From Appendix B,

$$k = 0.028\ \frac{Btu \cdot in.}{ft^2 \cdot sec \cdot {}^\circ F},$$

$$J_{th} = \left(0.028\ \frac{Btu \cdot in.}{ft^2 \cdot sec \cdot {}^\circ F}\right)\left(\frac{140^\circ F}{0.11\ in.}\right)\left(\frac{60\ sec}{1\ min}\right)$$

$$= 2100\ \frac{Btu}{ft^2 \cdot min}.$$

Comment. We shall learn later that stainless steel has a much lower thermal conductivity than other metals because it contains large quantities of alloying elements (Chapter 4). ◀

Table 2–7.1

THERMAL CONDUCTIVITY COEFFICIENTS (AT 68°F)*

Material	k English units, $\dfrac{Btu \cdot in.}{ft^2 \cdot sec \cdot {}^\circ F}$	k Metric units,** $\dfrac{joule \cdot cm}{cm^2 \cdot sec \cdot {}^\circ C}$
Aluminum	0.43	2.23
Brass (70 Cu–30 Zn)	0.24	1.24
Copper	0.77	3.99
Iron	0.14	0.72
Building brick	0.0012	0.0062
Window glass	0.0014	0.0072
Polyethylene	0.00065	0.0034
Polystyrene	0.00016	0.0008
Rubber	0.00025	0.0013

* See Appendix B for additional data.
** See comments with Example 2–1.1 regarding metric units.

ELECTRICAL PROPERTIES

2–8 ELECTRICAL RESISTIVITY (AND CONDUCTIVITY) $\boxed{m}$

In Example 2–1.2 we defined electrical conductivity σ as the reciprocal of resistivity ρ. The latter is more commonly given in tables of data (Appendix B) because the resistivity can be readily converted to resistance and other electric measurements. Electrical conductivity can also be defined as

$$\sigma = nq\mu, \tag{2–8.1}$$

where n is the number of charge carriers per unit volume, q is the charge per carrier, and μ is the *mobility* of the charge carriers. The units are:

$$\text{ohm}^{-1} \cdot \text{cm}^{-1} = \left(\frac{\text{carriers}}{\text{cm}^3}\right) \left(\frac{\text{coul}}{\text{carrier}}\right) \left(\frac{\text{cm/sec}}{\text{volt/cm}}\right). \tag{2–8.2}*$$

In metals and in those semiconductors in which electrons are the charge carriers, the charge per carrier is 1.6×10^{-19} coul, or 1.6×10^{-19} amp $\cdot$ sec. The mobility may be considered as the net, or *drift velocity* $\bar{v}$ of the carriers, which arises from the electric field $\mathscr{E}$, where the units are (cm/sec) and (volt/cm), respectively:

$$\mu = \bar{v}/\mathscr{E}. \tag{2–8.3}$$

Mobility is commonly expressed as $\text{cm}^2/\text{volt} \cdot \text{sec}$. The relationships of Eqs. (2–8.1), (2–8.2), and (2–8.3) will be particularly useful in Chapter 12, when we consider semiconductors.

Example 2–8.1. A 1-meter wire has a diameter of 1.0 mm and a resistance of 0.13 ohm at 20°C (68°F) and 0.18 ohm at 30°C (86°F). What are its conductivities at these temperatures?

Solution. From Eq. (2–1.1),

$$\sigma = \frac{1}{\rho} = \frac{l}{RA}. \tag{2–8.4}$$

At 20°C,

$$\sigma = \frac{100 \text{ cm}}{(0.13 \text{ ohm})(\pi/4)(0.1 \text{ cm})^2} = 98{,}000 \text{ ohm}^{-1} \cdot \text{cm}^{-1};$$

$$\rho = 10.2 \times 10^{-6} \text{ ohm} \cdot \text{cm}.$$

---

* The units check because coul = amp $\cdot$ sec, and volt = amp $\cdot$ ohm.

At 30°C,

$$\sigma = (98{,}000 \text{ ohm}^{-1}\text{cm}^{-1}) \left(\frac{0.13 \text{ ohm}}{0.18 \text{ ohm}}\right) = 71{,}000 \text{ ohm}^{-1} \cdot \text{cm}^{-1};$$

$$\rho = 14 \times 10^{-6} \text{ ohm} \cdot \text{cm}.$$

Comments. As a general rule, metals increase in resistivity (decrease in conductivity) as the temperature is increased. But nonmetals, at elevated temperatures, lose their extremely high resistivity.

Electrical resistivity is commonly reported as ohm · cm (or ohm · m). Ohm · in. is less commonly used. ◀

Example 2–8.2. The resistivity ρ of a semiconductor which has 10^{18} electron carriers per cm^3 is 2.0 ohm · cm. At what velocity $\bar{v}$ do the electrons move through the semiconductor when the voltage gradient is 7 millivolts per 0.11 mm?

Solution. From Eq. (2–8.1), we have

$$\mu = \frac{\sigma}{nq} = \frac{1}{\rho nq}$$

$$= \frac{1}{(2 \text{ ohm} \cdot \text{cm})(10^{18}/\text{cm}^3)(1.6 \times 10^{-19} \text{ amp} \cdot \text{sec})}$$

$$= 3.1 \text{ (cm/sec)/(volts/cm)}.$$

From Eq. (2–8.3),

$$\bar{v} = \left(3.1 \frac{\text{cm/sec}}{\text{volts/cm}}\right)\left(\frac{0.007 \text{ volt}}{0.011 \text{ cm}}\right) = 2 \text{ cm/sec}.$$

Comments. Recall from earlier science courses that

$$\text{volts} = \text{amp} \cdot \text{ohm}. \tag{2–8.5}$$

The velocity which was calculated is an average or *drift velocity*. The maximum velocity of the electrons is greater because they are deflected and reflected. ◀

2–9 RELATIVE DIELECTRIC CONSTANT ⓜ Ⓜ

Electrical insulators do not, of course, transport electric charges. However, they are not inert to an electrical field. We can show this by separating two electrode plates a distance d and applying a voltage V between them (Fig. 2–9.1a). The *electric field* $\mathscr{E}$ is the voltage gradient:

$$\mathscr{E} = V/d. \tag{2–9.1}$$

Under these conditions, when there is nothing between the plates, the charge density $\mathscr{D}_0$ on each plate is proportional to the field $\mathscr{E}$. For each volt/cm of field, there are 8.85×10^{-14} coulombs per cm^2 of electrode. (This charge density requires 550,000 electrons/cm^2 for each volt/cm, since each electron has 1.6×10^{-19} coulombs.)

If a material m is placed between the electrodes of Fig. 2–9.1, the charge density can be increased to $\mathscr{D}_m$ from the $\mathscr{D}_0$ just described. The ratio, $\mathscr{D}_m/\mathscr{D}_0$, is called the *relative dielectric constant* κ of the material which is used as a dielectric spacer between the electrodes:

$$\kappa = \mathscr{D}_m/\mathscr{D}_0. \tag{2–9.2}$$

The relative dielectric constant will be important to us in Chapters 9 and 10, in which we look at the *dielectric* properties of the plastics and ceramics that are used in capacitors. The relative dielectric constant is greater than 1.0 because the electrons and the positive and negative ions are displaced within the material.

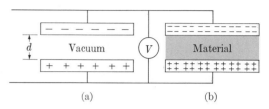

(a)   (b)

Fig. 2–9.1 Charge density $\mathscr{D}$ versus relative dielectric constant, κ. The presence of a material increases the charge density, $\mathscr{D}_m$, held by the capacitor plates in proportion to the relative dielectric constant, κ: $\mathscr{D}_m = \kappa\mathscr{D}_0$.

Example 2–9.1. A capacitor (Fig. 2–9.1a) with two parallel plates 1 cm × 2 cm each, receives a 2.25-volt potential difference between the electrodes. (a) How far apart must they be to produce a charge density of 10^{-12} coul/cm^2? (b) How many electrons accumulate on the negative plate under these conditions? No dielectric insulator is placed between these plates.

Solution

a) From the first paragraph of this section,

$$\mathscr{D}_0 = (8.85 \times 10^{-14})\,\mathscr{E}, \tag{2–9.3}$$

$$10^{-12}\ coul/cm^2 = (8.85 \times 10^{-14}\ coul/volt \cdot cm)(2.25\ volt/d),$$

$$d = 0.2\ cm.$$

b) No. of electrons $= \dfrac{1\ cm \times 2\ cm \times 10^{-12}\ coul/cm^2}{1.6 \times 10^{-19}\ coul/electron}$

$$= 1.25 \times 10^7\ electrons.$$

Comments. The dimensions of Eq. (2–9.3) are

$$coul/cm^2 = (coul/volt \cdot cm)(volt/cm).$$

Since a farad f is a (coul/volt), the proportionality term is 8.85×10^{-14} f/cm, when the space between the capacitor plates is empty. ◀

Example 2–9.2. The relative dielectric constants κ of a glass and of a plastic are 3.9 and 2.1, respectively. What voltage should be applied between electrodes which are separated with 0.13 cm of glass if one wants the same charge density $\mathscr{D}_m$ as would develop on another set of electrodes separated by 0.42 cm of plastic and 210 volts?

Solution

$$\kappa_{\mathrm{gl}}\mathscr{D}_0 = \mathscr{D}_{\mathrm{gl}} = \mathscr{D}_{\mathrm{pl}} = \kappa_{\mathrm{pl}}\mathscr{D}_0.$$

Since

$$\mathscr{D}_0 = (8.85 \times 10^{-14}\ \mathrm{coul/volt \cdot cm})\ \mathscr{E},$$

$$[\kappa(8.85 \times 10^{-14})(V/d)]_{\mathrm{gl}} = [\kappa(8.85 \times 10^{-14})(V/d)]_{\mathrm{pl}},$$

$$3.9\ V/0.13\ \mathrm{cm} = 2.1(210\ \mathrm{volts})/(0.42\ \mathrm{cm}),$$

$$V = 35\ \mathrm{volts}.$$

Comments. In calculating electric properties, in the United States as elsewhere, one usually uses metric units.

Equation (2–9.3) is often written as

$$\mathscr{D}_0 = \varepsilon_0\mathscr{E}, \tag{2–9.4}$$

and describes the charge density when there is nothing between the electrodes. The term ε_0, which is 8.85×10^{-14} coul/volt·cm, or 8.85×10^{-14} farad/cm, is called the *permittivity* of a vacuum. ◄

REVIEW AND STUDY

2–10 SUMMARY

The engineer must use materials which meet specific requirements of strength, thermal and electrical conductivity, fabricability, etc. Therefore he is interested in the various properties of materials; it is these properties which give him control over his designs and products. We have cited some—but far from all—such properties. The ones we have discussed, however, will enable us to discuss the features of internal structure which (a) provide resistance to mechanical alteration by applied stresses, (b) account for the more common characteristics of a material when it is thermally agitated, and (c) relate to the behavior of a material within an electric field. These features are important because the knowledge of them enables the design engineer to specify, and the materials engineer to produce, materials with optimum structures and characteristics for present and future products.

Materials recover from elastic strain after the stress is removed because the atoms retain their original neighbors. Plastic strain proceeds beyond the yield point to give permanent deformation. Highly ductile materials are tough in that they require considerable energy input before failure.

More thermal agitation exists at higher than at lower temperatures. This extra energy expands the structure. Thermal energy is transferred to cooler parts of the material by atom vibrations and (in metals) by electron conduction.

The diffusion of electrons accounts for the electrical conductivity of most conducting materials. The charges within an insulator do not migrate in an electric field, but are displaced, so that in addition to simply being "isolators," dielectrics have functional electric properties useful to the engineer.

Terms for Study

Breaking strength, BS

Brinell hardness number, BHN

Brittle

Charge density, $\mathscr{D}$

Charpy test

Dielectric

Drift velocity, $\bar{v}$

Ductility

Elastic

Electric field, $\mathscr{E}$

Electrical conductivity, σ

Electrical resistivity, ρ

Elongation

Expansion coefficient, α

Gage length

Hardness

Impact strength

Mechanical properties

Mobility

Modulus of elasticity

Plastic

Reduction of area, Red. of A

Relative dielective constant, κ

Rockwell hardness number, R

Strain, ε

Strain hardening

Stress, σ

Tensile strength, TS

Thermal conductivity, k

Thermal expansion, α

Toughness

Transition temperature

True stress, σ_{tr}

Yield strength, YS

Young's modulus, Y

STUDY PROBLEMS

2–1.1 The thermal expansion of a steel rod is 6.3×10^{-6} in./in. · °F. What is the thermal expansion in metric units?

Answer. 11.3×10^{-6}/°C

2–1.2 Copper has a resistivity of 1.7×10^{-6} ohm · cm. What is its conductivity in mhos/in.?

Answer. 1.5×10^{6} mhos/in.

2–1.3   A no. 14 wire has a resistivity of 1.7×10^{-6} ohm · in. and a diameter of 0.064 in. What is the resistance per foot?

Answer. 0.006 ohm

2–2.1   A steel rod must carry a load of 1500 lb with an elastic strain of no more than 0.0001 in./in. What is the minimum permissible cross-sectional area?

Answer. 0.5 in.^{2}

2–2.2   How much strain would a glass rod of the same dimensions of the steel rod in the previous problem have if it carried a 10,000-lb load?

Answer. 0.002 in./in. (or, more simply, 0.002)

2–3.1   The yield strength of a brass (70 Cu–30 Zn) is 30,000 psi. What is the maximum length to which a 96-in. wire made of this material can be stretched without permanent deformation?

Answer. 96.18 in.

2–3.2   The tensile strength of a certain steel is 78,000 psi. A bar 0.75 in. × 1.25 in. deforms to a cross section of 0.51 in. × 0.85 in. before it breaks with a load of 55,000 lb. (a) What is the maximum load the bar can carry? (b) What is the maximum stress before fracture? (c) What is the ductility?

Answer. (a) 73,000 lb (b) 127,000 psi (c) 54% Red. of area

2–3.3   (a) An 0.05-in. aluminum wire 25 ft long is drawn through an 0.035-in. diameter die. What is the new length? (b) The tensile strength of the wire before drawing was 16,000 psi; after drawing, it is 38,000 psi, but its reduction of area dropped from 75% to 15%. Explain these changes.

Answer. (a) 51 ft

2–4.1   By means of a 10-mm indenter, a 3000-kg Brinell load makes a 4.0-mm indentation into one steel (a), and a 3.2-mm indentation into another steel (b). Estimate how their tensile strengths compare. See Eq. (2-4.1).

Answer. (a) BHN = 225 (TS ≈ 112,000 psi)   (b) BHN = 360 (TS ≈ 180,000 psi)

2–5.1   A 1.2-lb steel ball had to fall 37 in. onto a sample of plate glass to break it. What height would a 200-g ball require to deliver the same impact?

Answer. 100 in.

2–5.2   What is the maximum impact energy measurable in the testing equipment of Fig. 2–5.2, given that 120° is the required starting position? (See Example 2–5.1 for dimensions.)

Answer. 75 ft · lb

2–6.1   A borosilicate glass gage block is 1.00000 in. at 80°F. What will its dimension be on a 66°F day?

Answer. 0.99998 in.

2–6.2 .A steel wire is stretched with a stress of 7000 psi at 65°F. To what temperature must it be heated to reduce the stress to 4000 psi, given that the length is held constant after the initial stretching?

Answer. 81°F

2–6.3 A welded steel rail is laid in place at 90°F, and anchored so that shrinkage cannot occur when the temperature drops to 30°F. How much stress develops in the rail?

Answer. +11,000 psi (tension)

2–6.4 A steel rim for a 48-in. locomotive wheel is heated to 600°F. At that temperature its diameter has an 0.05-in. clearance over the wheel. Assume that the wheel, at 60°F, is absolutely rigid. What stresses are developed in the rim after it is shrunk-fit onto the wheel?

Answer. +71,000 psi (tension)

2–7.1 Calculate the joules conducted through a plate of copper which is substituted for the stainless steel of Example 2–7.2.

Answer. 67,000 joules/cm^2 · min.

2–7.2 Using data from Appendix B, determine the ratio k/σ for each metal.

Answer. (The units of k/σ are joule · ohm/sec · °C.)   Al, 6.0×10^{-6}; brass, 7.4×10^{-6}; bronze, 8.0×10^{-6}; Cu, 6.8×10^{-6}; Fe, 7.1×10^{-6}; Pb, 6.5×10^{-6}; Mg, 8.6×10^{-6}; monel metal, 12×10^{-6}; Ag, 6.6×10^{-6}; steel, 8.3×10^{-6}

2–8.1 A small semiconductor cube (1 mm × 1 mm × 1 mm) has a conductivity of 10.3 ohm^{-1} · cm^{-1}. What is its resistance?

Answer. 0.97 ohm

2–8.2 Measurements indicate that the drift velocity of electrons in a 110-V heating element is 87 m/sec. It has an extended length of 27.5 in. What is the mobility of those electrons?

Answer. 5500 cm^2/volt · sec

2–8.3 There are 10^{17} electron carriers per cm^3 in a semiconductor which has an electron mobility of 1900 cm^2/volt · sec. What is its resistivity? Conductivity?

Answer. (a) 0.033 ohm · cm (b) 30.4 mho/cm

2–8.4 A wire with a resistance of 0.0025 ohm/ft carries a current of 1.28 amp. (a) How many electrons move through the wire per second? (b) What voltage is required to produce this current through a 2640-ft wire?

Answer. (a) 8×10^{18} electrons/sec (b) 8.5 volts

2–9.1 Two capacitor plates (3 cm × 2 cm each) are parallel and 0.22 cm apart with nothing between them. What voltage is required to produce a charge of 0.24×10^{-10} coul on the electrodes?

Answer. 10 volts

2–9.2 What relative dielectric constant will be required for the same capacitor characteristics, given that the spacing is reduced from 0.010 cm to 0.006 cm? The present dielectric material has a dielectric constant of 3.3.

Answer. 2.0

2–9.3 Two electrode plates of a capacitor are 10 mm × 5 mm each and are separated by a 1-mm-thick dielectric material with $\kappa = 80.2$. What voltage is required to force 10^{10} electrons from one electrode to the other?

Answer. 45 volts

3–1 SINGLE-COMPONENT METALS

We shall look first at materials which have only one kind of atom. Several of our common metallic products fall in this category: copper used in electric wiring, iron for car fenders, and aluminum used in kitchen utensils. Each of these metals is commercially pure, containing only minor amounts of other components not removed during production. Those metals *not* included in this chapter are the large majority of steels, brass, bronze, solder, most aluminum alloys, and the high-temperature alloys used in energy-conversion equipment.

Single-component metals—i.e., metals with only one kind of constituent—receive our attention first because we shall not have to consider composition as a variable. Further, every atom has only one kind of neighbor: an atom identical to itself.

When all the atoms are the same, we can readily calculate the number of atoms which are present per cubic centimeter (or any other volume). To do this (Example 3–1.1), we must recall that the atomic weight of an atom, in amu's (Fig. 1–2.2), is equal to the mass in grams of 6×10^{23} atoms. This number is called *Avogadro's number*, AN, and will be used frequently in the discussions and calculations of this text. As examples, from Fig. 1–2.2 we observe that it takes 63.54 g of copper to provide 6×10^{23} atoms of copper; and 26.98 g of aluminum to make 6×10^{23} atoms of aluminum; and 55.85 g of iron to make 6×10^{23} atoms of iron. Thus, since 1 cm³ of copper weighs 8.96 g, we can calculate that each cm³ has $(8.96 \text{ g/cm}^3)/(63.54 \text{ g}/6 \times 10^{23} \text{ atoms})$, or 0.85×10^{23} atoms of copper.

Atomic radii. The distance between atoms can be measured by x-ray diffraction. We shall not do that in this text, but the procedure is well established, so that we have access to tables such as Table 3–1.1. Although the analogy has shortcomings, it is convenient for us to envisage a *hard-ball model*, in which the atoms in simple metals are spherical. Thus, since the interatomic distance in copper is 2.556 Å, the radius of a copper atom is simply half that dimension, or 1.278 Å.

Table 3–1.1

INTERATOMIC DISTANCES AND ATOMIC RADII (METALS)

Metal	Structure	Distance, Å**	Radius, Å
Ag	fcc	2.888	1.444
Al	fcc	2.862	1.431
Au	fcc	2.882	1.441
Be	hcp	2.28*	1.14*
Cd	hcp	2.96*	1.58*
Co	hcp	2.50*	1.25*
Cr	bcc	2.498	1.249
Cu	fcc	2.556	1.278
Fe (α)	bcc	2.4824	1.2412
Fe (γ)	fcc	2.540	1.270
K	bcc	4.624	2.312
Li	bcc	3.038	1.519
Mg	hcp	3.22*	1.61*
Mo	bcc	2.725	1.362
Na	bcc	3.714	1.857
Ni	fcc	2.491	1.246
Pb	fcc	3.499	1.750
Pt	fcc	2.775	1.386
Ti (α)	hcp	2.93*	1.46*
Ti (β)	bcc	2.85	1.42
V	bcc	2.632	1.316
W	bcc	2.734	1.367
Zn	hcp	2.78*	1.39*
Zr	hcp	3.24*	1.62*

* The average dimension is given for hcp, since the radius varies somewhat with direction; i.e., the hard-ball atom is not quite spherical, but ellipsoidal.
** An angstrom, Å, is 10^{-8} cm ($= 3.937 \times 10^{-9}$ in.).

(One angstrom, Å, equals 10^{-8} cm. Although we shall use the simpler angstrom units whenever possible, at times it will be necessary to revert to centimeter units; for example, $R_{Cu} = 1.278 \times 10^{-8}$ cm.) The volume of a sphere with a radius of 1.278 Å is 8.7 Å^3 (or 8.7×10^{-24} cm^3). Realizing that space cannot be fully packed with spheres, we can quickly *estimate* that there may be slightly more than 10^{-23} cm^3 per copper atom, or slightly less than 10^{23} copper atoms per cm^3. In the preceding paragraph, we calculated 0.85×10^{23} as the more precise number.

Example 3–1.1. An aluminum wire is 30.48 cm long and 2.58 mm in diameter. How many atoms are present?

Solution

$$\text{No. of Al atoms} = \frac{(2.699 \text{ g/cm}^3)[\pi(0.129 \text{ cm})^2(30.48 \text{ cm})]}{(26.98 \text{ g}/6 \times 10^{23} \text{ atoms})}$$

$$= 0.96 \times 10^{23} \text{ atoms.}$$

Comments. There are 0.6×10^{23} Al atoms/cm^3 compared to 0.85×10^{23} Cu atoms/cm^3, indicating that $R_{Al} > R_{Cu}$. This is verified by data in Table 3–1.1. ◄

Example 3–1.2. Assume spherical copper atoms with a hard-ball radius of 1.278 Å. We know from the previous discussion that there are 0.85×10^{23} atoms of copper per cm^3. What is the *atomic packing factor* (APF) for copper; i.e., what fraction of the volume is occupied by the ball-like atoms?

Solution

$$\left(\frac{\text{Volume of atoms}}{\text{Total volume}}\right) = \frac{(0.85 \times 10^{23} \text{ atoms})(4\pi/3)(1.278 \times 10^{-8} \text{ cm})^3/\text{atom}}{1 \text{ cm}^3}$$

$$= 0.74 \text{ cm}^3/\text{cm}^3;$$

$$\text{APF} = 0.74.$$

Comments. The value of 0.74 is the highest atomic packing factor possible for spheres of only one size. (See Section 3–3.)

Our hard-ball model of atoms simply notes that we cannot force two copper atoms with their charged nuclei and accompanying electrons closer together than 2.556 Å. However, an *uncharged* particle, such as a neutron, can move through this spherical volume, indicating that open space remains within the atom. ◄

3–2 CUBIC METALS

Fcc metals. When a metal solidifies, the atoms arrange themselves into regular arrays which we call *crystals*. This occurs because there can be stronger bonding between atoms if there is a regular arrangement than if there is a random atomic coordination. Since bonding

Fig. 3–2.1 Coordination number CN $= 12$. The center atom (dashed circle) is surrounded by twelve atoms of the same size.

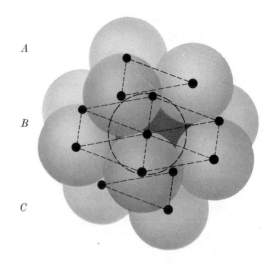

exists because of the attraction between atoms, many metals arrange themselves so that each atom is touching, or is coordinated with, 12 neighboring atoms. We say then that the *coordination number*, CN, is 12. As shown in Fig. 3–2.1, 12 is the maximum number of spheres which can surround a given sphere when all of them are the same size. When this coordination of atoms is extended in a regular pattern of three dimensions, they form a cubic array of atoms (Fig. 3–2.2). This particular array, or *crystal lattice*, turns out to be one of the most common arrangements found in metals, or for that matter in all materials. It is called *face-centered cubic*, *fcc*, because equivalent *lattice points* exist, not only at the corner of each cube, but also at the center of each face. Familiar metals which have an fcc structure are aluminum, calcium, iron (at elevated temperatures), nickel, copper, silver, platinum, gold, and lead.

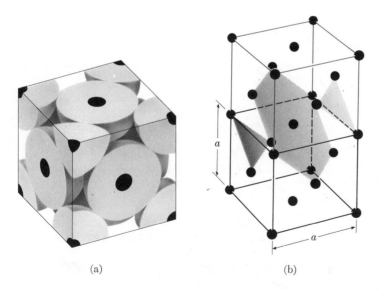

(a)   (b)

Fig. 3–2.2 Face-centered cubic (fcc). (a) Hard-ball model with atoms at cube corners and face centers. There are four "balls" per cube (8/8 + 6/2). (b) Correlation between Fig. 3–2.2 and Fig. 3–2.1.

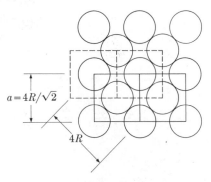

$a = 4R/\sqrt{2}$

$4R$

Fig. 3–2.3 Equivalent positions. In the fcc, the face-centered position is fully equivalent to the corner position in every respect. In fact, the reference corner could be shifted from one to the other.

The fcc lattice is highly symmetric. Each edge a of the *unit cell* shown in Fig. 3–2.2 is equal in dimensions. Each angle between edges is 90°. Each atom has *twelve* neighbors. Finally, the center of each face has a lattice site which is *equivalent* to the corner site. This means that we could have picked an atom which is now at the face-centered position and used it as a cube corner, and in doing so we would have developed a new fcc unit cell identical to the first. You can verify this by extending the front face of Fig. 3–2.2(a) for three or four unit cell distances in each direction and comparing corner locations with face-centered locations (Fig. 3–2.3).

An examination of the unit cell of Fig. 3–2.2(a) reveals that each unit cell contains the equivalent of four atoms (6 half-atoms on face centers plus 8 eighth-atoms at cube corners). From this, we can calculate the hard-ball volume for each unit cell as $4(4\pi R^3/3)$, the volume of four spheres. Furthermore, the edge a of the unit cell is $4R/\sqrt{2}$, where R is the atomic radius (Fig. 3–2.3). The volume of the unit cell is therefore $32R^3/\sqrt{2}$. Since we saw in the last section that the APF is the ratio of hard-ball volumes to total volume, we can calculate that value as $\pi/(3\sqrt{2})$, or 0.74. This, of course, is the value we obtained by other means in Example 3–1.2; it is also the answer to Study Problem 3–1.2.

Bcc metals. Figure 3–2.4 shows a second coordination arrangement commonly found in metals. As we would expect, this one is called a *body-centered cubic*, *bcc*, lattice. Familiar metals which have this lattice include sodium, potassium, titanium (at elevated temperatures), chromium, iron (at normal temperatures), and tungsten. This lattice, like the fcc lattice, is highly symmetric. Each side of the unit cell is equal in dimension, a. The angle between each pair of edges is 90°. However, the bcc lattice differs from the fcc lattice, in that each atom has only *eight* neighbors; CN = 8. Finally, only the center lattice position is fully equivalent to the corner positions.

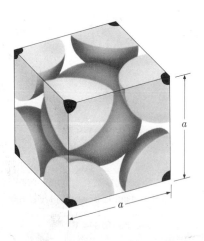

a

a

Fig. 3–2.4 Body-centered cubic (bcc). Hard-ball model, with atoms at cube corner and body center. There are two "balls" per cube (8/8 + 1).

Although bcc metals are somewhat less common than fcc metals, they are fully as important in our engineering products. For example, the fact that iron changes from face-centered cubic when it is at high temperatures to body-centered cubic as it is cooled provides the basis for heat-treating steels, which in turn gives us the wide range of properties described in the first section of Chapter 1. These structures will be important to us in our study of materials.

Example 3–2.1. Determine the APF of a bcc metal lattice.

Solution. There are two atoms per unit cell with R as the hard-ball radius. From Fig. 3–2.4, the long diagonal of the unit cell is

$$4R = \sqrt{a^2 + a^2 + a^2},$$

$$a_{bcc} = 4R/\sqrt{3}; \qquad\qquad\qquad (3\text{–}2.1)$$

$$APF = \frac{2(4\pi R^3/3)}{(4R/\sqrt{3})^3} = 0.68.$$

Comments. Note that the APF of metals is independent of the radius of their atoms.

    The APF of bcc metals is less than that of fcc metals because each atom has only eight rather than twelve neighbors.

    Cubic unit cells are *equiaxed* because their dimensions are equal in the three principal directions. ◀

Example 3–2.2. Nickel has an fcc lattice with one nickel atom at each lattice point. Calculate its density.

Solution. Basis: 1 unit cell. Density ρ is mass per unit volume, g/cm^3. With four nickel atoms per fcc unit cell, and since 6×10^{23} atoms equal 58.71 g (Fig. 1–2.2), each unit cell has $4(58.71/6 \times 10^{23})$ g. From Fig. 3–2.2(a),

$$a_{fcc} = 4R/\sqrt{2}. \qquad\qquad\qquad (3\text{–}2.2)$$

and since

$$R = 1.246 \text{ Å (Table 3–1.1)}, \qquad a_{Ni} = 3.525 \text{ Å}.$$

$$\rho = \frac{(4 \text{ atoms})(58.71 \text{ g}/0.6 \times 10^{24} \text{ atoms})}{(3.525 \times 10^{-8} \text{ cm})^3}$$

$$= 8.95 \text{ g/cm}^3.$$

Comments. Density measurements give 8.90 g/cm^3. Calculations for other cubic metals give additional verification that our fcc and bcc lattice models are correct. ◀

Example 3–2.3. Pure iron changes from bcc to fcc at 1670°F (910°C). Will iron expand or contract when it changes from bcc to fcc as it is heated past that temperature? *Assume* that iron has the same radii in both the bcc and fcc structures.

Solution. Basis: 4 atoms = 2 bcc unit cells = 1 fcc unit cell.

$$V_{bcc} = 2a_{bcc}^3 = 2(4R/\sqrt{3})^3 = 24.6R^3.$$

$$V_{fcc} = (4R/\sqrt{2})^3 = 22.6R^3.$$

Therefore

$$V_{fcc}/V_{bcc} = 22.6R^3/24.6R^3 = 0.92.$$

Thus there is an 8% volume *contraction during heating* as iron is heated through this transformation temperature and changed from bcc Fe to fcc Fe.

Comments. We would have anticipated this, since fcc metals have a greater APF than bcc metals, 0.74 versus 0.68, respectively.

Close measurements reveal that the bcc iron has a slightly smaller radius than fcc, so that the volume contraction is less than the 8% calculated above. ◀

3–3 NONCUBIC METALS

Some important metals are not cubic. These include magnesium, titanium (at normal temperatures), cobalt, zinc, zirconium, and cadmium, all of which are hexagonal and have the same APF as fcc metals; that is, 0.74. Figure 3–2.1 is presented again in Fig. 3–3.1(a), along with a subtle variation in Fig. 3–3.1(b). In each case the atoms have twelve immediate neighbors. However, in part (b) the lower plane of atoms has been rotated 60°. This subtle difference leads to a

Fig. 3–3.1 Coordination arrangements. (a) CN = 12 in fcc. (b) CN = 12 in hcp.

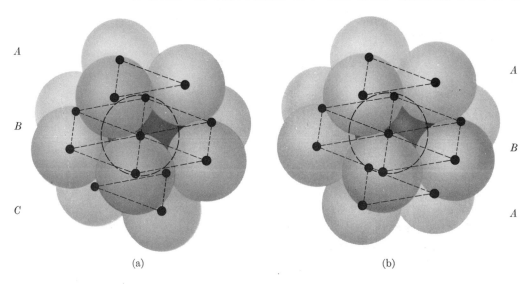

(a)   (b)

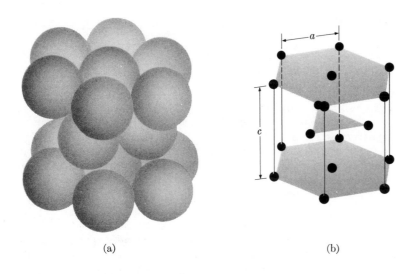

(a)   (b)

Fig. 3–3.2 Hexagonal close-packed (hcp). (a) Hard-ball model. (b) Stacking arrangement. Each atom has 12 close neighbors.

hexagonal lattice (Fig. 3–3.2) rather than a cubic lattice. Because it has the densest packing possible for equal-sized spheres, it is commonly called *hexagonal close-packed, hcp*. (Likewise, fcc is sometimes called *cubic close-packed, ccp*.)

In some respects it is unfortunate that hcp metals have the subtle differences which cause them to crystalize in the hexagonal system rather than as an fcc crystal, because, as we shall see later, plastic deformation is much more restricted in hexagonal systems than in cubic metals. Magnesium, which has a density of only 1.74 g/cm³, would be used much more extensively if it could more readily be formed into wire, rods, beams, etc. Likewise, one of the problems encountered in fabricating zirconium for nuclear reactor applications arises from its restricted plasticity.

A few metals—for example, tin and uranium—are neither cubic nor hexagonal. Each has rather specialized uses as a pure metal, so we shall not be required to give our attention to their structures in this introductory book.

Example 3–3.1. Titanium, which is hcp at normal temperatures, has a radius of approximately 1.46 Å and an APF of 0.74. Calculate the density ρ of titanium.

Solution. Basis: 1 atom.

$$\text{APF} = 0.74 = \frac{(4\pi/3)(1.46 \times 10^{-8} \text{ cm})^3}{x \text{ cm}^3}$$

where x is the total "space" per atom;

$x = 17.6 \times 10^{-24}$ cm³/atom,

$$\rho = \frac{47.9 \text{ g}/0.6 \times 10^{24} \text{ atoms}}{17.6 \times 10^{-24} \text{ cm}^3/\text{atoms}} = 4.5 \text{ g/cm}^3.$$

Comments. Atoms of hexagonal metals are usually not quite spherical. For example, titanium is slightly "squashed" in one direction, with the largest radius equal to 1.47 Å and the shortest radius equal to 1.45 Å. Except for magnesium and titanium, which are slightly compressed, most other hexagonal metals are distorted by being slightly elongated in the hexagonal prism direction. ◀

3–4 IMPURE METALS

In some metals impurities are undesirable; in other metals they are intentionally added as alloying elements. Furthermore, even commercially pure metals have 0.01 to 1 a/o of other kinds of atoms present. In this section we shall pay particular attention to *where* the impurity atoms are located, and what percentage of them may be present. In Chapter 5 we shall look at alloys in which impurities are added intentionally.

Solid solutions. Our concept of solutions from high school chemistry involves liquids. For example, salt dissolves in water and loses its identity. Water absorbs the sodium and chlorine ions of the salt within the structure of the liquid. A similar situation can exist in solids. We shall observe in Chapter 4 that brass is fcc copper in which some ($\simeq 30\%$) of the copper atoms have been replaced by zinc atoms. As shown in Fig. 3–4.1, the fcc structure accommodates the zinc atoms because they are approximately the same size as the copper atoms ($R_{Zn} = 1.39$ Å versus $R_{Cu} = 1.278$ Å), and they have somewhat similar tendencies to release electrons. We call this structure a *substitutional solid solution.* As a rule of thumb, the ability of two dissimilar elements to coexist in a substitutional solid solution is very limited if their radii differ more than 12–15% in size.

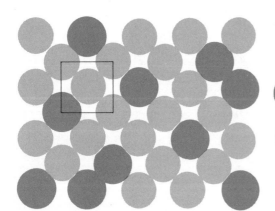

Zinc

Copper

Fig. 3–4.1 Substitutional solid solution (zinc in copper). The substitution of zinc atoms for copper atoms does not alter the fcc structure of the metal.

A structure of one element may also dissolve atoms of a second element to form an *interstitial solid solution*. If the impurity atom is very small compared to the host atoms, it can reside in the vacant spaces (or interstices) among the large atoms. Carbon atoms dissolve in fcc iron in this manner at elevated temperatures (Fig. 3–4.2). However, the carbon atoms are large enough so that there is some crowding. As a result, it is impossible to dissolve carbon atoms into more than 5–10% of the interstices at any one time. The interstices in bcc iron, although they are more numerous than they are in fcc iron, are smaller; thus an interstitial solid solution of iron and carbon at normal temperatures is extremely limited. (See Section 5–4.) In Chapter 6 we shall see that heat treatment of steels involves the dissolving of carbon in fcc iron at high temperatures, followed by various cooling processes in which the carbon atoms come out of solid solution as the iron changes to a body-centered crystal structure.

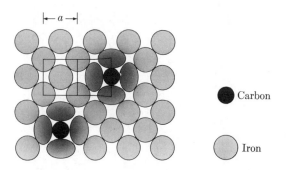

Fig. 3–4.2 Interstitial solid solution (carbon in fcc iron). Although the carbon atoms are smaller than the iron atoms, they are larger than the holes, which necessitates some crowding.

Example 3–4.1. Iron may contain nitrogen as an interstitial impurity. Analysis of one sample shows 0.05 w/o nitrogen; what is the atom percent nitrogen?

Solution. Basis: 100,000 amu = 99,950 amu Fe = 50 amu N. Using the atomic weights of Fig. 1–2.2, we have

$$\frac{99,950 \text{ amu}}{55.85 \text{ amu/Fe atom}} = 1790 \text{ Fe atoms};$$

$$\frac{50 \text{ amu}}{14.01 \text{ amu/N atom}} = 3.5 \text{ N atoms.}$$

Therefore

a/o N = (3.5/1793)100 = 0.2 a/o.

Comments. Only boron, carbon, nitrogen, and sometimes oxygen are small enough to dissolve interstitially into iron. Other common alloying elements, such as nickel, manganese, and chromium, replace the iron atoms in the bcc or fcc structure to form substitutional solid solutions. ◀

Example 3–4.2. Assume no crowding in fcc iron at high temperature. What is the radius of the largest atom which can be introduced interstitially? (The radius of iron at elevated temperatures is 1.29 Å; a = dimension of unit cell.)

Solution. Refer to Fig. 3-4.2.

$$a = 4(1.29 \text{ Å})/\sqrt{2} = 3.65 \text{ Å,}$$

$$r_{\text{hole}} = [3.65 \text{ Å} - 2(1.29 \text{ Å})]/2 = 0.54 \text{ Å.}$$

Comments. A commonly accepted value for the radius of a carbon atom *in iron* is 0.7 Å. Thus there must be some local strain of the iron lattice around each carbon atom (Fig. 3–4.2).

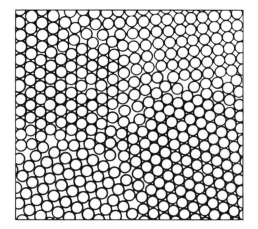

Fig. 3–5.1 Grains and grain boundaries. Each grain is an individual crystal. The grain boundaries are paths for diffusion; they affect deformation, and they modify properties in other ways. Therefore they are important in understanding materials. (Clyde Mason, *Introductory Physical Metallurgy.* Metals Park, O.: American Society for Metals)

Fig. 3–5.2 Growth of grains from liquid. Grains grow from various nuclei, producing a grain boundary where they meet.

The 2.58 Å between the centers of the iron atoms is larger than given in Table 3–1.1 because (1) there is thermal expansion, and (2) there is an increase in radius when there are 12 neighbors (fcc) instead of 8 neighbors (bcc). ◀

3–5 GRAINS AND GRAIN BOUNDARIES ⓔ

Only in special cases does an engineering product contain just one crystal. Usually, a metal such as a copper wire is composed of many crystals packed together to fill space. Small parts of the metal, called *grains*, are individual crystals in which all the atoms are in a lattice of one specific orientation. An adjacent grain of a single-component material has the same crystal lattice but a different orientation so that there is a zone of mismatch where the two join (Fig. 3–5.1). This mismatched zone, called a *grain boundary*, markedly affects many of the characteristics of metals.

Most materials are *polycrystalline*, i.e., they have many grains, because crystals grow simultaneously from various nuclei. As they grow they encounter other growing crystals until all the material is in one or another of the grains (Fig. 3–5.2).

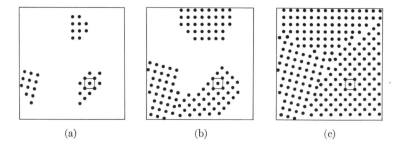

(a)   (b)   (c)

As shown in Fig. 3–5.1, the grain boundary is less dense than the grain proper. This means that other atoms, e.g., carbon in iron, can diffuse more readily through the grain boundary than in the bulk of the crystal. At normal temperatures the grain boundary interferes with plastic deformation; therefore *fine-grained* materials are generally stronger than coarse-grained ones. [At elevated temperatures, on the other hand, the grain boundary facilitates *creep*, which we shall encounter in Section 4–10; therefore it is desirable to have coarse-grained metals for applications such as jet turbine blades (Fig. 3–5.3).]

The atoms at grain boundaries naturally have fewer neighbors than atoms within the grains. (Cf. Fig. 3–5.1.) Since energy is *required* to

Fig. 3–5.3 Coarse-grained metal in turbine blades. These blades are used at high temperatures; therefore the grain-boundary area is minimized by the development of coarse grains in processing. In contrast, grain boundaries increase the strength at low temperatures. (Courtesy M. E. Shank, Pratt and Whitney)

pull atoms apart, or *released* as they come together, the less densely packed atoms along the grain boundaries have *more* energy than those within the grains. This means that the atoms along the boundary are not held as tightly, and are more reactive, than those within the grain. Corrosion, for example, progresses more rapidly along grain boundaries.

Grain size and grain boundary area. ◎ Figure 3–5.4(a) shows a polished and etched "section" of metal. The grain boundaries were corroded by the etchant; as a result they do not reflect light directly back into the microscope (Fig. 3–5.4b) and they appear as dark lines. Since

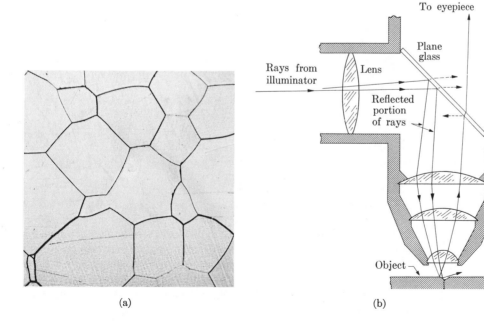

(a)   (b)

Fig. 3–5.4 Photomicrograph of metal. (a) Molybdenum (× 250), polished and etched. (Courtesy O. K. Riegger) (b) The grain boundaries of (a) are dark because, after being etched, they do not reflect light through the microscope. (Bruce Rogers, *The Nature of Metals*, Metals Park, O.: American Society for Metals)

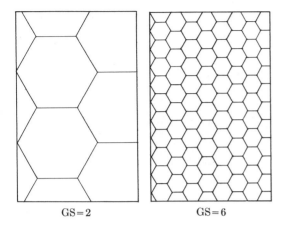

GS=2   GS=6

Fig. 3–5.5 ASTM grain size nets. (a) GS # 2 has 2 grains per in^2 at ×100. (b) GS # 6 has 32 grains per in^2 at ×100. The number of grains per in^2 at ×100 is 2^{n-1}, where n is the GS number.

the grain size and grain boundary area of a metal are important, the materials engineer has devised ways of indexing grain size. The American Society for Testing and Materials (ASTM) has established the reference nets shown in Fig. 3–5.5 with *grain size numbers* (GS #) for comparison with observation through a microscope at × 100, that is, when the linear dimension is magnified 100 times. We shall refer to these grain size numbers many times in subsequent chapters.

The *grain boundary area* can be readily calculated. Consider the circle in Fig. 3–5.6, which is shown at × 250 and therefore is 0.025 in. in circumference. If that circle is randomly placed on the photomicrograph, the number of times per unit length which it intercepts the grain boundary, P_L, is simply equal to one-half the boundary area per unit volume, S_V, or

$$S_V = 2P_L \tag{3–5.1}$$

We shall not prove this, but will use it in the following example.

Example 3–5.1 ◎.

a) Determine the grain boundary area per cubic inch in the molybdenum of Fig. 3–5.6.
b) What is the ASTM grain size number?

Solution

a) The circumference of the 2.00-in. circle intercepts 11 boundaries at × 250. Therefore P_L equals 11/(2.00π in./250) = 440/in.:
$S_V = 2(440/\text{in.}) = 880$ in.^{2}/in.^{3} (= 350 cm^2/cm^3).
b) At × 250, the 2.3 × 2.3 in.^{2} area contains 17 grains, i.e., (corners × $\frac{1}{4}$) + (edges × $\frac{1}{2}$) + inside. Therefore, at × 100, where the area would be 0.92 × 0.92 in.^{2}, there would be approximately 20 grains/in.^{2}, or 2^{5+-1}. Thus GS # $\cong$ 5. (See Fig. 3–5.5 legend.)

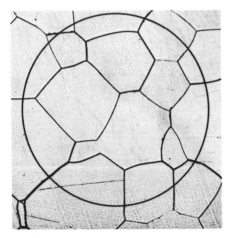

Fig. 3–5.6 Calculation of grain boundary area. See Example 3–5.1 (× 250) and Fig. 3–5.4.

Comments. The circle of Fig. 3–5.6 must be at least three times as large as the largest grain in order to get a good statistical sample of P_L. Otherwise, a random sampling procedure is required.

Often we speak of grain size in terms of a typical "diameter." Thus the grains in Fig. 3–5.6 could be said to have an average grain size diameter of ~ 0.07 mm (after the magnification is taken into account). Note, however: (a) the grains are not spherical; and (b) all the grains do not have the same "diameter." This latter difference may arise in part because we are "catching only the top corner" of some of the grains, and in part because the grains do vary in size. In general, grain "diameters" do not offer a fully satisfactory measure of either the average size of the grains or the grain boundary area. ◄

REVIEW AND STUDY

3–6 SUMMARY

Solid metals are crystalline, i.e., the atoms have a repeating pattern. Most metallic materials which contain only one type of element crystallize into one of three types of lattices: fcc, bcc, and hcp. We can calculate the density of these metals with considerable accuracy. The fcc and hcp metals are close-packed, but have different symmetries, and therefore markedly different deformation characteristics.

Metals, like other solids, can dissolve impurity atoms into their structures. If the impurity atoms are small, they may distribute themselves interstitially among the metal atoms. The interstitial solution of carbon in iron will be the basis for studying heat treatment of steel in later chapters. If the impurity atom is very similar in size to the predominant atom, it may dissolve by substituting for one of the regular atoms in the crystal. Many alloys are designed to have this characteristic because solid solutions modify the properties of metals.

All engineering metals are polycrystalline, which means that they are composed of a number of grains, a grain being an individual crystal. The boundaries between the grains markedly affect the behavior of metals.

Terms for Study

Atomic packing factor, APF	Grains
Atomic radius, r or R	Grain boundary
Avogadro's number, AN	Grain boundary area, S_v
Body-centered cubic, bcc	Grain size number, GS #
Component	Hard-ball model
Coordination number, CN	Hexagonal close-packed, hcp
Crystal	Lattice point
Crystal lattice	Polycrystalline
Cubic crystals	Solid solution, interstitial
Density, ρ	Solid solution, substitutional
Equiaxed	Symmetry
Face-centered cubic, fcc	Unit cell

STUDY PROBLEMS

3–1.1 Calculate the number of atoms per cm^3 of aluminum.

Answer. $6 \times 10^{22}/cm^3$.

3–1.2 What is the atomic packing factor for fcc aluminum?   *Answer.* 0.74

3–1.3 (a) Calculate the number of atoms per cm^3 of chromium. (b) What is the atomic packing factor?   *Answer.* (a) $8.3 \times 10^{22}/cm^3$ (b) 0.68

3–2.1 Calculate the density of lead (fcc).

Answer. 11.4 g/cm^3 (versus 11.34 g/cm^3 by experiment)

3–2.2 Calculate the density of chromium (bcc).

Answer. 7.25 g/cm^3 (versus 7.19 g/cm^3 by experiment)

3–2.3 Measurements indicate that if fcc iron could be cooled quickly enough so that it did not transform from fcc to bcc, then its fcc unit cell dimension, a_{fcc}, would be 3.59 Å. Compare the resulting density ρ with that of bcc iron which has a value for a_{bcc} of 2.861 Å.

Answer. $\rho_{fcc} = 8.1$ g/cm^3, $\rho_{bcc} = 7.9$ g/cm^3 (vs. 7.87 g/cm^3 by experiment)

3–3.1 Assume that the density of magnesium is 1.74 g/cm^3. Calculate the volume of a unit cell of hcp magnesium which contains six atoms.

Answer. 140 $Å^3$ (versus 139 $Å^3$ by experiment)

3–3.2 The hexagonal unit cell of zinc (Fig. 3–3.2a) is 4.94 Å in height and 2.665 Å along each edge of the base. (a) How many atoms are there per unit cell? (b) What is the volume of the unit cell? (c) Calculate the density of zinc.

Answer. (a) 6 (b) 9.1×10^{-23} cm^3 (c) 7.17 g/cm^3 (versus 7.135 by experiment)

3–4.1 Impure copper contains 0.4 a/o oxygen. What is the weight percent?

Answer. 0.1 w/o

3–4.2 Copper contains a trace (0.08 w/o) of silver. What is the atomic percent Ag?

Answer. 0.047 a/o Ag

3–4.3 How much would the edge of a bcc iron unit cell be expanded if a carbon atom ($r_C = 0.7$ Å) were forced into a position between the two corner atoms ($R_{Fe} = 1.24$ Å)?

Answer. 1.02 Å (or ~ 35 l/o)

3–5.1 (a) Assume that the GS #6 of Fig. 3–5.5 represents a 2-D cut through a polycrystalline solid. Determine the corresponding grain boundary area. (b) Repeat for GS #2.

Answer. (a) ~ 1460 $in.^2/in.^3$ (b) ~ 380 $in.^2/in.^3$

3–5.2 (a) How many grains are there per square inch (at $\times 100$) for GS #8? (b) GS #5? (c) GS #2?

Answer. (a) 128 (b) 16 (c) 2

CHARACTERISTICS
OF
METALLIC
PHASES

CHAPTER 4

4–1 SINGLE-PHASE METALS

In this chapter we shall consider the characteristics of those metals which have only one structural arrangement of atoms, i.e., they are single-phase. The behavior of these metals depends on their properties, plus (1) the magnitude and directions of applied forces, (2) electric fields, (3) temperature and thermal gradients, and (4) other chemical and radiation conditions of the environment. Thus this chapter will pay attention not only to properties as they relate to structure, but also how properties and structure vary with processing and/or service conditions. In Chapters 5 and 6 we shall discuss multiphase alloys.

Commercial single-phase alloys. In Section 3–1 we noted that copper electric wire, sheet metal in car fenders, and aluminum in kitchen utensils are single-component metals. They are copper, iron, or aluminum, respectively, with only unavoidable minor fractions of other elements present. If special attention is directed to reducing impurities during production, we speak of metals as being *commercially pure.*

The word *alloy* is used when additional components are intentionally added to a metal. Many commercial alloys involve only one phase because the second component is dissolved into a solid solution (Section 3–4). *Brass* is the most familiar of these alloys. The usual kind of brass (70–30) contains 70 w/o Cu and 30 w/o Zn. Like pure copper, brass is fcc, because the zinc atoms simply substitute for the copper atoms; however, when 40% of zinc is added, the composition exceeds the solubility limit. We shall look at such alloys in Chapter 5.

Brass is attractive to the design engineer because (1) the solid solution of copper and zinc is stronger than copper alone, and (2) zinc is cheaper than copper. Therefore, he can simultaneously specify a stronger and cheaper alloy for those applications in which copper and brass are equally suitable otherwise. (Although brass is stronger than pure copper, the solid solution of zinc in copper has a much lower conductivity than pure copper. Therefore we never find the cheaper brass used for electrical wiring.)

Other common single-phase alloys include (a) Cu-Ni alloys, (b) *bronze*, which is copper plus tin,* (c) *sterling silver*, which is silver plus copper, and (d) stainless steel which is either bcc (Fe + Cr) or fcc (Fe + Cr + Ni). The copper-nickel alloys may be mixed in all ratios and still retain only one phase because the two metals are very similar (both are fcc, both can form divalent ions, and their radii are 1.278 Å and 1.243 Å, respectively). The remainder of the metals cited above are like brass in that there is a solubility limit of the minor metal in the alloy.

Example 4–1.1. Nickel costs approximately twice as much per pound (or per gram) as copper. Determine the cost ratio of a rod made of pure copper 0.5 in. in diameter and a similar rod made of monel (70Ni-30Cu).

Solution. Data are available in Appendix B. Basis: 100 cm³ of metal (= 890 g Cu and 880 g monel). Let P = price of copper per gram and $2P$ = price of nickel per gram.

Cost of copper = $890P$.

Cost of monel = $(0.3)(880)(P) + (0.7)(880)(2P)$

$$= 1496P.$$

Cost ratio = 0.595.

Comments. The relative cost of nickel, copper, and other metals varies in this country with the supply and demand of each. ◄

MECHANICAL BEHAVIOR

4–2 ELASTIC DEFORMATION OF METALS ⓔ

The bonding forces holding metal atoms together are strong. Therefore it takes considerable tensile stress to produce even a minor amount of strain. The ratio of tensile stress to tensile strain in the linear elastic range is called *Young's modulus*, Y:

$$Y = \sigma/\varepsilon. \tag{2–2.3}$$

This modulus ranges from 2,000,000 psi for lead to 57,000,000 psi for tungsten (Table 4–2.1).

An examination of Table 4–2.1 reveals that there is a close correlation between Young's moduli and the melting temperatures. This

---

Table 4–2.1

ELASTIC MODULI (YOUNG'S MODULI) OF CUBIC METALS

Metal	Melting temperature, °C (°F)	Average elastic modulus,* psi
Tungsten, W	3410 (6170)	57×10^6**
Molybdenum, Mo	2610 (4730)	47
Chromium, Cr	1875 (3407)	36
Iron, Fe (bcc)	1537 (2798)	30
Nickel, Ni	1453 (2647)	30
Copper, Cu	1083 (1981)	16
Aluminum, Al	660 (1220)	10
Lead, Pb	327 (621)	2

* The elastic modulus varies with orientation. For example, in bcc iron, the cube diagonal has a modulus of 41×10^6 psi; the cube edge has a modulus of only 18×10^6 psi. The average figures of this table are appropriate for polycrystalline metals.
** Multiply by 6900 to obtain newtons/m².

---

* Unless specifically designated otherwise, bronze is a Cu-Sn alloy. An Al-bronze is a copper-aluminum alloy, and a Mn-bronze is basically a Cu-Mn alloy.

should be expected, since a more strongly bonded material not only requires a higher stress to produce strain, but also a higher temperature to produce the thermal agitation that destroys the crystal lattice, i.e., to produce melting. [The correlation between elastic modulus Y and melting temperature T_m, shown in Table 4–2.1, is much better for metals than for other materials, in which a variety of bonds may be present. Therefore one should hesitate to predict the modulus elasticity for polymers and ceramics on the basis of their melting points alone.]

Compressive moduli. Values of compressive Young's moduli, where both the stress and the strain have negative signs, are, for metals, essentially identical to values obtained in tension. The atoms are pulled together naturally by attractive forces until they encounter equally strong repulsive forces. In the small strain range encountered in metals ($< \pm 0.01$ in./in.), the forces required for additional compression are equivalent to the forces required for comparable tensile strain. Thus, in calculations for metals, we do not need to concern ourselves with different modulus values in compression and tension.

Effect of temperature. As long as the structure does not change, the elastic modulus decreases at higher temperatures (Fig. 4–2.1). The discontinuity in the curve for iron is due to the change of iron from bcc to fcc at 1670°F (910°C). Since the atoms of fcc iron have twelve neighbors rather than eight, as in bcc iron, the fcc structure is more rigid (has a higher modulus) and responds less to applied stresses. In fact, the normal decrease in moduli with increased temperatures may be explained in part by the thermal expansion, which allows more strain for a given stress (lower modulus).

Poisson's ratio. A tensile elongation produces an accompanying elastic contraction in the two lateral directions (and a compression in one direction produces an accompanying elastic expansion in the other two directions, Fig. 4–2.2). If an elastic strain involved absolutely no change in volume, the two lateral strains would be equal to one-half the tensile strain (but of the opposite sign). However, when we elongate a material in one direction, the cross-sectional area is not reduced quite as much proportionately. Therefore the ratio of strains, $-\varepsilon_x/\varepsilon_z$, is always less than 0.5. This ratio is called *Poisson's ratio*, v, and ε_z is the strain in the direction of applied load. The two lateral strains, ε_x and ε_y, are equal if the stress is applied in only one direction,

$$v = -\varepsilon_x/\varepsilon_z. \qquad (4\text{–}2.1)$$

Under tension the signs are

$$v = -(-)/(+);$$

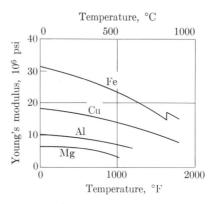

Fig. 4–2.1 Elastic modulus (Young's) versus temperature. In general, the elastic modulus of metals decreases at elevated temperatures. Exceptions occur only when there is a change in structure. (A. G. Guy, *Elements of Physical Metallurgy*, Reading, Mass.: Addison-Wesley)

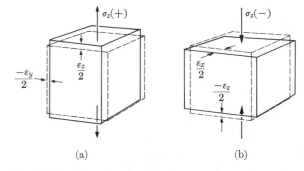

Fig. 4–2.2 Poisson's ratio. This is the ratio between lateral strain and axial strain (Eq. 4–2.1). With no volume change, $v = 0.5$. Normally the volume changes slightly, so $v < 0.5$.

Table 4–2.2

POISSON'S RATIO OF CUBIC METALS

Metal	Poisson's ratio
Tungsten	0.28
Iron and steel	0.28
Copper	0.35
Aluminum	0.34
Lead	0.4
Brass	0.37

and with compression, where ε_z is negative,

$$v = -(+)/(-).$$

Therefore Poisson's ratio is always positive. Typical values are shown in Table 4–2.2.

Shear modulus. ◎ To date we have considered only stresses which are axial, i.e., the opposing forces are perpendicular to the area on which they act. In many engineering applications there are shear forces in which the opposing forces have components which are parallel to the area of interest. As shown in Fig. 4–2.3, a *shear stress* τ produces a strain angle α. The *shear strain* γ is defined in terms of that angle,

$$\gamma = \tan \alpha, \tag{4–2.2}$$

and the amount of elastic strain is initially proportional to the amount of elastic shear stress,

$$G = \tau/\gamma. \tag{4–2.3}$$

This shear modulus G is different from Young's modulus Y; however, the two are related as follows:

$$Y = 2G(1 + v). \tag{4–2.4}$$

Since Poisson's ratio v is normally between 0.25 and 0.5 (Table 4–2.2), the value of G is between 0.4 and 0.33 of Y.

Example 4–2.1. A precisely machined steel rod has a specified diameter of 0.7321 in. (1.860 cm). It is to be loaded longitudinally with a force of 24,400 lb. How much will its diameter d change? Its area A?

Solution

$$A_0 = (\pi/4)(0.7321 \text{ in.})^2 = 0.42095 \text{ in.}^2,$$

$$\varepsilon = \sigma/Y = (F/A)/Y.$$

$$= \frac{(24,400 \text{ lb})/(0.42095 \text{ in.}^2)}{30,000,000 \text{ psi}} = 0.002 \text{ in./in.}$$

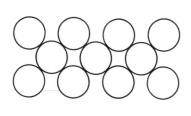

(a)

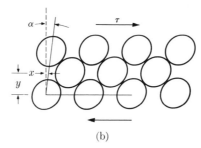

(b)

Fig. 4–2.3 Elastic shear strain. (a) No strain. (b) Shear strain. The strain is elastic as long as the atoms maintain their original neighbors.

$\varepsilon_x = -(0.28)(0.002) = -0.00056.$

$d_f = 0.7321 - (0.00056)(0.7321) = 0.7321 - 0.0004 = 0.7317.$

$\Delta A = 0.42095 \text{ in.}^2 - 0.42048 \text{ in.}^2 = 0.00047 \text{ in.}^2 \ (\cong 0.1 \text{ area percent}).$

Comments. This change in area is smaller than the indicated accuracy of the load; therefore a correction does not need to be made to determine the actual stress. However, if there was a reason for specifying a ± 0.0001 in. dimension, the engineer should point out that the contraction exceeds that limit. ◄

Example 4–2.2. Refer to Example 2–6.1, but assume that Y drops from 16,000,000 psi, at 68°F (20°C) to 15,500,000 psi at 122°F (50°C). What is the change in total length ΔL if the stress remains constant at 24,000 psi while the temperature rises 54°F?

Solution. Elastic strain at 68°F:

$\Delta L/L = +24,000 \text{ psi}/16,000,000 \text{ psi} = +0.00150 \text{ in./in.}$

(Zero load at 68°F to zero load at 122°F) + elastic strain at 122°F:

$\Delta L/L = (9 \times 10^{-6}/°F)(+54°F) + (24,000 \text{ psi}/15,500,000 \text{ psi})$

$= 0.00049 + 0.00154 = 0.00203 \text{ in./in.}$

Therefore

$\Delta L = (0.00203 - 0.00150 \text{ in./in.})(62 \text{ in.}) = 0.033 \text{ in.}$

Comments. Designs for many high-temperature applications—e.g., turbine blades (Fig. 3–5.3)—must take into account both thermal expansion and the change in Young's modulus. ◄

4–3 PLASTIC DEFORMATION OF METALS

A crystal may be viewed as collections of parallel planes containing layers of atoms (Fig. 4–3.1). Plastic deformation commonly occurs by the slip of one plane over another. The plane does not slip all at once, but with the localized motion shown schematically in Fig. 4–3.2,

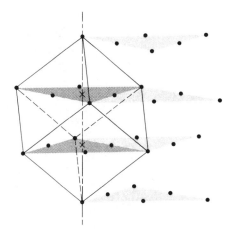

Fig. 4–3.1 Crystal planes. The atoms have a regular arrangement on each plane. The planes are stacked and may slide over one another by shear stresses.

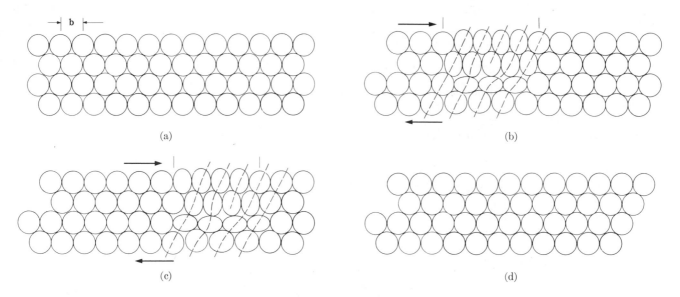

(a)

(b)

(c)

(d)

Fig. 4–3.2 Plastic deformation by shear. The edges of four planes are shown. Slip of this type does not occur simultaneously along the total slip plane. Rather, the shear force produces an imperfection, called a *dislocation,* which moves along the slip plane.

where the edge rows of four planes are shown responding to shear forces. Note that in the shear process only a portion of the atoms are displaced at any one time, and that in effect extra atoms are squeezed into the upper rows. The crystal is not perfect in this region. We call this type of imperfection an *edge dislocation.*

Slip is very nonuniform within a polycrystalline solid for several reasons: (1) When tensile (or compressive) stresses are applied, they must be resolved as shear stresses onto the slip planes (Fig. 4–3.3). The greatest shear stress develops at 45° to the direction of tension; no shear stress is developed on planes parallel or perpendicular to the direction of tension. Thus slip can occur on only some of the crystal planes within a metal grain. (2) Some grains are oriented more favorably for deformation than others. (3) Plastic deformation results from dislocation movements, and occurs with lower applied stress on crystal planes which are farthest apart and in directions in which the dis-

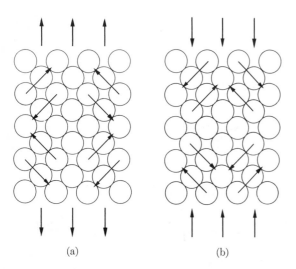

(a)   (b)

Fig. 4–3.3 Resolved shear stresses. A tensile stress (a), or a compressive stress (b), also provides shear stresses which are at a maximum at 45°.

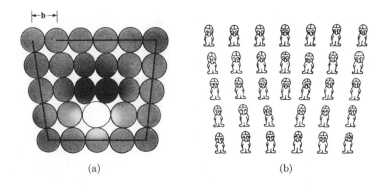

Fig. 4–3.4 (a) Edge dislocation. Strain accompanies an edge dislocation with compression (dark) on one side and tension (light) on the other. (b) For fun! (Courtesy L. S. Darken in the style of L. Matteoli)

placement step, b, is shortest. In other words, slip occurs preferentially on the most densely packed planes (most atoms per unit area), and in the most densely packed directions of those planes (most atoms per unit length).

Strained zones of compression and tension accompany a dislocation (Fig. 4–3.4). Here the atoms are not separated at the equilibrium distance listed in Table 3–1.1, the distance found elsewhere in the crystal. These distortions require extra energy. This fact will help us explain, in the next section, why solid solutions are stronger than pure metals.

Example 4–3.1. Refer to Fig. 4–3.5, which shows three different sets of planes commonly referred to in an fcc crystal. Assume that the lattice is that of copper, with $a = 3.61$ Å. What are the perpendicular distances p_{010}, p_{110}, and $p_{\bar{1}11}$ between the sets of planes labeled (010), (110), and ($\bar{1}$11), respectively?

Fig. 4–3.5 Planes (low-index) in an fcc metal. (a) (010), (b) (110), (c) ($\bar{1}$11).

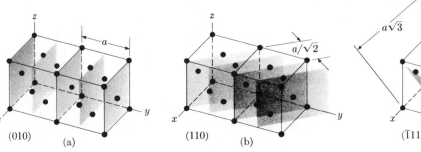

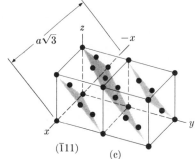

Solution

$p_{010} = a/2 = 1.805$ Å,

$p_{110} = a\sqrt{2}/4 = 1.278$ Å,

$p_{\bar{1}11} = a\sqrt{3}/3 = 2.08$ Å.

Comments. Note that the (010) planes cut the *y*-axis only. The (110) planes cut the *x*- and *y*-axes at equal distances from the origin, but are parallel to the *z*-axis. The (111) planes cut all three axes at equal distances from the origin. We call these notations for planes *Miller indices*.

In fcc metals, slip occurs almost exclusively on (111)-type planes, which have greater interplanar spacings than the (010) and (110)-type planes. There are *four* (111)-type planes with various orientations around the origin.

On a (111)-type plane, the shear stress is lowest in the direction of the *face diagonal* of the cube. In those directions, the atoms touch (cf. Fig. 3–2.2), and therefore have the shortest displaced distance *b* for dislocations (cf. Fig. 4–3.2). Each of the four (111)-type planes has *three* of these easy directions (each with a plus and a minus sense). Consequently an fcc crystal has 4 × 3, or twelve, possible slip combinations. This is one reason why fcc metals are among the most ductile of materials. ◀

Example 4–3.2. What is the copper atom packing, i.e., atoms/cm², on the (111)-type planes shown in Fig. 4–3.5?

Solution no. 1. There are 4 atoms per unit cell, which is a $(3.61 \times 10^{-8}\text{cm})^3$ cube. Therefore, one calculates 0.85×10^{23} atoms/cm³ to match the figure we obtained in Section 3–1.

From Example 4–3.1, the interplanar spacing $p_{\bar{1}11}$ is 2.08 Å. Thus there are 0.48×10^8 planes per centimeter.

$$\text{Atoms/cm}^2 = \frac{0.85 \times 10^{23} \text{ atoms/cm}^3}{0.48 \times 10^8/\text{cm}} = 1.8 \times 10^{15} \text{ atoms/cm}^2.$$

Solution no. 2. The ($\bar{1}11$) plane cuts each unit cell as an equilateral triangle which is $a\sqrt{2}$ (or 5.1 Å) on each side. If one draws a sketch, it can become apparent that this triangle contains $(\frac{3}{6} + \frac{3}{2})$, or two atoms:

$$\text{Atoms/cm}^2 = \frac{2 \text{ atoms}}{\frac{1}{2}[5.1 \text{ Å}][0.866 \times 5.1 \text{ Å}]} = 1.8 \times 10^{15} \text{ atoms/cm}^2.$$

Comments. The (010) and (110) planes of copper have 1.5×10^{15} atoms/cm² and 1.1×10^{15} atoms/cm², respectively. Slip occurs most readily on those planes with the highest atom packing, as well as in directions with closest atom contact. ◀

Example 4–3.3. ◎ A force of 100 pounds is directed along the *x*-axis of Fig. 4–3.5(b). What is the resolved force in the direction of the body diagonal?

Solution

$$\frac{\text{Diagonal force}}{\text{Force along axis}} = \text{cos of angle between directions};\qquad (4\text{--}3.1)$$

Diagonal force $= (100\ \text{lb})(a/a\sqrt{3}) = 58\ \text{lb}.$

Comments. If we were to calculate a resolved *stress*, we would have to take into account not only the resolved force, but also the resolved area. (See Schmid's law in other textbooks on materials science.) ◄

4–4 STRENGTHENING OF METALS

We usually want metals to be as strong as possible. There are three widely used methods of strengthening single-phase metals: (1) strain hardening, (2) solution hardening, and (3) grain-size control.

Strain hardening. In Section 2–3 and Fig. 2–3.1, we observed that added strength (and hardness) develops in ductile materials when they are deformed into the plastic range. We use the term *cold work* because this hardening usually comes about during the processing, or "working," of metals which have not been heated. Figure 4–4.1(a) shows how the hardness of iron, copper, and 70-30 brass is affected by the amount of cold work (%CW), as measured by the change in cross-sectional area A.

$$\%\text{CW} = \left[\frac{A_o - A_f}{A_o}\right] 100. \qquad (4\text{--}4.1)$$

Dislocations are generated in materials as they are plastically deformed; indeed dislocations are necessary for plastic deformation. However, as more and more dislocations are accumulated in the metal, they become entangled in each other, requiring a greater stress to

Fig. 4–4.1 Strain hardening. Cold work (a) increases the hardness, (b) increases the strength, and (c) decreases the ductility.

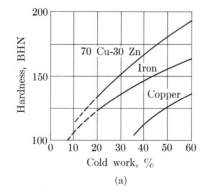 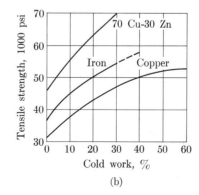 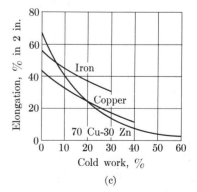

(a)   (b)   (c)

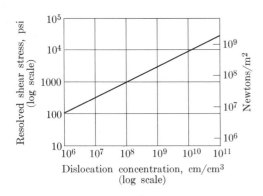

Fig. 4–4.2 Shear stress versus dislocation density (copper). Slip occurs through dislocation *movements*. Therefore a high shear stress is required for deformation when (1) *no* dislocations are present, or (2) the numbers of dislocations are such that they become *entangled*. In the latter cases strain hardening (cold work) results. [Adapted from H. Wiedersick, "Hardening Mechanisms and Theory of Deformation," *J. Metals,* **16,** (1964) page 425.]

keep them moving. Figure 4–4.2 shows how the shear strength and the dislocation content are related. Figure 4–4.1(b) indicates the tensile strength in terms of the amount of cold work. Strain hardening resulting from cold work may be removed by *annealing* (Section 4–9).

Solution hardening. A solid solution is stronger than a pure metal. This is shown in Fig. 4–4.3, and accounts for the use of brass, bronze, and many single-phase, two-component alloys as engineering materials.

The strengthening is accomplished because a larger *or* a smaller atom reduces the strained region around a dislocation (Fig. 4–4.4). Recall from the previous section that extra energy is required for these strained regions. Thus if a dislocation encounters an impurity atom, the strained volume is reduced. It resists moving beyond the impurity atom because the strained region would have to be enlarged again. Consequently a larger stress is required for continued deformation.

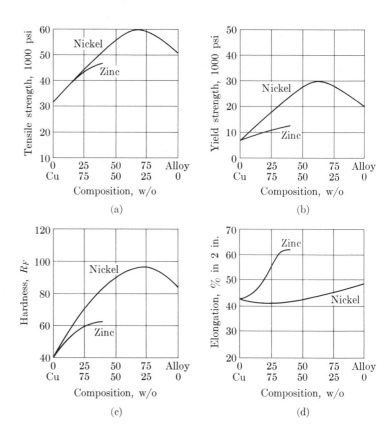

Fig. 4–4.3 Solution hardening (copper alloys). Impurity atoms such as Zn, Ni, or Sn serve as anchor points for dislocations. Therefore a greater shear stress is required.

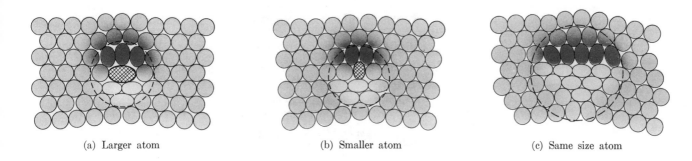

(a) Larger atom   (b) Smaller atom   (c) Same size atom

Either component of a two-component alloy is strengthened by the addition of the second component. (See the Cu-Ni alloys of Fig. 4–4.3.) This means that some intermediate compositions have the maximum strength. In the Cu-Ni alloy system, this is *monel* metal at 70Ni-30Cu. This maximum would probably be between 40 and 50% zinc in Cu-Zn alloys, except for the fact that the solubility limit for zinc in the fcc brass structure is reached first.

Effect of grain size on strength. (e) Plastic deformation occurs by slip along crystal planes. A grain boundary is a discontinuity where the slip plane ends; the planes in the next grain are oriented differently (Fig. 4–4.5). Therefore a grain boundary interferes with movements of dislocations along the slip plane. Fine-grained materials are thus stronger than coarse-grained materials because their grain boundary area is larger (Fig. 4–4.6). This relationship applies at normal service temperatures. We shall observe in Section 4–10 that grain boundaries facilitate creep at elevated temperatures.

Fig. 4–4.4 Impurity atoms and dislocations. An odd-sized atom, larger *or* smaller, decreases the strain around a dislocation. As a result, an increased shear stress is required to move the dislocation beyond the impurity atom.

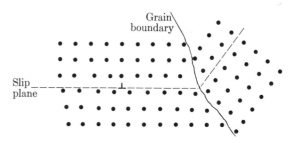

Fig. 4–4.5 Misorientation across a grain boundary. The grain boundary interferes with dislocation movements. Therefore, at normal temperatures, at which the dislocation cannot leave its original plane, fine-grained materials are stronger than coarse-grained ones.

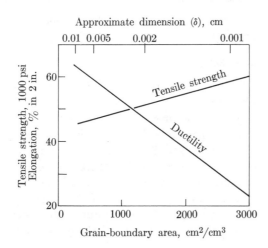

Fig. 4–4.6 Mechanical properties versus grain-boundary area (annealed 70-30 brass).

Example 4–4.1. A copper wire must have a tensile strength of at least 41,000 psi, a ductility of at least 20% elongation (2 in.), and be 0.08 in. in diameter. Propose the necessary processing.

Solution. From Fig. 4–4.1, we observe the following requirements.
For TS > 41,000 psi: CW > 14%.
For elongation > 20% (2 in.): CW < 25%.
Use 20% CW to obtain final diameter $D_f = 0.08$ in.

$$0.20 = \frac{(\pi/4)(D_o)^2 - (\pi/4)(0.08 \text{ in.})^2}{(\pi/4)(D_o)^2} ;$$

$D_o = 0.09$ in.

(Select a 0.09-in. wire, then cold work it 20% to 0.08 in.)

Comments. We determine a specification "window" for an allowable working range from Fig. 4–4.1. In the absence of other requirements, we choose a percent CW near the center. This permits some latitude in processing. In the next example, there will be an economic incentive to work close to one side of the "window."

The initial wire with a diameter of 0.09 inch must be annealed so that it contains no strain-hardening. We shall discuss this in Section 4-9. ◀

Example 4–4.2. Specifications for a product which you manufacture call for a Cu-Ni alloy with a tensile strength of at least 54,000 psi and a yield strength of at least 20,000 psi. The company you own buys six tons of alloy per year for this product. What kind of alloy would you purchase?

Solution. From Fig. 4-4.3, we determine the alloy as follows.
For TS > 54,000 psi: 45% Ni ↔ 90% Ni.
For YS > 20,000 psi: 30% Ni ↔ 100% Ni.

Use 45% nickel (55% copper); this alloy meets both specifications and is cheaper than an alloy with a higher percentage of nickel.

Comments. If you are uncertain about the relative cost of copper and nickel, think of the change you have in your pocket. ◀

4–5 MECHANICAL FAILURE OF METALS ⓔ

Of course the engineer designs his products so that they will *not* be likely to fail. However, failures do occur—sometimes because service conditions become more severe than anticipated (e.g., lamp posts are not, and probably should not, be designed to withstand the impact of a careening car). Sometimes the failure follows deterioration in service (e.g., a worn rail is subject to fracture unless it is replaced). Sometimes fracture is a result of a designer's insufficient knowledge about service stresses. Finally, sometimes failure arises because quality control was not absolutely 100% perfect during the processing of the metal. Whatever the cause of failure, the engineer can control and/or prevent it better if he understands its causes and nature. Details of design and control of metal failure must of necessity be presented in texts which are specifically focused on such problems. However, we shall take time in this section to cite the two prime causes of mechanical failure: *fracture* and *fatigue*. In Section 4–10 we shall also briefly discuss *creep*.

Fracture. Some metals break with little or no plastic deformation. Others have enough ductility to deform considerably before a final abrupt failure. We shall use the terms *brittle fracture* and *ductile fracture* to describe these two types of failure. [In rare instances at normal temperatures, and somewhat more commonly at elevated temperatures, a metal may have a ductility of 100% Red. of A., i.e., plastic deformation continues and no fracture occurs prior to complete *rupture*.]

Almost universally the engineer prefers ductile fracture to brittle fracture in the materials he uses. (1) When there is ductile fracture, inspection can reveal failure in progress before final fracture occurs. Under many conditions, corrections can be made. (2) Energy is consumed by ductile fracture. Thus the energy that accompanies the impact type of failure may be dissipated.

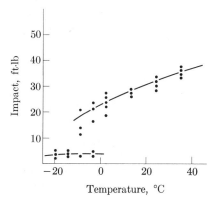

Fig. 4–5.1 Brittle-ductile transition (Charpy test). At lower temperatures, a brittle fracture starts, and propagates faster than plastic deformation can occur. The transition temperature is lowered through metallurgical control of the grain size. (Adapted from N. A. Kahn and E. A. Imbembo, American Society for Testing and Materials)

Almost all fcc metals fail ductilely. This is not always true for bcc metals, which have a *transition temperature* below which they fracture brittlely, i.e., with only slight consumption of energy. The energy absorption during failure may drop by a factor of 4 or more at a transition temperature interval of only 10–20°F (Fig. 4–5.1). During World War II, before the importance of transition temperatures was understood, and when fully welded ships were first used, some of the steels used had transition temperatures above 0°C, the temperature of winter ocean travel. A crack once started at some point of stress concentration, e.g., a corner of a hatchway, could proceed until it ran out of ship. The results were obviously disastrous. Nowadays we know that design plays a big role in avoiding such catastrophic failures. Furthermore, it has been learned that *fine-grained* steels have transition temperatures which are lower than those of coarse-grained steels. Through the use of these two options, the problem just described for ships is now under general control. However, modern cryogenic applications continue to demand attention to this design consideration.

Fatigue. ◎ Man has long been aware that certain materials fail if they remain in service a long time. The layman assumed that the metal got "tired," so he began to speak of fatigue failure.

The most familiar type of fatigue occurs with cyclic stresses.* We are aware that a thin sheet can be broken if we repeatedly bend it back and forth. More subtle but equally important stress cycles occur in a loaded rotating shaft or a vibrating plate. In one position the surface of the shaft is in tension; 180° later (0.017 sec later at 1800 RPM), it is on the other side and in compression. Under such conditions, 10^6 or 10^8 stress cycles can be accumulated during a relatively short period of service time.

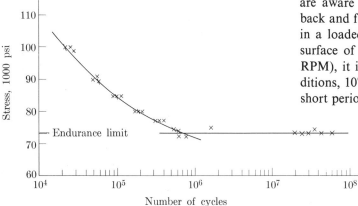

Fig. 4–5.2 *S–N* curve (4340 steel, hot-worked bar stock). *S–N* = permissible stress versus the number of cycles before failure. (Adapted from M. F. Garwood, H. H. Zurburg, and M. A. Erickson, American Society for Metals)

The stress which a material can tolerate under cyclic loading is much less than it is under static loading. The yield strength can be used only as a guide because, as Fig. 4–5.2 shows, the tolerable stress decreases as the required number of cycles increases. Were it not for the fact that an *endurance limit* (the stress below which fatigue does not occur in reasonable times) exists for many materials, the uncertainty in design would be great.

Fatigue arises because each half-cycle produces minute strains which are not fully recoverable. When these strains are added together, they produce local plastic strains which are sufficient to reduce ductility in the cold-worked areas (cf. Fig. 4–4.1c) so that submicroscopic cracks are formed. A crack serves as a notch which concentrates stresses until finally complete fracture occurs.

---

* Static fatigue can occur with constant stresses in glass and in metals by *stress corrosion*, where there is a slow destructive reaction with the surrounding environment.

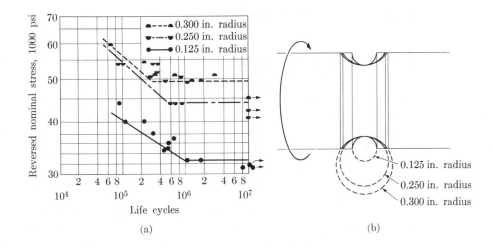

(a)   (b)

Fig. 4–5.3 *S–N* curves for filleted test samples (cf. Fig. 4–5.4). A smaller radius of curvature produces a high stress concentration and therefore a lower endurance limit. (Adapted from M. F. Garwood, H. H. Zurburg, and M. A. Erickson, American Society for Metals)

These strains are localized along slip planes, at grain boundaries, and around surface irregularities caused by compositional or geometric defects. The influence of geometric irregularities (notches) is illustrated in Fig. 4–5.3 and Table 4–5.1. The three sets of data in Fig. 4–5.3 are all for identical steels; however, a notch of $\frac{1}{8}$ in. serves to reduce the endurance limit by 25–30%. Likewise, better surface finishes increase the endurance limit because there are fewer surface irregularities on which stresses can concentrate.

Any design which leads to stress concentrations can lead to premature fatigue failure. This is illustrated schematically in Fig. 4–5.4, which shows that the use of fillets is recommended by the mechanical engineer to avoid sharp corners.

Table 4–5.1

SURFACE FINISH VERSUS ENDURANCE LIMIT*

Type of finish	Surface roughness, micro-inches	Endurance limit, psi
Circumferential grind	16–25	91,300
Machine lapped	12–20	104,700
Longitudinal grind	8–12	112,000
Superfinished (polished)	3–6	114,000
Superfinished (polished)	0.5–2	116,750

* SAE 4063 steel, quenched and tempered to R_C 44. (Adapted from M. F. Garwood, H. H. Zurburg, and M. A. Erickson, American Society for Metals.)

  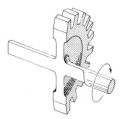

(a) Poor design   (b) Better design   (c) Better design

Fig. 4–5.4 Design of fillet. The use of generous fillets is recommended in mechanical engineering design. It should be observed that (c) is a better design than (a), even with some additional material removed. Of course, if too much metal is removed, failure may occur by mechanisms other than fatigue.

ELECTRICAL BEHAVIOR

4–6 RESISTIVITY OF METALS ⓜ

Metals are conductors, and therefore have relatively low electrical resistivity. As discussed in Section 1–2, this is one of the definitive characteristics of metals. Table 4–6.1 lists the resistivity values for a number of common single-phase metals. You will quickly note that resistivities increase at higher temperatures, and that alloys have higher resistivities than their pure components. These observations, which are summarized in Fig. 4–6.1, are explained by the way electrons move through metals.

Since electrons move in a wavelike fashion they can proceed a great distance (long mean-free path) in a perfectly regular crystal structure, before they are deflected or reflected. Interference with their movement along the potential gradient is increased by any factor which makes the crystal structure less perfect. Thermal vibrations introduce lattice irregularities in proportion to the temperature increase. Therefore at higher temperatures we see a direct reduction in conductivity (increase in resistivity). Plastic deformation by cold work also introduces irregularities because it produces dislocations, with accompanying lattice strains (Fig. 4–3.4). Not surprisingly, therefore, cold-working increases the resistivity, as shown by the dashed lines in Fig. 4–6.1. Finally, a solute atom, e.g., tin in fcc copper, causes local irregularities because the tin atoms are not precisely the same size as the host atoms. Furthermore, such a tin atom differs electrically from copper by having more protons, more electrons, and a different valence for ionization.

Table 4–6.1

ELECTRICAL RESISTIVITY OF SINGLE-PHASE METALS

Metal	Resistivity, microohm · cm*	
	68°F (20°C)	212°F (100°C)
Aluminum	2.7*	3.0
Copper	1.7	2.0
Brass (70 Cu–30 Zn	6.2	7.0
Bronze (95 Cu–5 Sn)	9.5	—
Constantan (55 Cu–45 Ni)	49.0	49.0
Nickel	6.8	7.4
Gold	2.3	2.6
Iron	9.7	10.2
Lead	20.6	20.9
Silver	1.6	1.9
Tin (white)	12.8	13.1

* 2.7 microohm · cm = 2.7 × 10^{-6} ohm · cm.

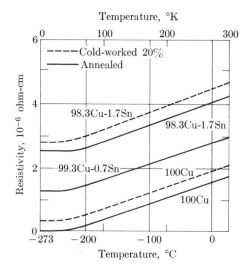

Fig. 4–6.1 Resistivity variables. The resistivity increases with increased temperature, increased solid solution, and cold work.

Thus, as an electron moves through a bronze into the vicinity of a tin atom, it may be deflected by the irregular electric field around the tin. Again this leads to a shorter mean free path and some scattering and back-tracking during conduction. Solid solutions have higher resistivities, or lower conductivities, than their pure components (Fig. 4–6.2).

The total resistivity ρ of a metal may be expressed as the sum of the thermal resistivity ρ_T, the solid solution resistivity ρ_X, and the strain-hardening resistivity ρ_s:

$$\rho = \rho_T + \rho_X + \rho_s. \tag{4–6.1}$$

Example 4–6.1. The resistivity of a heating element for a toaster triples between room temperature and service temperature, where it draws 550 watts (at 110 volts). What is the initial amperage when the toaster is turned on?

Solution. Since watts = (volts)(amps) = (volts)^2/ohms, the hot resistance may be determined.

$$R_{hot} = (110)^2/550 = 22 \text{ ohms.}$$

The dimensions of the element do not change significantly. Therefore the resistivity ρ and resistance R change together, and thus with $\rho_{hot} = 3\rho_{cold}$, $R_{hot} = 3R_{cold}$.

$$R_{cold} = 7.3 \text{ ohms.}$$

Amperes initially = 110 volts/7.3 ohms = 15 amps.

Comments. With the high initial wattage, the element heats quickly, increasing the resistance and cutting back the current. ◀

Example 4–6.2. A copper wire has a resistance of 0.5 ohm per 100 ft. Consideration is being given to the use of a 75Cu-25Zn brass wire instead of a copper wire. What would the resistance be if the brass wire were the same size?

Solution

$$R_{75\text{-}25} = \rho_{75\text{-}25}(L/A) = (1/\sigma_{75\text{-}25})(L/A),$$

$$R_{Cu} = \rho_{Cu}(L/A) = (1/\sigma_{Cu})(L/A).$$

Based on Fig. 4–6.2, where $\sigma_{75\text{-}25} = 0.3\sigma_{Cu}$:

$$\frac{R_{75\text{-}25}}{R_{Cu}} = \frac{\sigma_{Cu}}{0.3\sigma_{Cu}}$$

$$R_{75\text{-}25} = R_{Cu}/0.3 = \frac{(0.5 \text{ ohm}/100 \text{ ft})}{0.3} = 1.7 \text{ ohm}/100 \text{ ft.}$$

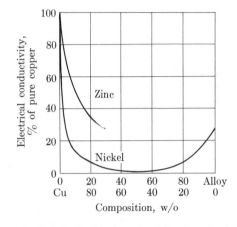

Fig. 4–6.2 Electrical conductivity of solid solution alloys. The conductivity is reduced when nickel is added to copper, and when copper is added to nickel, because the impurity atoms shorten the paths of the electrons between deflections.

Comments. If a wire with low resistance were the only specification, one could afford to pay more than three times as much for copper wire as for brass wire. ◀

Example 4–6.3 ◎ Based on data in Fig. 4–6.1 and in Appendix B, estimate the resistivity of an annealed 70-30 brass at −100°C.

Solution. With no cold work, Eq. (4–6.1) becomes

$$\rho = \rho_T + \rho_X.$$

The resistivities of copper and 70-30 brass are 1.7×10^{-6} ohm·cm and 6.2×10^{-6} ohm·cm, respectively, at 20°C (Appendix B). Therefore $\rho_X = 4.5 \times 10^{-6}$ ohm·cm from the zinc addition to the copper. From Fig. 4–6.1, ρ_T for copper at −100°C is 0.9×10^{-6} ohm·cm.

$$\rho_{70\text{-}30} = 0.9 \times 10^{-6} \text{ ohm·cm} + 4.5 \times 10^{-6} \text{ ohm·cm}$$

$$= 5.4 \times 10^{-6} \text{ ohm·cm at } -100°C.$$

Comments. The value of the solid solution resistivity ρ_X depends on the atom fractions, X and $(1 - X)$,* of the components as follows:

$$\rho_X = y_X(X)(1 - X). \tag{4-6.2}$$

From the above data, y_X for Cu-Zn alloys becomes 21×10^{-6} ohm·cm. The value of ρ_X for an 85Cu-15Zn brass would be 2.7×10^{-6} ohm·cm and the value of ρ becomes 4.4×10^{-6} ohm·cm, which is about 2.5 times as great as that for pure copper. Compare that resistivity figure with the conductivity figure given in Fig. 4–6.2. ◀

THERMAL BEHAVIOR

4–7 THERMAL PROPERTIES OF METALS

Melting. ◎ In Chapter 3 we described the regular pattern of a crystal structure. At the melting point, a material loses the crystal pattern and its long-range order. In liquid metal, interatomic distances between close neighbors are somewhat variable (Fig. 4–7.1). The closest average approach is not much greater than in the crystalline solid; however, almost all metals have some decrease in atomic packing factor when they melt, which causes an increase in volume.

When metals are in a molten state, more energy is required to maintain the atomic disorder. Hence we observe a *heat of fusion* which is required for melting. In general, the more refractory a metal is—

Table 4–7.1

HEATS OF FUSION OF METALS

Metal	Melting temperature, °C (°F)	Heat of fusion, joule/mole*
Tungsten, W	3410 (6170)	32,000
Molybdenum, Mo	2610 (4730)	28,000
Chromium, Cr	1875 (3407)	21,000
Titanium, Ti	1668 (3035)	21,000
Iron, Fe	1537 (2798)	15,300
Nickel, Ni	1453 (2647)	17,900
Copper, Cu	1083 (1981)	13,500
Aluminum, Al	660 (1220)	10,500
Magnesium, Mg	650 (1202)	9,000
Zinc, Zn	419 (787)	6,600
Lead, Pb	327 (621)	5,400
Mercury, Hg	−38.9 (−38)	2,340

* joule/6×10^{23} atoms. It requires 4.18 joules to heat 1 g of water 1°C. To obtain cal/mole, divide joule/mole by 4.18.

* Since the atomic weights of copper and zinc are 63.5 and 65.4 amu, respectively, the atom fraction is approximately the weight fraction.

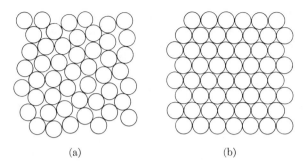

(a)   (b)

Fig. 4–7.1 Liquid and solid structures. A liquid metal (a) has no long-range order. The packing factor is somewhat less than in a (111) plane of an fcc crystalline solid (b). The *average* interatomic distance between first neighbors in the liquid is slightly greater than that in a crystal because there is no structural order.

i.e., the higher a melting point it has—the higher its heat of fusion (Table 4–7.1).

Thermal expansion. We observed in Section 4–2 that there is a direct correlation between the moduli of elasticity and the melting temperatures of metals. The explanation was that a more strongly bonded material not only requires a higher stress to produce strain, but also requires a higher temperature to introduce the thermal agitation that produces a melting of the crystals. We may make one more comparison. A change in temperature introduces less extensive agitation per degree rise in a material in which the atoms are tightly bonded than in one in which the bonds between atoms are weak. This leads us to expect that metals with high melting points will have lower coefficients of thermal expansion than metals with low melting points. This is indeed the case, as Table 4–7.2 shows. Consequently there is an inverse relationship between thermal expansion and Young's modulus.

Table 4–7.2

THERMAL EXPANSION OF METALS

Metal	Melting temperature, °C (°F)	Young's modulus, psi*	Linear thermal expansion at 20°C (68°F)	
			per °C	per °F
Tungsten, W	3410 (6170)	57×10^6	4.6×10^{-6}	2.5×10^{-6}
Iron, Fe	1537 (2798)	30	11.7	6.5
Copper, Cu	1083 (1981)	16	16.5	9.2
Aluminum, Al	660 (1220)	10	22.6	12.5
Lead, Pb	327 (621)	2.2	29.3	16.3
Mercury, Hg	− 38.9 (− 38)	—	40**	22**

* Multiply by 6900 to obtain newtons/m^2.
** At − 40°C (− 40°F).

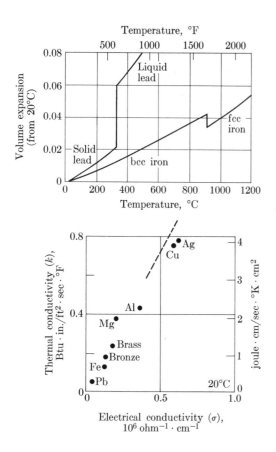

Fig. 4–7.2 Thermal expansion versus temperature (iron and lead). The discontinuities occur at phase changes. At 621°F (327°C), lead melts. At 1670°F (910°C), iron changes from bcc to fcc, a denser phase.

Thermal expansion is not always uniform. Figure 4–7.2 shows a discontinuity in the expansion curve for iron where the phases change from bcc to fcc at 1670°F (910°C). Likewise lead, and all closely packed materials, undergo a major change in volume at the melting point. Furthermore, the volume expansion coefficient is greater for the liquid than for the crystalline solid. In the liquid, the atoms do not have to maintain a specific center of vibration, as they do in a crystal. Therefore the added temperature agitates the atoms more accordingly.

Thermal conductivity. ⓜ ◎ In Section 2–7 we commented that in metals there is a direct relationship between electrical conductivity σ and thermal conductivity k, because the electrons carry much of the thermal energy down the temperature gradient. Figure 4–7.3 shows the correlation for common metals. The ratio k/σ is called the *Wiedemann-Franz ratio*; it has a value of about 7.5×10^{-6} at normal temperatures, when k is expressed in joule·cm/sec·°C·cm^2 and σ is expressed in ohm^{-1}·cm^{-1} (see study problem 2–7.2). Temperature, alloy content, and cold work affect thermal conductivity in the same manner and for the same reasons as they affect electrical conductivity (cf. Fig. 4–7.4 with 4–6.2).

Fig. 4–7.3 Thermal conductivity versus electrical conductivity (20°C or 68°F). In metals, electrons conduct both heat and charge; therefore the ratio k/σ is similar for all metals.

Fig. 4–7.4 Thermal conductivity versus solid solution. Compare this with Fig. 4–6.2. In each case, impurity atoms reduce conductivity.

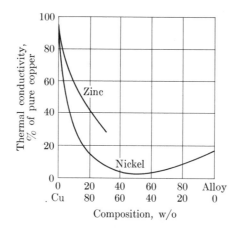

Not all the heat is transferred through a metal by electrons. Just as in other solids, some heat is transferred by lattice vibrations. However, these vibrations also move in a wavelike manner, so that they are reflected and scattered by many of the same crystal irregularities that scatter electrons. Thus each of the two mechanisms parallels the other's behavior.

Example 4–7.1. Cubes of iron, copper, and aluminum are each heated from 68°F (20°C) to 122°F (50°C) between two rigid supports so that they can't expand. Which will develop the greatest stress?

Solution. We use data from Appendix B, and calculate as follows.

$$\Delta l/l = 0 = \sigma/Y + \alpha_l \,\Delta T,$$

$$\sigma = -Y\alpha_l \,\Delta T.$$

$$\sigma_{Fe} = -(30 \times 10^6 \text{ psi})(6.5 \times 10^{-6}/°F)(54°F) = -10,500 \text{ psi}.$$

$$\sigma_{Cu} = -(16 \times 10^6 \text{ psi})(9 \times 10^{-6}/°F)(54°F) = -7800 \text{ psi}.$$

$$\sigma_{Al} = -(10 \times 10^6 \text{ psi})(12.5 \times 10^{-6}/°F)(54°F) = -6750 \text{ psi}.$$

Comments. It may be easier to analyze the above problem if you imagine free expansion from 68°F to 122°F. Then consider that a stress is applied to return the metal to its original dimensions between the supports. ◀

Example 4–7.2. Suppose that you are asked to design a product made of brass. Specifications call for tensile strength at room temperature of at least 43,000 psi and thermal conductivity of at least 1.3 joule · cm/°C · sec · cm^2 (that is, > 0.25 Btu · in./°F · sec · ft^2). How large is the permissible zinc composition range of the brass?

Solution. From Appendix B, $k_{Cu} = 4.0$ joule · cm/°C · sec · cm^2; therefore k must be equal to, or greater than 0.325 k_{Cu}.
    From Fig. 4–7.4, for $k \geq 0.325 k_{Cu}$, Zn $\leq$ 28 w/o.
    From Fig. 4–4.3, for TS $\geq$ 43,000 psi, Zn $\geq$ 26 w/o.

Comments. The use of room-temperature specifications is common, even though the material is to be used at some other temperature. When the alert engineer designs on this basis, he will have made allowances for the difference in property values as the temperature is raised (or lowered). ◀

4–8 DIFFUSION

Atoms move within solids. Not only does thermal energy cause them to vibrate, but under appropriate conditions they can also jump to neighboring lattice sites. High temperatures favor atom jumps, or *diffusion*, more strongly than any other factor. In the first place, at high temperatures the atoms are vibrating more energetically than at low temperatures and are more likely to jump to neighboring vacant sites.

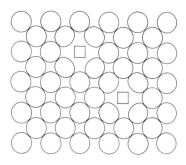

Second, at high temperatures the lattice develops more atom *vacancies* (Fig. 4–8.1) because a significant fraction of the atoms are actually able to move to grain boundaries and to the surface, leaving atomic voids behind. Finally, small atoms such as carbon can move readily through the interstices of the lattice.

Diffusion coefficient. Let us consider the diffusion of carbon through iron in a little more detail. *Carburization* is a metallurgical process of heating iron to form the fcc structure, then diffusing the small carbon atoms into the metal.* As carburization proceeds, there are more carbon atoms near the surface than there are deep in the metal, so that there is a net diffusion inward. The number of carbon atoms which move inward per cm^2 per sec, J, is proportional to the concentration gradient $(C_2 - C_1)/(x_2 - x_1)$ as sketched in Fig. 4–8.2:

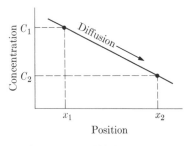

Fig. 4–8.2 Concentration gradient. (Cf. Fig. 2–7.1.) The net flow of atoms is down the concentration gradient, $(C_2 - C_1)/(x_2 - x_1)$.

$$J = -D\left(\frac{C_2 - C_1}{x_2 - x_1}\right). \tag{4–8.1}$$

The term D is called the *diffusion coefficient*. The units for Eq. (4–8.1) are

$$\frac{\text{atoms}}{cm^2 \cdot \text{sec}} = -\left(\frac{cm^2}{\text{sec}}\right)\left(\frac{\text{atoms}/cm^3}{cm}\right), \tag{4–8.2}$$

and the negative sign is present because the atoms always have a net movement *down* the concentration gradient.

Table 4–8.1 shows diffusion coefficients for different atoms in iron and other common metals. Note that: (1) The diffusion coefficient increases quite markedly with an increase in temperature. (2) A small atom like carbon ($r = 0.7$ Å) has a higher diffusion coefficient through iron than larger atoms do. (3) Atoms have a higher diffusion coefficient, and therefore diffuse more readily through bcc iron than through fcc iron (the APF is lower in bcc iron). (4) Nickel diffuses more readily through fcc copper than through fcc iron because the Fe-Fe bonds are stronger than the Cu-Cu bonds. (Compare their melting points and elastic moduli.)

---

* Subsequent quenching provides the steel with a very hard surface (Chapter 6).

Table 4–8.1

DIFFUSION COEFFICIENTS

Diffusion couple	Temperature, °C (°F)	Diffusion coefficient D, cm²/sec
Copper in copper	900 (1652)	3.6×10^{-10}
Copper in copper	500 (932)	10^{-14}
Copper in silver	500 (932)	10^{-13}
Copper in aluminum	500 (932)	5×10^{-11}
Carbon in fcc iron	900 (1652)	10^{-7}
Carbon in bcc iron	900 (1652)	4×10^{-6}
Carbon in bcc iron	800 (1472)	1.8×10^{-6}
Carbon in bcc iron	700 (1292)	6×10^{-7}
Carbon in bcc iron	600 (1112)	1.8×10^{-7}
Carbon in bcc iron	500 (932)	4×10^{-8}
Thorium in tungsten	900 (1652)	10^{-22}
Iron in fcc iron	900 (1652)	6×10^{-14}
Iron in bcc iron	900 (1652)	4×10^{-11}
Nickel in fcc iron	900 (1652)	2×10^{-13}
Nickel in copper	900 (1652)	2.5×10^{-11}

Example 4–8.1. There is one carbon atom per 10 unit cells at the surface of fcc iron. One millimeter behind the surface, there is one carbon atom per 30 unit cells. At 1650°F (900°C), how many carbon atoms diffuse down the gradient per second through each cm²?

Solution. Recall from Example 3–4.2 that the radius of an fcc iron atom at elevated temperatures is 1.29 Å, so that the lattice constant is 3.65 Å. Concentrations are:

$C_2 = 1 \text{ carbon}/30(3.65 \times 10^{-8} \text{ cm})^3 = 7 \times 10^{20}/\text{cm}^3;$

$C_1 = 1 \text{ carbon}/10(3.65 \times 10^{-8} \text{ cm})^3 = 21 \times 10^{20}/\text{cm}^3.$

Using the diffusion coefficient D from Table 4–8.1, we have

$J = -D(C_2 - C_1)/(x_2 - x_1)$

$= -(10^{-7} \text{ cm}^2/\text{sec})(7 \times 10^{20}/\text{cm}^3 - 21 \times 10^{20}/\text{cm}^3)/(0.1 \text{ cm})$

$= 1.4 \times 10^{15} \text{ carbon atoms/cm}^2 \cdot \text{sec}.$

Comments. Since the cross-sectional area of an fcc iron unit cell is about 14 Å² (or 1.4×10^{-15} cm²), there are approximately 2 carbon atoms moving through each unit cell each second. The interior of a heated solid is definitely not a static, inert structure, but subject to continual change. ◀

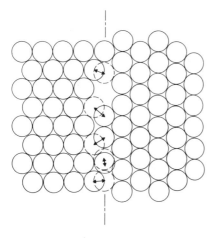

Fig. 4–9.1 Movements of atoms across grain boundaries. As an atom vibrates, it may jump back and forth across the boundary many times per second. This causes the boundary to shift slightly. However, if the boundary is approximately straight, there is insignificant net movement of the boundary.

4–9 GRAIN GROWTH AND ANNEALING

Grain growth. ⓔ Atoms move relatively easily across a grain boundary, where the APF is low (Fig. 4–9.1). At high temperatures at which they have considerable thermal energy, the atoms adjacent to the boundary may jump back and forth from one side to the other as many as 10^{10} times per second.* If the boundary is curved, the atoms move less readily from the concave to the convex surface than vice versa, because, on the average, the atoms on the concave surface associate with more neighbors than the atoms on the convex surface do. This difference exposes "second-layer" atoms for movement from the convex to the concave surface and gives a net shift of the boundaries *toward* their centers of curvature (Fig. 4–9.2a). Inside a material, the centers of grain-boundary curvatures are within the smaller grains and outside the larger grains. As a result, there is a net growth because the smaller grains are progressively eliminated (Fig. 4–9.2b).

This process occurs more readily at high than at low temperatures because the atoms have more energy to jump across the boundary. Thus we can readily see evidence of grain growth as the temperature increases (Fig. 4–9.3). This is not to say that grain growth does not occur at lower temperatures; it occurs appreciably more slowly, however.

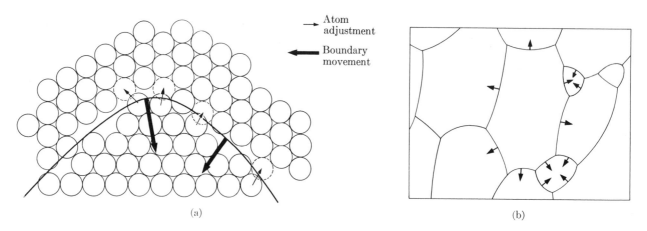

→ Atom adjustment

⬅ Boundary movement

(a)   (b)

Fig. 4–9.2 Movement of curved boundaries. An atom on the concave side is held more firmly than an atom on the convex side because, on the average, it has more neighbors. Thus the boundary shifts toward its center of curvature, reducing the total boundary area.
(a) Individual atoms. (b) Total grains.

---

* The frequency of vibration of an atom within a solid is about 10^{13} per second. This means that, with each oscillation of an atom which lies next to a grain boundary, there is about one chance out of 1000 that it will break free and jump across the boundary (provided that that boundary is straight).

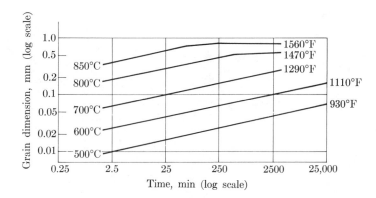

Fig. 4–9.3 Grain growth (brass). Note that the dimensions increase as much between 2.5 and 25 minutes at 600 °C as between 250 and 2500 minutes at 500 °C. (J. E. Burke, *Metals Technology*)

Grain growth is an irreversible process. The only way the grain size can be *refined*—i.e., reduced—is to destroy the original grains, as such, and start over again. One such refinement procedure is through recrystallization and annealing (which we shall discuss in the next two paragraphs). Another method is through *transformation*. For example, when pure iron that has been heated above 910°C (1670°F) to form large fcc grains is cooled to lower temperatures, it transforms to bcc crystals which are initially small grains.

Recrystallization. A heavily cold-worked or strain-hardened metal contains considerable strain energy. This was shown schematically as the strain zone adjacent to the dislocations (Fig. 4–3.4). Strain energy is also identified with the $\sigma\varepsilon$-area under the stress-strain curve (Fig. 2–3.1). At high temperatures, at which atoms move more readily, this strain energy can be eliminated* by the nucleation and growth of new strain-free grains (Fig. 4–9.4). We call this process *recrystallization*. It is used commercially to refine—i.e., make smaller—the average grain size of metals.

Annealing. Recrystallization provides a means of grain-size refinement, mentioned earlier in this section. The importance of recrystallization lies in the fact that in the very early stages (even before it is detectable microscopically) the lattice strains are relieved, and the material's properties, which were altered during cold work, revert to earlier values. Thus we see (Fig. 4–9.5) that a marked softening, or *annealing*, occurs in a 65Cu-35Zn brass around 300°C ($\sim$ 570°F). It is at this temperature

---

* Any material is more stable with lower energy, and will spontaneously change in that direction. Water runs downhill to decrease the potential energy. A cup of hot coffee cools off to a lower level of thermal energy. A condenser discharges to reduce stored electrical energy.

Fig. 4–9.4 Recrystallization (brass, × 40). The deformed brass of (a) is gradually replaced by new, strain-free grains which have few dislocations. This softens the metal (Cf. Fig. 4–9.5). (J. E. Burke, General Electric Co.)

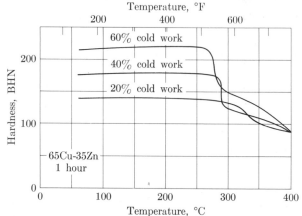

Fig. 4–9.5 Softening during annealing (65Cu-35Zn brass). The more severely cold-worked brass (60%) softens and recrystallizes at lower temperatures than a brass with 20% cold work, 280°C versus ~320°C (ASM data).

that recrystallization starts. Note that with more cold work (60% versus 20%) the *recrystallization temperature* is lower. When there is more strain energy, there is a greater driving force for the atoms to rearrange themselves into annealed, strain-free crystals. Also note that, above the recrystallization temperature, there is added annealing— i.e., added softening—because of extra grain growth. (See Figs. 4–9.5 and 4–4.6.)

Recrystallization involves rearrangements of atoms, or diffusion that is very local. Recrystallization occurs less readily in metals that are strongly bonded. In fact, there is a direct correlation between recrystallization temperature and melting temperature. Tungsten, which melts at 3410°C (> 6000°F), recrystallizes at about 1200°C (~ 2200°F), whereas lead, with a melting temperature of 327°C (620°F) recrystallizes near 0°C. We cannot cite an exact recrystallization temperature for a given metal because, as we have already seen in Fig. 4–9.5, recrystallization temperature depends on the amount of cold work. It also depends on time; the atomic rearrangements which occur in one hour at 300°C still occur at 290°C, but require a longer time. In general, however, the recrystallization temperature is between one-third and one-half of the absolute melting temperature. In any event, a metal strengthened by strain-hardening (Section 4–4) loses its strength if in service it encounters temperatures above the recrystallization temperature.

Example 4–9.1. A 70-30 brass plate is to be 0.25 in. thick, have a tensile strength of at least 50,000 psi, and an elongation (2 in.) of at least 28%. Indicate how this plate should be rolled from a 0.50-in.-thick plate. (Assume negligible change in width during rolling.)

Solution. Based on Fig. 4–4.1, the brass must have between 4 and 16% cold work. A direct reduction of thickness from 0.50 to 0.25 in. would introduce 50% cold work. This would miss the specification. Therefore, use ~ 10% cold work as final step after previously annealing. Thus roll the plate from 0.50 in. to 0.28 in. thickness; then anneal it. As the final step, cold-roll the 0.28-in.-thick plate 10.7% to 0.25 in.

Comments. If the cold-rolling were done without annealing, the rolling from 0.50 in. to 0.28 in. would require 44% cold work. In such a case, the brass would have a ductility of only 5% elongation (2 in.). Under some conditions this lack of ductility could cause cracking during deformation. When that is a problem, it might take two or three cold-work-anneal cycles to attain the 0.28-in. dimensions. We shall observe later that the brass could also be hot-worked above the recrystallization temperature (Section 6–2). ◀

4–10 CREEP ⓔ ◎

At high temperatures, even low stresses produce a slow plastic deform-ation called *creep*. The *creep rate* may be only a fraction of a percent a day; however, in a steam boiler, or similar application, steel tubing may be in service at 1150°F (620°C) for several years. The steam pressure at these temperatures is very high, so the steel is never free from stresses. Thus even a 0.01% strain per day of service would not be tolerable.

Creep has the characteristic of a high initial rate, followed by a long period of "steady-state creep," i.e., a constant creep rate (Fig. 4–10.1). Finally if there is eventual necking of the metal, the rate may be accelerated again before final failure. The steady-state creep is more critical than the initial creep, because it continues over such a long period of time to high strain values.

Higher temperatures or higher stresses increase the slope of the ε-versus-t curve; conversely lower creep rates accompany lower stresses and lower temperatures.

Creep rates are faster in fine-grained materials because the grain boundary (1) contributes directly to creep as a result of its lower packing factor, and (2) serves as a "sink" into which dislocations can move, thus avoiding the dislocation "tangles" which retard slip (Section 4–4). This means that grain boundaries weaken metals at high temperatures, which of course is opposite to the situation at low temperatures, at which fine-grained materials are stronger (Fig. 4–4.6).

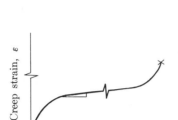

Fig. 4–10.1 Creep. This slow rate of strain is critical in high-temperature applications, particularly if long-time service is involved, e.g., power-plant boiler tubes. The slope of the extended center portion of the curve is called the steady-state creep rate.

REVIEW AND STUDY

4–11 SUMMARY

The characteristics of single-phase metals arise from their crystal structures and microstructures. An analysis of plastic deformation as being dislocation movements along slip planes enables us to rationalize two of the common strengthening procedures: strain hardening and solid-solution hardening.

Irregularities in crystal structure such as thermal agitation, solid solution impurities, and dislocations from cold work interfere with electron movements in metals. This leads to a reduction of the electrical (and thermal) conductivity.

Thermal behavior is closely related to bond strength. High-elastic-modulus metals have higher melting points and lower thermal ex-pansion. As the temperature is increased, atoms move more and more readily within the metal. This appears as diffusion down a concentration gradient, grain growth, annealing, and creep, all of which affect the behavior of a metal in processing or in service.

Terms for Study

Alloy	Miller indices (hkl)
Annealing	Plastic deformation
Brass	Poisson's ratio, v
Brittle fracture	Polycrystalline
Bronze	Recrystallization
Carburization	Recrystallization temperature
Cold work, CW	Resistivity
Creep	solid solution, ρ_X
Creep rate	strain-hardening, ρ_s
Diffusion	thermal, ρ_T
Diffusion coefficient, D	Shear modulus, G
Dislocation	Shear strain, γ
Displacement distance, $\mathbf{b}$	Shear stress, τ
Ductile fracture	Single-phase metals
Edge dislocation	Slip
Elastic deformation	Solution hardening
Endurance limit	Sterling silver
Fatigue	Strain hardening
Fracture	Transition temperature
Grain growth	Wiedemann-Franz ratio, k/σ
Heat of fusion	Young's modulus, Y
Mean free path	

STUDY PROBLEMS

4–1.1 What price ratio should copper and zinc have to produce a 10% cost advantage for 70-30 brass over pure copper on a pound-for-pound basis?

Answer. $\$_{Zn}/\$_{Cu} = 0.67$

4–2.1 Silver is not listed in Table 4–2.1; however, make an estimate of its Young's modulus.

Answer. ~ 13,000,000 psi, based on melting points and Young's moduli of aluminum and copper (versus 11,000,000 psi by experiment)

4–2.2 What percent elastic volume change occurs in a brass when it is loaded under compression to its yield strength of 48,000 psi?

Answer. −0.1 v/o

4–2.3 A shear stress of 7300 psi produces a shear angle of 0.1°. Other experiments indicate that a 7300-psi tensile stress gives a strain of 0.0006 in./in. What is the Poisson's ratio?

Answer. 0.45

4–3.1(a) Based on the comments following Example 4–3.1, sketch the (001) and the (100) planes onto Fig. 4–3.5(a). (b) Compare p_{100} and p_{001} with p_{010}, and the atom packing on the (100) plane with that on the (010) plane.

4–3.2 What is the packing of atoms per cm^2 on the (110) plane of bcc chromium ($r = 1.249$ Å)?

Answer. $1.7 \times 10^{15}/cm^2$

4–3.3 Compare the interplanar spacings p_{010}, p_{110}, and p_{111}, for bcc iron ($a = 2.86$ Å). (Modify the sketches of Fig. 4–3.5 in an appropriate manner.)

Answer. $p_{010} = 1.43$ Å, $p_{110} = 2.02$ Å, $p_{111} = 0.83$ Å

4–3.4 How much force should be applied in a face-diagonal direction on the bottom side of the cube to produce a one-pound force resolved in a direction parallel to the x-axis?

Answer. 1.41 lb

4–4.1 A 70-30 brass wire, originally 87 in. long and without cold work, is drawn through a circular die which elongates it to 127 in. What is its hardness after drawing?

Answer. BHN 155

4–4.2 Copper, iron, or a 70-30 brass may be used for a support rod. Cold working is acceptable. Specifications call for TS = ≥ 40,000 psi; elongation (2 in.) ≥ 30%. Make a selection and indicate the reasons for your choice.

4–4.3 A pure iron sheet is to be cold-rolled from 0.14 to 0.12 in. thick (with negligible change in width). (a) What will the ductility of the iron be after cold-rolling? (b) The tensile strength?

Answer. (a) 40% elongation (2 in.) (b) TS = 47,000 psi

4–4.4 Repeat Problem 4–4.3 for cold-drawing a wire from a diameter of 0.14 to 0.12 in.

Answer. (a) 32% elongation (2 in.) (b) 53,000 psi

4–5.1 Examine closely three examples of metal failure. Check for evidence of plastic deformation before fracture.

4–5.2 Sometimes it is assumed that if failure does not occur in a fatigue test in 10^8 cycles, the stress is below the endurance limit. A test machine is connected directly to a 1740 rpm electric motor. How long will it take to log that number of cycles?

Answer. 40 days

4–6.1 A brass alloy is to be used in an application which must have a tensile strength of more than 40,000 psi and an electrical resistivity of less than 5×10^{-6} ohm·cm (resistivity of copper $= 1.7 \times 10^{-6}$ ohm·cm). What percent zinc should the brass have?

Answer. 15 to 20% zinc

4–6.2 The electrical resistivity of pure copper is 1.7×10^{-6} ohm·cm at 20°C and 2.4×10^{-6} ohm·cm at 120°C. That of a 98Cu-2Ni alloy is 4.1×10^{-6} ohm·cm at 20°C. Estimate the resistivity of (a) pure copper at 45°C, (b) 98Cu-2Ni at 120°C, (c) 96Cu-4Ni at 20°C, and (d) 96Cu-4Ni at 95°C.

Answer. (a) $\sim 1.9 \times 10^{-6}$ ohm·cm, (b) $\sim 4.8 \times 10^{-6}$ ohm·cm, (c) $\sim 6.4 \times 10^{-6}$ ohm·cm, (d) $\sim 7 \times 10^{-6}$ ohm·cm

4–6.3 A 96Cu-4Ni alloy has a resistivity of 6.4×10^{-6} ohm·cm at 20°C. Refer to the comments in Example 4–6.3. Estimate the electrical resistivity of a 50Cu-50Ni alloy.

Answer. $\sim 30 \times 10^{-6}$ ohm·cm

4–6.4 A wire whose diameter is 0.039 in. has a resistance of 0.10 ohm/ft. It is cold-drawn to 0.031 in. in diameter and then has a resistance of 0.20 ohm/ft. What fraction of the initial conductivity remains after cold-working?

Answer. $\sigma_{CW} = 0.79\ \sigma_{initial}$

4–7.1 Refer to Fig. 4–7.2. Determine the atomic packing factors for lead just above and just below its melting point. (Use the room temperature radii as the hard-ball size.)

Answer. At 327°C, $(APF)_s = 0.725$; $(APF)_l = 0.698$

4–7.2 An iron wire 0.1-in. in diameter has a 0.01-in. coating of aluminum surrounding it. It is stress relieved at 300°F, then allowed to cool to 70°F. (a) Which metal will contain a tensile (+) stress? (b) What stresses will be in each metal?

Answer. $\sigma_{Fe} = -5,300$ psi, $\sigma_{Al} = +12,000$ psi (assume that the aluminum does not deform plastically)

4–7.3 Refer to the aluminum of Example 4–7.1. What is the volume change as the metal is heated under restraint? [$\alpha_v \simeq 3\alpha_l$ from Eq. (2–6.3), and v is 0.35 from Table 4–2.2.]

Answer. +0.18 v/o

4–7.4 The designer has a choice of a brass or a Cu-Ni alloy to meet room-temperature specifications of hardness $\geq R_F 80$, TS $\geq 40,000$ psi, and thermal conductivity ≥ 0.10 cal·cm/°C·sec·cm² ($\geq 10\%\ k_{Cu}$). Pick a suitable composition.

Answer. 85Ni-15Cu

4–7.5 Basing your answer on Fig. 4–6.1, estimate the thermal conductivity of a 20% cold-worked 98.3Cu-1.7Sn alloy at 0°C.

Answer. ~ 1.7 joule·cm/°C·sec·cm²

4–8.1 An iron bar is plated with nickel and heated to 1650°F (900°C) so that there is 30 a/o nickel at 0.01 cm below the surface and 20 a/o nickel at 0.02 cm below the surface. What is the atom flux per cm^2 per sec? (Assume fcc iron.)

Answer. 0.18×10^{12} Ni atoms/cm$^2 \cdot$ sec (or one atom per second for the cross section of $\sim$ 5500 unit cells)

4–8.2 What is the ratio of diffusion fluxes of copper in aluminum and copper in silver at 500°C (930°F) when the concentration gradient is (10^{18} Cu atoms/cm^3)/cm?

Answer. $(J_{\text{Cu in Al}}/J_{\text{Cu in Ag}}) = 500$

4–9.1 Compare the length of time required to grow grains of brass from a dimension of 0.02 mm to 0.1 mm at 500°C and at 600°C (see Fig. 4–9.3).

Answer. $t_{500°C} = \sim 100{,}000$ min ($\sim$ 70 days), $t_{600°C} = \sim 2500$ min ($\sim$ 42 hr)

4–9.2 An iron rod (0.42 in. in diameter) is to be drawn through a series of dies to a 0.165-in. diameter, at which time it must have a ductility of at least 40% elongation (in 2 inches) and a tensile strength of at least 45,000 psi. What diameter is required prior to the last cold-drawing?

Answer. CW $\cong$ 12%; therefore the diameter = 0.176 in.

4–10.1 Other things being equal, which will have the lowest creep rate: (a) steel in service with a high tensile stress and low temperature, (b) steel in service with a low tensile stress and high temperature, (c) steel in service with a high tensile stress and high temperature, or (d) steel in service with a low tensile stress and low temperature? Why?

4–10.2 Design considerations could permit a pressure tube to have 3l/o strain during a year's service. What maximum creep rate is tolerable when reported in the normal %/hr figure?

Answer. 3.4×10^{-4} %/hr

MICROSTRUCTURES OF ALLOYS

5–1 PHASES ⓔ

A tumbler partially filled with ice water includes four different structures. The tumbler has the structure of glass. Although of the same composition, the structures of the water and the ice are different. Finally, the air above the ice water has its own structure, admittedly not very dense, and with no molecule-to-molecule coordination. Each of these structures is called a *phase*. A *phase boundary* appears as a discontinuity in structure and/or composition within a material.

There is never more than one gaseous phase. It is capable of receiving any and all components that have the thermal energy to evaporate.

Several insoluble liquid phases may be in contact, e.g., oil and water, mercury and water, or liquid solder and a rosin flux. Often, however, liquid phases have the ability to accommodate large amounts of solute within their structures, because these structures are non-crystalline and do not have the close-packed requirements of a crystal lattice.

There are many solid phases. We have already discussed some of these: fcc iron, bcc iron, hcp titanium, fcc nickel, and fcc copper, to name a few. All of these, and essentially all solid metallic phases, are crystalline. (This broad statement will not be true for polymer and ceramic phases, discussed in later chapters.) The numbers of crystalline phases certainly run into many thousands, and may be almost limitless. There are many because, when atoms are arranged in close-packed structures, there are only limited possibilities for solid solution. It is not uncommon to find metal products with more than one solid phase (Fig. 5–1.1).

In Chapter 4 we discussed the characteristics of single-phase alloys. In this chapter we shall look at those alloys which have more than one phase. Among multiple-phase metals, *steels* are iron-base alloys containing various amounts of a carbide phase. Ordinary *solder*

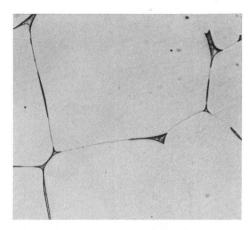

Fig. 5–1.1 Metallic phases (bismuth in copper) (× 190). The larger light areas are fcc grains and are essentially pure copper. The darker intergranular areas are rich in bismuth and remained liquid long after the copper solidified. (Photomicrograph by C. S. Smith; courtesy AIME)

contains lead-rich and tin-rich phases. *Diecasting alloys* commonly contain a major amount of zinc, plus selected intermetallic compounds. Hardenable aluminum alloys commonly contain, as a second component, copper, which collects in the second phase during controlled heat-treating processes.

One of the important concepts of this chapter is that of a *solubility limit*, i.e., the maximum amount of a second component which can be added to a phase without *supersaturation*. Many of our metal-processing operations (Chapter 6) involve a *solution treatment*, followed by cooling in which the solubility limit is exceeded. Separation (i.e., *precipitation*) follows, sometimes slowly.

Example 5–1.1. A syrup may contain only 67% sugar (33% water) at 68°F (20°C), but 83% sugar at 212°F (100°C). One pound of sugar and 0.25 pound water are mixed and boiled until all the sugar is dissolved. During cooling, the solubility limit is exceeded so that (with time), excess sugar separates from the syrup. If equilibrium is attained, what is the weight ratio of syrup to sugar at 68°F?

Solution

Basis: 1.00 lb sugar + 0.25 lb H_2O = 1.25 lb.

At 68°F, amount of syrup = 0.25 lb H_2O/(0.33 lb H_2O/lb syrup)

$$= 0.757 \text{ lb syrup}$$

$$= 0.493 \text{ lb excess sugar.}$$

Comment. All the sugar is dissolved at the higher temperature, to give a single phase of syrup. When the mixture cools, the *solubility limit* for an 80 sugar–20 water composition is reached at 87°C (189°F). Typically, however, some *supercooling* is encountered before the excess phase (in this case, sugar) starts to separate. In fact, in this case, supercooling can proceed to room temperature so that the start of separation may be delayed considerably. Similar supersaturation commonly occurs in metals and other materials. ◀

5–2   TWO-PHASE ALLOYS [e]

Sterling silver is an example of alloys which can have more than one phase. This metal is of course the basis for much of the world's currency. It contains ∼92.5 w/o silver and ∼7.5 w/o copper (and traces of other elements, notably lead). This proportion is not without reason. It is near the practical limit at which copper can be added to fcc silver while still maintaining a single phase. Because the single phase is maintained, the copper does not degrade the luster of the silver. True, the copper is a diluent; however, like the addition of zinc to copper in brass, the addition of copper to silver strengthens the

Fig. 5–2.1 Solid solubility. (a) Copper in silver. (b) Silver in copper.

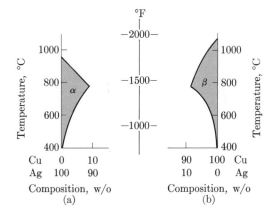

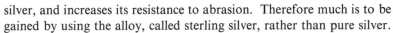

silver, and increases its resistance to abrasion. Therefore much is to be gained by using the alloy, called sterling silver, rather than pure silver.

The solubility limit of copper in solid fcc silver varies with temperature (Fig. 5–2.1a). For an alloy of 7.5% Cu (92.5% Ag), there is a temperature range of about 50°C (90°F), in which *all* the copper enters the fcc silver phase. This takes place between 760°C and 810°C (1400°F to 1490°F). If we were to add more copper, this range would decrease rapidly, so that, in production, control problems might arise during the necessary solution treatment.

If sterling silver is cooled rapidly, its solubility limit is exceeded and it becomes a supersaturated solid solution at room temperature. Of course room temperature is a relatively cold temperature compared to the above-mentioned solution-treating temperature, and the atom diffusion is very slow; so slow, in fact, that the solid solution may remain supersaturated for an indefinite period of time because the fcc silver cannot reject the copper atoms. This can be good from our point of view because we now have the benefit of solution hardening that would disappear if the copper became completely separated.

We have been focusing on a silver-rich Ag-Cu alloy, but the same situation exists in a copper-rich Cu-Ag alloy. Copper will dissolve up to 8 w/o silver within its fcc lattice (Fig. 5–2.1b). As before, the solubility limit is highly temperature sensitive. Shortly we shall see the reason for our reverse plotting of Fig. 5–2.1(b) and the apparent coincidence that produces a maximum solubility just under 800°C for either copper or silver alloys. First, however, we shall note that both silver and copper are fcc. The difference in their sizes ($r_{Ag} = 1.444$ Å; $r_{Cu} = 1.278$ Å) precludes extensive solid solution at low temperatures. Only as the temperature is raised does the solubility limit become significant. We shall be referring to an Ag-rich fcc solid solution and to a Cu-rich fcc solid solution. For simplicity, let us call the former α and the latter β. Note that α (and β) is the name of a phase, and does not indicate a specific composition. Conversely, the term α does not imply silver alone, but a solid solution, admittedly Ag-rich.

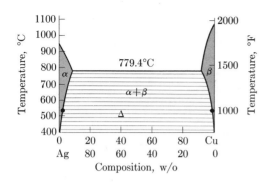

Fig. 5–2.2 Solid-phase regions of the Ag-Cu system.

The Ag-Cu system. Figure 5–2.2 connects the two parts of Fig. 5–2.1. In each of the solid-solution regions only one phase exists, α at the left and β at the right. Beyond the solubility limits, a second phase is required to accommodate all the silver and copper. Thus with increasing amounts of copper at 1000°F (540°C), we find first only α (0 to 2.5 w/o Cu, or 100 to 97.5 w/o Ag). Beyond 2.5% copper, the excess copper must appear in the β phase and produce a two-phase, α-β alloy. Not until 98.5 w/o Cu (1.5 w/o Ag) does the single-phase alloy occur again. Likewise, a 97.5Ag–2.5Cu alloy is a single-phase alloy above 1000°F (540°C) and a two-phase alloy below that temperature. Note that for alloys with compositions in the α + β region, there is more copper than the α can hold, and more silver than the β can hold. Therefore both α and β have compositions at their respective solubility limits. *Whenever two phases are in equilibrium within a two-phase composition–temperature region, the solubility limit curves at each side define their respective compositions.* Thus a 60Ag–40Cu alloy, Δ, at 1000°F (540°C) has the phase compositions shown by the dots on the solubility curves of Fig. 5–2.2.

Eutectics. You are aware that H_2O and NaCl, when in solution together, remain liquid below 32°F, the freezing point of water. [This is much below the 800°C (1472°F) freezing point of NaCl.] Likewise, an Ag-Cu alloy remains liquid below the melting temperature of either pure silver (960°C), or pure copper (1083°C). The lowest liquid temperature possible for Ag-Cu alloys is 779°C (1435°F). This is called the *eutectic temperature.* Any alloy which can remain fully molten to

Fig. 5–2.3 The Ag-Cu system. The greatest solid solubility occurs at the eutectic temperature. (ASM *Handbook of Metals*, Metals Park, O.: American Society for Metals)

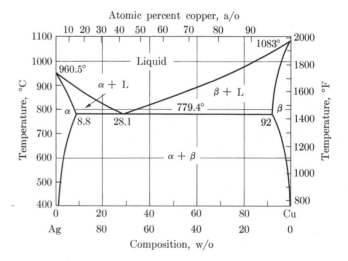

this temperature has the *eutectic composition*, in this case 71.9Ag–28.1Cu. The inclusion of this information with the information presented in Figs. 5–2.1 and 5–2.2 completes the Ag-Cu *phase diagram* (Fig. 5–2.3).

Before we discuss the quantitative aspects of phase diagrams, let us note the appearance of the eutectic in a phase diagram. A liquid with a eutectic composition solidifies to produce two solid phases. In an Ag-Cu alloy, this *eutectic reaction* is:

$$\text{Liq } (28.1\% \text{ Cu}) \xrightarrow{\text{cooling}} \alpha \ (8.8\% \text{ Cu}) + \beta \ (92\% \text{ Cu}). \qquad (5\text{–}2.1)$$

On a phase diagram the eutectic appears as a characteristic "V-slot" at the bottom of the liquid field and above an area containing two solid phases. [Cf. 71.9Ag–28.1Cu at 779.4°C in Ag-Cu (Fig. 5–2.3) with 38.1Pb–61.9Sn at 183°C in Pb-Sn (Fig. 5–2.4), with 95.7Fe–4.3C at 2065°F (1130°C) in Fe-C (Fig. 5–4.2), and with 32Al–68Mg at 437°C in Al-Mg (Fig. 5–7.1).] In a solidified eutectic, the two solid phases are commonly intimately mixed because they form simultaneously from the final liquid. The mixture is called a *eutectic microstructure*.

Phase diagrams. Diagrams such as Fig. 5–2.3 are very useful to the engineer and technician. Consider the Pb-Sn phase diagram in Fig. 5–2.4. (1) From it, as with the Ag-Cu diagram, one can identify the stable *phase(s)* for any composition and temperature. (2) The phase *compositions* can be determined. In a single-phase field, the phase analysis matches the alloy composition, for example, 90Pb–10Sn at

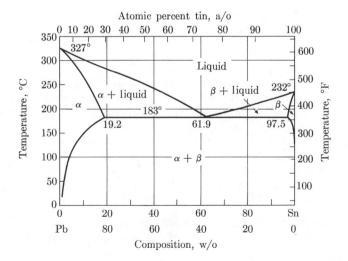

Fig. 5–2.4 The Pb-Sn system. The eutectic composition (∼ 60 Sn–40 Pb) is used for low-melting solders. (ASM *Handbook of Metals*, Metals Park, O.: American Society for Metals)

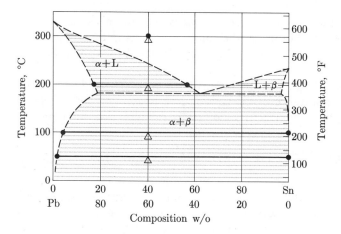

Fig. 5–2.5 Temperatures and compositions for Example 5–2.1.

200°C.   In a two-phase field, the compositions are defined by the solubility-limit curves of the two phases.   (3) The equilibrium *amounts* of each phase may also be determined.   This will be discussed in Section 5–3.

Example 5–2.1. A solder containing 60Pb–40Sn is melted (a) at 300°C (572°F) and cooled (b) to 200°C (392°F); (c) to 50°C (122°F).   Cite the equilibrium phase(s) and their compositions at each of the three temperatures.

Solution. Refer to Fig. 5–2.5.

a) 300°C: Only liquid   60Pb–40Sn
b) 200°C: α   82Pb–18Sn
            +
            liquid   43Pb–57Sn
c) 50°C: α   98Pb–2Sn
            +
            β   100Sn

Comments. The curves at the lower edges of the liquid field in Fig. 5–2.4 are also solubility-limit curves.   The one to the left is the solubility-limit curve for lead in the liquid phase.   The curve to the right is the solubility-limit curve for tin in the liquid phase.

A common—but not universal—procedure is to label the solid phases as α, β, γ, etc., across the phase diagram from left to right.   Thus in the Pb-Sn system (Fig. 5–2.4), α is a lead-rich phase; in the Cu-Zn system (Fig. 5–2.6), α is a brass, a copper-rich phase.   This procedure seldom, if ever, causes confusion arising from the multiplicity of α phases. ◀

Example 5–2.2. Refer to Fig. 5–2.6 for the Cu-Zn phase diagram.   Note that a 70Cu–30Zn brass is single-phase, α, at all temperatures below 925°C.

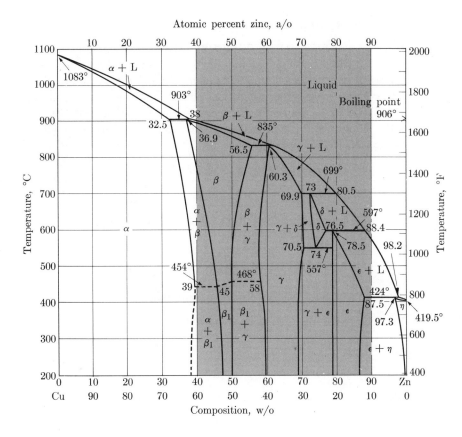

Atomic percent zinc, a/o

Fig. 5–2.6 The Cu-Zn system. The common brasses possess the single-phase α structure, a solid solution of zinc in copper. Most commercial alloys fall in the unshaded areas. (ASM *Handbook of Metals*, Metals Park, O.: American Society for Metals)

Since this is a copper-rich Cu-Zn solid solution which is contiguous to pure copper, we know that α is fcc.

Cite the phase(s) and their compositions for a 26Cu–74Zn alloy at (a) 600°C (1112°F); (b) at 525°C (977°F).

Solution

a) Only δ, therefore 26Cu–74Zn
b) γ, which is   30Cu–70Zn
   and ε, which is   21.5Cu–78.5Zn

Comments. Don't panic because of the apparent complexity of phase diagrams, such as the one for the Cu–Zn system. This problem is included to indicate that even in the more formidable looking areas, one can cite the phase(s) and compositions by focusing on the small area only.

Fortunately the commonly used alloys are found in simpler-looking areas of phase diagrams. Brass is used extensively; the 26Cu–74Zn alloy has extremely rare applications. ◀

5–3 LEVER RULE ⓔ

As indicated in Example 5–2.1, a 60Pb–40Sn solder contains Liq + α at 200°C, with each phase containing 57% tin and 18% tin respectively. From this we can calculate the *amount of* α and the *amount of liquid*. Assume that some solder weighs ten pounds and contains four pounds of tin. Let L be the amount (or pounds) of liquid, so that (10 − L) is the amount of α. Now make a materials balance, or an accounting of the total tin in the two phases.

Total tin = 4 lb = 0.57L lb + 0.18(10 − L) lb.

Thus

L = 5.65 lb liquid = 0.565 of the 10-lb sample,

and

10 − 5.65 = 4.35 lb α = 0.435 of the 10-lb sample.

Alternatively, we could have made a materials balance for lead.

Total lead = 6 lb = 0.43L lb + 0.82(10 − L) lb.

Again

L = 5.65 lb liquid

and

10 − 5.65 = 4.35 lb α.

These calculations are consistent with Fig. 5–2.5. At 200°C, the 60Pb–40Sn is somewhat nearer the composition of liquid than the composition of α; thus we would expect to have more liquid than α. Our calculations show $\alpha/L = 0.435/0.565 = 0.77$.

Now let us generalize our calculations of the *amount* of each of two phases (for instance, the amount of x and y), calling these *amounts* X and Y in an alloy consisting of metals A and B. If the alloy has a *composition* C_A and C_B (read at the bottom of the phase diagram), where C_A is the fraction of A, then there are $[C_A(X + Y)]$ pounds of A in the total of the two phases.

Next let us find the total amount of metal A in a second way: by determining how much A there is in each of the two phases that make up the alloy. The fraction of A in phase x is C_{A_x}; and in y, the fraction is C_{A_y}. So the number of pounds of A in the two phases is $[C_{A_x}(X)$ pounds $+ C_{A_y}(Y)$ pounds$]$. Equating these, we have

$$C_A(X + Y) = C_{A_x}(X) + C_{A_y}(Y). \tag{5–3.1}$$

Rearranging, we see that the ratio of X to Y is

$$\frac{X}{Y} = \frac{C_{A_y} - C_A}{C_A - C_{A_x}}. \tag{5–3.2}$$

Let us picture this in Fig. 5–3.1(a) where X and Y are the amounts of α and liquid, respectively. We let A be lead and B be tin, and work with the lead fractions.

Filling in Eq. (5–3.2) with data for lead from Fig. 5–3.1,

$$\frac{\alpha}{L} = \frac{0.43 - 0.60}{0.60 - 0.82} = 0.77,$$

which also equals 0.435/0.565, as we previously calculated. We may envision a lever (Fig. 5–3.1b) with the fulcrum at the overall composition which is C_A and C_B (in this case 60Pb and 40Sn), and the necessary pounds of α at C_{A_α} and liquid at C_{A_L} to balance each other. As drawn, there must be more pounds of liquid than α, which of course checks our previous calculations.

Equation (5–3.2) is a statement of the *lever rule* [or, as it is sometimes called, the *inverse lever rule*, since the value of X is proportional to $(C_{A_y} - C_A)$, the *opposite* end of the lever]. Equation (5–3.2) may be rearranged algebraically to correspond to Fig. 5–3.1, giving

$$\frac{X}{X + Y} = \frac{C_{A_y} - C_A}{C_{A_y} - C_{A_x}}. \tag{5–3.3}$$

This gives the fraction of the total alloy that is X. Likewise the fraction of the total alloy that is Y is

$$\frac{Y}{X + Y} = \frac{C_{A_x} - C_A}{C_{A_x} - C_{A_y}}. \tag{5–3.4}$$

These latter two versions are more commonly used than Eq. (5–3.2).

Example 5–3.1. Five pounds of a 90Mg–10Al alloy (Fig. 5–7.1) are melted and cooled slowly to 500°C (932°F). By making a material balance on the magnesium, determine the amount of ε which is present.

Solution. Let E = pounds ε, and (5 − E) = pounds liquid.

$$(0.90)(5 \text{ lb}) = (0.92)(E) + (0.76)(5 - E),$$

$$(0.90)(5 \text{ lb}) - (0.76)(5) = 0.92E - 0.76E,$$

$$E/5 \text{ lb} = (0.90 - 0.76)/(0.92 - 0.76) \tag{5–3.5}$$

$$= \text{fraction } \varepsilon,$$

$$E = 4.38 \text{ lb } \varepsilon \text{ (and 0.62 lb L)}.$$

Comments. Observe that Eq. (5–3.5) is an application of the lever rule. ◀

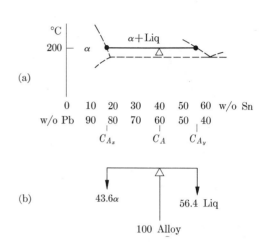

Fig. 5–3.1 (a) 200°C Pb-Sn isotherm. (Cf. Fig. 5–2.5 and Example 5–2.1b.) (b) Inverse lever rule applied to a 60Pb–40Sn alloy composition at 200°C.

Example 5–3.2. Determine the amount of α and β in a 60Pb–40Sn alloy at 100°C (assuming equilibrium).

Solution. Refer to Fig. 5–2.5.

$$\frac{\beta}{\alpha + \beta} = \frac{0.04 - 0.40}{0.04 - 1.00} = 0.375\beta.$$

$$\frac{\alpha}{\alpha + \beta} = \frac{1.00 - 0.40}{1.00 - 0.04} = 0.625\alpha.$$

Comments. The above calculation was based on the tin fractions. We could have made the same calculation based on lead fractions.

$$\frac{\beta}{\alpha + \beta} = \frac{0.96 - 0.60}{0.96 - 0} = 0.375\beta.$$

The phrase "assuming equilibrium" will be implied in the remainder of this chapter, since a phase diagram is, *per se*, an equilibrium diagram. In Chapter 6 we shall discuss nonequilibrium conditions in metals. ◀

Example 5–3.3. Five pounds of a 20Ag–80Cu alloy are completely melted and then cooled. How many pounds of β are there at each 100°C interval from 1000°C (1832°F) down to 400°C (752°F)?

Solution. Using the silver fraction, calculate as follows.

1000°C: All liquid $\hspace{4.5cm}$ = 0 lb β

900°C: $5[\beta/(\beta + L)] = (5\ \text{lb})(0.39 - 0.20)/(0.39 - 0.07)$ = 2.97 lb β

800°C: $5[\beta/(\beta + L)] = (5\ \text{lb})(0.66 - 0.20)/(0.66 - 0.08)$ = 3.96 lb β

700°C: $5[\beta/(\beta + \alpha)] = (5\ \text{lb})(0.94 - 0.20)/(0.94 - 0.05)$ = 4.16 lb β

600°C: $5[\beta/(\beta + \alpha)] = (5\ \text{lb})(0.965 - 0.20)/(0.965 - 0.02)$ = 4.05 lb β

500°C: $5[\beta/(\beta + \alpha)] = (5\ \text{lb})(0.98 - 0.20)/(0.98 - 0.01)$ = 4.02 lb β

400°C: $5[\beta/(\beta + \alpha)] = (5\ \text{lb})(0.99 - 0.20)/(0.99 - \text{nil})$ = 3.99 lb β

Comments. We would have obtained the same answers had we used the copper fraction.

A plot of the weight fraction β versus temperature is shown in Fig. 5–3.2. Note the discontinuity at the eutectic temperature (780°C or 1436°F).

In Fig. 5–3.3 (for study problem 5–3.4), the fractions of all phases are present in one figure. ◀

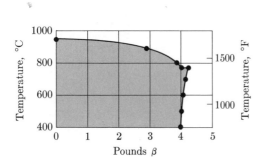

Fig. 5–3.2 Phase amount versus temperature (β in a 20Ag–80Cu alloy). (See Example 5–3.3.)

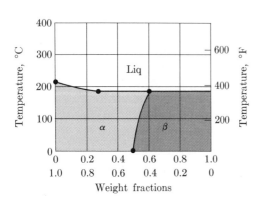

Fig. 5–3.3 Phase fractions versus temperature (50Pb–50Sn). (See study problem 5–3.4.)

5–4 Fe-C SYSTEM [e]

We shall give special attention to the Fe-C phase diagram, since it serves as the basis for our knowledge about the internal structure of steels. First, however, let us review several facts about iron itself.

Ferrite (α). At normal temperatures, iron is bcc. Because of this, it can dissolve very little carbon interstitially, but it *can* dissolve significant amounts of chromium, silicon, tungsten, and molybdenum by substitution. Two terms are used synonymously to label this phase: α and *ferrite*.

Austenite (γ). At elevated temperatures iron is fcc. It can dissolve more carbon interstitially than bcc iron can (Fig. 5–4.1). It can also dissolve by substitution large amounts of nickel and manganese—elements which prefer the fcc structures. Here, also, two labels are used interchangeably: γ and *austenite*.

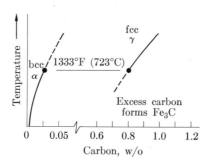

Fig. 5–4.1 Solubilities of carbon in ferrite (bcc) and austenite (fcc). Although the packing factor is higher in austenite, the interstices which are present in austenite (γ) are larger (and fewer) than in ferrite (α).

Carbide. Iron and carbon form the compound Fe_3C (= 93.33 w/o Fe–6.67 w/o C), sometimes called *cementite*. This compound is hard and brittle and will play an important role in our examination of steels. In steel, however, there are usually other alloying elements, such as Cr, Mo, and Mn, which also enter into the carbide phase. Therefore the carbide is seldom pure Fe_3C, or cementite. As a result we shall simply use the term *carbide*, abbreviated by a boldface **C**.

The Fe-Fe₃C phase diagram.* The Fe-C system has a eutectic at 2065°F (1130°C) and 95.7Fe–4.3C (Fig. 5–4.2). Since austenite, which is the stable iron-rich phase, can dissolve 2 w/o C at that temperature,

---

*   Our attention will be focused on iron that is below 2500°F (1370°C). At higher temperatures, fcc iron reverts to bcc iron to give what we call δ-ferrite. However, we shall not be involved with δ-ferrite in this text.

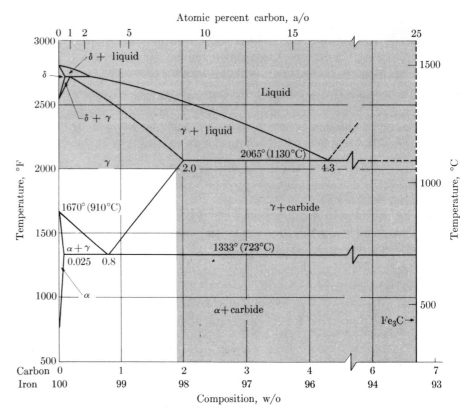

Fig. 5–4.2 The Fe-C system. The unshaded part of the phase diagram will be of prime interest to us. Note the eutectic-like appearance of the eutectoid region at 99.2Fe–0.8C (1333°F or 723°C).

the eutectic reaction during cooling is:

$$\text{Liq } (4.3\text{C}) \rightarrow \gamma \ (2.0\text{C}) + \text{carbide } (6.67\text{C}). \tag{5–4.1}$$

Pure bcc iron transforms, on heating, from ferrite to austenite at 1670°F (910°C). However, this temperature is decreased when carbon is present so that we may find γ (or austenite) stable as low as 1333°F (723°C). See Fig. 5–4.2. This temperature occurs where the solubility curves for carbon in γ and iron in γ cross, specifically, 99.2Fe–0.8C. A steel of this composition transforms during cooling by the reaction:

$$\gamma \ (0.8\% \ \text{C}) \rightarrow \alpha \ (0.025\% \ \text{C}) + \text{carbide } (6.67\% \ \text{C}). \tag{5–4.2}$$

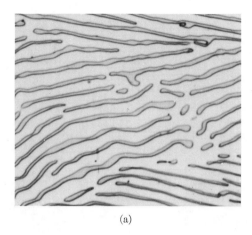

(a)

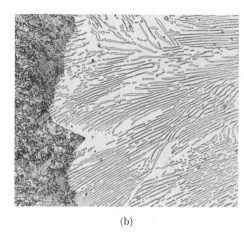

(b)

Eutectoid reactions. All *eutectic* reactions involve a liquid and two solid phases. Thus Eqs. (5–2.1) and (5–4.1) may be stated generally as:

$$L_2 \xrightleftharpoons[\text{heating}]{\text{cooling}} S_1 + S_3, \tag{5–4.3}$$

where the composition of the liquid is between the compositions of the two solid phases (hence the choice of subscripts).

The reaction of Eq. (5–4.2) is very much like the eutectic reaction, with one major exception: All phases are solid.

$$S_2 \xrightleftharpoons[\text{heating}]{\text{cooling}} S_1 + S_3. \tag{5–4.4}$$

We call this a *eutectoid* reaction (literally, eutectic-like). It has the eutectic-appearing V-notch in a phase diagram; also the composition of the single higher-temperature phase is between the compositions of the two lower-temperature phases. The reaction of Eq. (5–4.2) is the principal reaction during the heat-treatment of steels. It will be sufficiently useful that it is recommended that the reader become fully familiar with the details of the unshaded part of Fig. 5–4.2.

Pearlite. The Fe-C eutectoid reaction involves the simultaneous formation of ferrite and carbide from austenite of eutectoid composition. Since the α and C form simultaneously, they are intimately mixed in a characteristic lamellar microstructure which we call *pearlite* (Fig. 5–4.3). It may be formed in almost any steel which is cooled slowly from the austenite range into the [ferrite + carbide] temperature range.

Fig. 5–4.3 Pearlite. (a) The lamellar mixture of ferrite (lighter) and carbide (darker) forms from austenite of eutectoid composition (0.80 C) (×2500). (J. R. Vilella, U.S. Steel Corporation). (b) Pearlite formation (×1500). The pearlite grew from right to left in a temperature gradient before being quenched. The carbon has to diffuse to the carbide and leave the ferrite, which can hold no more than 0.025 w/o C. (B. L. Bremfitt and J. R. Kilpatrick, Bethlehem Steel Corporation)

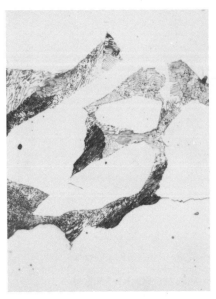  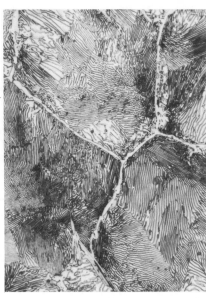

(a) 0.20% C   (b) 0.5% C   (c) 1.2% C

Fig. 5–4.4 Ferrite plus pearlite in Fe-C alloys (cf. Fig. 5–4.2) (×500). (a) A 0.20% C alloy contained about 75% α + 25% γ immediately prior to transformation. Since that austenite produced pearlite ($\gamma \to \alpha + C$), the microstructure is 75% proeutectoid ferrite plus 25% pearlite. (b) A 0.50% C alloy had more austenite (γ) than proeutectoid ferrite (α) before transformation. Therefore the major part of it is now pearlite. (c) A 1.2% C alloy contained carbide prior to transformation. The proeutectoid carbide formed at the grain boundaries of the former austenite grains. (U.S. Steel Corporation)

Pearlite is a specific mixture of ferrite and carbide. The two phases are layer-like or lamellar, and they must be of eutectoid origin. This distinction is important, since [α + C] may also form by reactions other than eutectoid reactions. With other origins, the microstructures are not lamellar, and have markedly different properties which we shall discuss later.

Since pearlite comes from austenite of eutectoid composition, the amount of pearlite present is equal to the amount of eutectoid austenite transformed. Figure 5–4.4(a), for example, shows that an 0.20% C steel is approximately three-fourths ferrite (white) and one-fourth pearlite (gray with lamellae barely resolvable at ×500). A 0.50% C steel (Fig. 5–4.4b) is mostly pearlite because it had more than 60% γ just prior to the eutectoid reaction.

Example 5–4.1. Determine the amount of γ and α in a 3-lb steel casting which contains 0.4% C as it is cooled to (a) 1600°F (871°C), (b) 1335°F (724°C), (c) 1331°F (722°C).

Solution

1600°F:	All γ		3.0 lb γ	0	lb α
1335°F:	γ(0.8% C) + α(0.025% C)		1.45 lb γ	1.55 lb α	
1331°F:	α(0.025% C) + C(6.67% C)		0	lb γ	2.83 lb α

Comments. The 1.45 lb of γ at 1335°F transforms to 1.45 lb of pearlite below the eutectoid temperature; this pearlite contains 1.28 lb α and 0.17 lb carbide. The steel thus contains 1.55 lb of *proeutectoid ferrite*, i.e., ferrite which formed before the eutectoid reaction, and 1.28 lb of *eutectoid ferrite*, which formed from the eutectoid reaction. ◀

Example 5–4.2. Show the phase fractions and compositions of ferrite, austenite, carbide, and liquid in an Fe-C alloy of 3.0 w/o carbon (97 w/o Fe) as they vary with temperature.

Solution

Above 2350°F	All liquid	
At	2065 (+) °F	0.435 liquid (4.3% C)
	1130 (+) °C	
		0.565 γ (2.0% C)
At	2065 (−) °F	0.785 γ (2.0% C)
	1130 (−) °C	
		0.215 C (6.67% C)
At	1333 (+) °F	0.625 γ (0.80% C)
	723 (+) °C	
		0.375 C (6.67% C)
At	1333 (−) °F	0.45 C (6.67% C)
	723 (−) °C	
		0.55 α (0.025% C)

See Fig. 5–4.5.

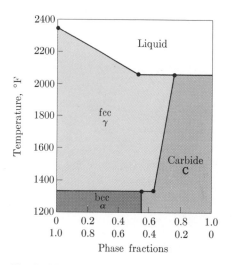

Fig. 5–4.5 Phase fractions versus temperature (97Fe–3C). See Example 5–4.2.

Comments. When cooling continues to room temperature under equilibrium conditions, even the minor one-fortieth of 1% of carbon will leave this ferrite. ◀

Example 5–4.3. Figure 5–4.6 is a photomicrograph of a steel containing only iron and carbon. It was cooled slowly through the eutectoid temperature and thus contains only ferrite and pearlite. Estimate the carbon content, X. (The densities of α and pearlite are 7.87 and 7.82 g/cm³, respectively.)

Solution. Approximately 45 v/o of the figure is pearlite and 55 v/o proeutectoid ferrite.

Basis: 1 cm³ of metal = 0.55 cm³ α + 0.45 cm³ pearlite

$$= 4.33\ g\ \alpha\ +\ 3.52\ g\ pearlite$$

$$= 7.85\ g\ alloy.$$

Total carbon = carbon in α + carbon in pearlite

$$(X\ w/o)(7.85\ g) = (0.025\ w/o)(4.33\ g) + (0.80\ w/o)(3.52\ g)$$

$$X = 0.37\ w/o\ C.$$

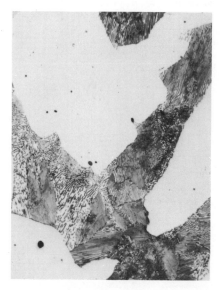

Fig. 5–4.6 Microstructure of Fe-C alloy (×500). See Example 5–4.3 (cf. Fig. 5–4.4).

Comments. There are various methods for estimating the volume percent (v/o). Probably the easiest is to take a piece of semitransparent graph paper, lay it at random over the photomicrograph, and count the grid intersections which lie in each constituent. ◀

5–5 PROPERTIES OF TWO-PHASE ALLOYS ⓔ

The properties of selected ferrous and nonferrous alloys are listed in Tables 5–6.3 through 5–6.7, and in Appendix B. Greatly expanded lists of similar data can be found in handbooks such as the *Metals Handbook*, published by the American Society for Metals. However, we shall not try to duplicate those data fully here. It will be useful, however, to examine some of the factors which affect properties in two-phase alloys. Most obvious among these are (1) the amount of each phase, (2) the size of the minor phase, and (3) the shape of the minor phase.

Properties versus amounts of phases. We can calculate the density of a two-phase mixture if we know the volume percent and the density of each phase. In fact, we did this in Example 5–4.3, when we found that 1 cm³ of metal weighed 7.85 g. As a formula, this calculation is

$$\rho_m = f_1\rho_1 + f_2\rho_2, \qquad (5\text{--}5.1)$$

where ρ_m is the density of the material and ρ_1 and ρ_2 are the densities of the two phases. The terms f_1 and f_2 represent volume fractions of the two phases. We call Eq. (5–5.1) a *mixture rule*, which in this case is relatively simple.

Fig. 5–5.1 Properties versus amounts of phases (annealed steels). (a) Hardness (read left) and strength (read right). (b) Ductility (read left) and toughness (read right).

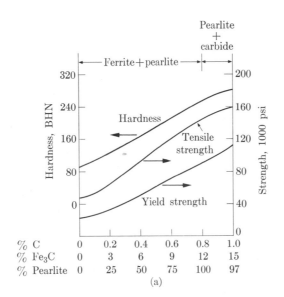
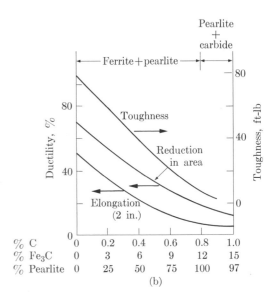

When other properties are involved, mixture rules are less simple. Rather than making calculations, it is easier for us just to observe the relationship of properties versus amounts of phases. Figure 5–5.1 shows (1) the hardness, (2) the strength, (3) the ductility, and (4) the toughness of annealed Fe-C alloys. The carbide is present in pearlite. In view of our description of the carbide phase in Section 5–4 as being hard and brittle, the data of Fig. 5–5.1 are just what we would expect. We should note, however, that 100% Fe_3C would give us a very weak material ($\sim$5000 psi) because of the cracking tendency of the brittle carbide. In fact, white cast iron, which has upwards of 40% carbide, has very restricted use as a load-bearing material because of its extreme brittleness and lack of strength.

Properties versus phase size. When steels are cooled slowly, the austenite transforms more slowly to $[\alpha + \mathbf{C}]$, and forms much coarser pearlite than normal. Figure 5–5.2 compares the hardness for Fe-C alloys with coarse pearlite versus fine pearlite. For a given composition, fine pearlite has more (though admittedly thinner) layers of carbide and of ferrite than coarse pearlite does. Slip occurs less readily in the finer pearlite, with a resulting increase in hardness.

Properties versus phase shape. We said in Section 5–4 that pearlite is lamellar, and that the nonlamellar mixtures of $[\alpha + \mathbf{C}]$ have markedly different properties. Figure 5–5.3 repeats our earlier photomicrograph of pearlite for comparison with *spheroidite*. Both have the same composition (0.8 w/o C), and therefore the same volume fraction of carbide. In spheroidite, unlike the situation in pearlite, the more nearly "sphere-like" particles of carbide were formed by extended heating below the eutectoid temperature, at 1250°F ($\sim$675°C).

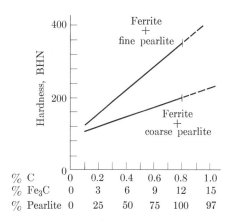

% C	0	0.2	0.4	0.6	0.8	1.0
% Fe₃C	0	3	6	9	12	15
% Pearlite	0	25	50	75	100	97

Fig. 5–5.2 Properties versus phase size. The coarser pearlite was obtained by slower cooling.

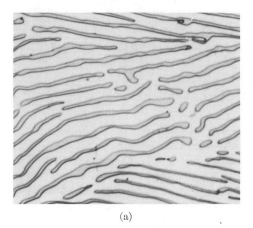

(a)

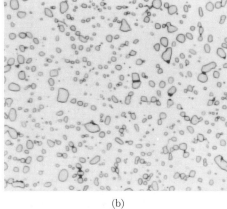

(b)

Fig. 5–5.3 Microstructures of eutectoid steels (0.80 % C) ($\times$ 2500). (a) Pearlite formed by transforming austenite (γ) of eutectoid composition. (b) Spheroidite formed by tempering at 1300°F. (Courtesy U.S. Steel Corporation)

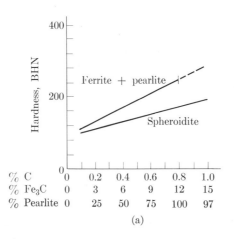
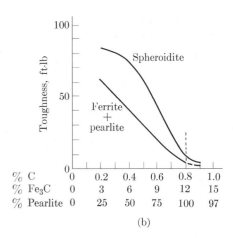

Fig. 5–5.4 Properties versus grain shape (cf. Fig. 5–5.3).
(a) Hardness. (b) Toughness.

Figure 5–5.4 compares the hardness and toughness properties. The difference is significant. The hardness of [ferrite + pearlite] is higher than that of spheroidite because the two-dimensional carbide lamellae provide more reinforcement to the soft ferrite matrix than do the equiaxed carbide particles in spheroidite. In contrast, the toughness of spheroidite is greater than pearlite-containing metal. A progressing crack can follow carbide layers through pearlite, but in spheroidite it must traverse the tougher ferrite.

Example 5–5.1. Refer to Fig. 5–6.3(b), which contains ferrite and graphite (dark spheres). Estimate the density of this alloy (97Fe–3C) if the densities of the two phases are 7.8 and 2.0 g/cm^3, respectively. Use the grid method to obtain approximately 12 v/o graphite (see comments on Example 5–4.3).

Solution

$$\rho = (0.88)(7.8 \text{ g/cm}^3) + (0.12)(2.0 \text{ g/cm}^3)$$

$$= 7.1 \text{ g/cm}^3.$$

Comments. Equation (5–5.1) is independent of shape and size of the phases when we use it to determine density or specific heat. This particular mixture rule is usually not applicable when we are calculating other properties.

The carbon in this alloy is not in the form of Fe_3C, as in steels; rather it is present in the form of graphite. This difference occurs in cast irons, particularly if there is a high silicon content. We shall see in Section 5–6 that different cast irons have significantly different properties, depending on the shape and amount of this graphite. ◀

5-6 COMMERCIAL ALLOYS (e)

The largest single group of commercial alloys are the *steels*, which are basically alloys of iron and carbon, plus selected amounts of other elements. They are also commonly referred to as ferrous alloys. Among the *nonferrous alloys* are the convenient but nonrigorous categorizations of copper-base alloys, light metals (aluminum and magnesium), white metals (zinc, lead, tin, cadmium, etc.), titanium, and superalloys.

Steels. The role of carbon in steel is sufficiently important so that identification procedures commonly indicate the carbon content. For example, a 1040 steel has 0.40 w/o carbon, while a 4068 steel has

Table 5-6.1

AISI AND SAE LOW-ALLOY STEELS

AISI or SAE number	Composition
10xx	Plain carbon steels
11xx	Plain carbon (resulfurized for machinability)
13xx	Manganese (1.5–2.0%)
23xx	Nickel (3.25–3.75%)
25xx	Nickel (4.75–5.25%)
31xx	Nickel (1.10–1.40%), chromium (0.55–0.90%)
33xx	Nickel (3.25–3.75%), chromium (1.40–1.75%)
40xx	Molybdenum (0.20–0.30%)
41xx	Chromium (0.40–1.20%), molybdenum (0.08–0.25%)
43xx	Nickel (1.65–2.00%), chromium (0.40–0.90%), molybdenum (0.20–0.30%)
46xx	Nickel (1.40–2.00%), molybdenum (0.15–0.30%)
48xx	Nickel (3.25–3.75%), molybdenum (0.20–0.30%)
51xx	Chromium (0.70–1.20%)
52xx	Chromium (1.30–1.60%)
61xx	Chromium (0.70–1.10%), vanadium (0.10%)
81xx	Nickel (0.20–0.40%), chromium (0.30–0.55%), molybdenum (0.08–0.15%)
86xx	Nickel (0.30–0.70%), chromium (0.40–0.85%), molybdenum (0.08–0.25%)
87xx	Nickel (0.40–0.70%), chromium (0.40–0.60%), molybdenum (0.20–0.30%)
92xx	Silicon (1.80–2.20%)

xx Carbon content, 0.xx w/o.
All steels have 0.50 ± w/o manganese, unless stated otherwise.

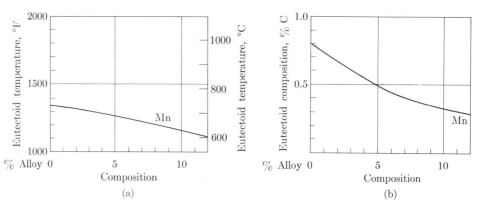

Fig. 5–6.1 Eutectoid shift in steel caused by addition of manganese. Manganese additions lower both (a), the eutectoid temperature, and (b) the eutectoid carbon content.

0.68 w/o carbon. The last two numbers of AISI and SAE coded steels (Table 5–6.1) indicate the *points* of carbon present, i.e., hundredths of one percent (0.68 w/o = 68 points). There has to be a small working range; thus a 0.40 carbon steel may have 38–43 points of carbon. The highest carbon content in normal steels is about 1.00 w/o in the 52xxx series, in this case 52100.

The term *plain-carbon steel* is used when there is a minimum of alloying elements other than carbon. Plain-carbon steels make up a major portion of the steel produced, because most of our structural steels fall in this category.

Manganese, nickel, chromium, and molybdenum are the more common alloying elements added to steels. Large tonnages of steels are made with up to 5 w/o additions. These are commonly called *low-alloy steels*, as contrasted to high-alloy steels (discussed below). Table 5–6.1 indicates the nomenclature scheme for plain-carbon and low-alloy steels. Low-alloy steels are widely used when heat-treating is required. We shall see in Chapter 6 that the temperatures and times used in heat treatment depend on both the alloy content and the carbon content. Therefore the notation procedure of Table 5–6.1 is highly informative to the process metallurgist.

Steels for severe service conditions commonly have a high alloy content. *High-alloy steels* include the *stainless steels*, which, because of their chromium and nickel contents, are designed to be resistant to corrosive conditions. *Tool steels* commonly have unusually high levels of those alloying elements which form carbides, because hard, stable carbides retain their hardness under the severe stress and the locally high temperatures encountered in metal-removal operations (machining). The common "carbide formers" are tungsten, chromium, molybdenum, and vanadium. *High-temperature steels* must possess resistance to oxidation and creep when they are used in applications involving destructive service temperatures. Strength at high

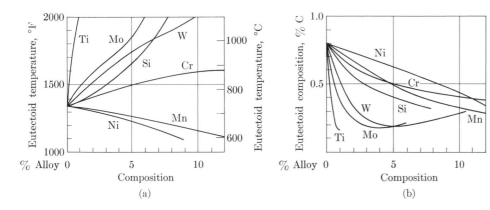

Fig. 5–6.2 Eutectoid shifts in steel caused by addition of alloying elements. (a) Shift in temperature of the eutectoid. (b) Shift in the carbon content of the eutectoid. (Adapted from ASM data.)

temperatures is best attained by adding considerable quantities of chromium, nickel, and molybdenum for solution hardening (Section 4–4). The first two of these, plus cobalt, also reduce oxidation rates. Since stainless steels, tool steels, and high-temperature steels generally have high alloy contents, they fall outside the nomenclature pattern of Table 5–6.1.

The steels discussed in Section 5–4 were Fe-C alloys with no other alloying elements present. In alloy steels, both the carbon and the iron atoms coordinate with other types of atoms in addition to iron atoms. Therefore we should not be surprised that the eutectoid is altered from 0.80% C and 1333°F (723°C). Figure 5–6.1 shows the shift which takes place in the eutectoid when manganese is added. All alloying elements lower the eutectoid composition to values less than 0.8 w/o carbon (Fig. 5–6.2). Manganese and nickel lower the eutectoid temperature, while chromium, tungsten, silicon, molybdenum, and titanium raise the eutectoid temperature.

Since the eutectoid composition is changed by the addition of alloying elements, the composition and amount of the pearlite is modified. For example, 2.0 w/o manganese lowers the eutectoid composition to 0.65 w/o C (Fig. 5–6.1b). Thus a 0.40 w/o C steel will contain 60 w/o γ (rather than approximately 50 w/o) immediately above the eutectoid temperature of approximately 1320°F (715°C). (See Example 5–6.1.)

Cast iron. ◎ When carbon is a major alloying element (3.0–3.5 w/o), the temperature of 100% liquid drops to below 2300°F (1260°C) as compared to 2800°F for pure iron and ~2750°F for most steels (Fig. 5–4.2). This reduced temperature makes high-carbon alloys very adaptable for castings. The liquid metal fills the mold readily and reacts very little with the mold surface. Therefore it is used extensively as *cast iron*.

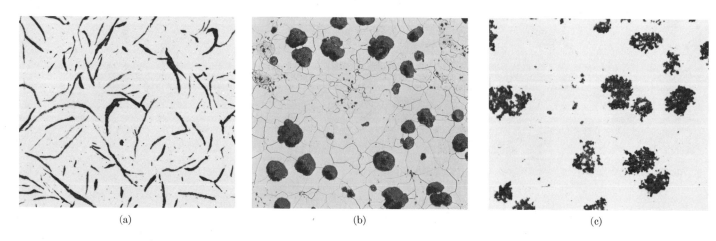

(a)   (b)   (c)

Fig. 5–6.3 Cast iron microstructures. (a) Gray cast iron (×100). The dark constituents are graphite flakes. (Courtesy J. E. Rehder) (b) Nodular cast iron (×100). Carbon is present as spheroids in the iron. As a result this material is more ductile than gray cast iron. (G. A. Colligan, Dartmouth) (c) Malleable iron (×100). Carbon is present as graphite clusters. These were formed by carbide dissociation (Fe$_3$C → 3Fe + graphite) within the solid iron. (American Society for Metals)

When carbon is at this high level and 2–3% silicon is also present, the Fe$_3$C dissociates to *graphite* and γ (or to graphite and α below the eutectoid temperature):

$$Fe_3C \xrightarrow{\;>1333°F\;} 3\,Fe_\gamma + C_{gr}, \qquad (5\text{–}6.1)$$

$$Fe_3C \xrightarrow{\;<1333°F\;} 3\,Fe_\alpha + C_{gr}. \qquad (5\text{–}6.2)$$

Of course, the γ and α are saturated with carbon. When the graphite forms within cast iron during solidification without special control, it appears in the microstructure as flakes (Fig. 5–6.3a). Obviously this weakens the alloy (and gives a gray rather than metallic-colored fracture, hence the label *gray iron*). In some applications, however, castability is more important than strength. Furthermore, the graphite flakes can even be advantageous where vibrations are a problem because they give a "damping capacity" to the metal. Such cast irons are used extensively in base castings of large machines.

With special control—e.g., through the addition of small amounts of magnesium—the graphite can be made to form nodules or spheres (Fig. 5–6.3b) rather than flakes. The resulting metal, called *nodular iron*, is more ductile than the gray cast iron simply because the graphite provides less opportunity for stress concentrations. Finally, those cast irons which contain <2% Si form graphite slowly and solidify as (γ + carbide). It is possible to have a *white cast iron* (no graphite), but a cast iron which will, on extended reheating, react according to Eqs. (5–6.1) and (5–6.2). When the graphite forms in the solid casting, its microstructure is that shown in Fig. 5–6.3(c). This early form of cast iron was the first cast iron to exhibit some ductility (or malleability); hence the name *malleable iron*.

It is worth pausing a moment to compare the microstructures of
Fig. 5–6.3. If each cast iron is heat-treated to complete the reaction of
Eq. (5–6.2), and thereby produce only ferrite and graphite, their
ductilities may be compared with that of a ferritic steel (Table 5–6.2).
The effects depend not only on the presence of graphite, but also on the
shape of the graphite flakes. Other data for cast irons are given in
Table 5–6.3.

Table 5–6.2

DUCTILITY OF FERRITIC ALLOYS

Iron type	Percent elongation (2 inches)	See Figure
Ferritic steel	40	—
Ferritic nodular iron	15 to 20	5–6.3(b)
Ferritic malleable iron	15±	5–6.3(c)
Ferritic gray iron	1	5–6.3(a)

Table 5–6.3

TYPICAL MECHANICAL PROPERTIES OF VARIOUS IRON–CARBON ALLOYS*

	Alloy	Tensile strength, lb/in^2	Yield strength, lb/in^2	Elongation in 2 inches, %	Modulus of rupture, lb/in^2
← Increasing cost (rough estimate for finished part)	Ferritic gray cast iron	25,000	–	0.5	50,000
	Pearlitic gray cast iron	45,000	–	0.5	65,000
	Nodular cast iron	80,000	50,000	15–20	–
	Malleable cast iron	55,000	35,000	18	–
	Cast steel (as cast)	74,000	34,000	19	
	Cast steel (normalized)	76,000	38,000	24	
	Machine steel, 0.2% C	50,000	27,000	35	
	Hardened and tempered 0.45% C steel	100,000	65,000	28	
	Wrought iron	50,000	30,000	30	
	Ingot iron (ferrite)	45,000	30,000	40	
	Cold drawn 0.80% C steel	200,000	150,000	5	
	Spheroidized 1.0% C steel	80,000	45,000	30	
	Hardened and tempered 1.0% C steel	250,000	200,000	1	

* From *Physical Metallurgy for Engineers*, by A. G. Guy, Reading, Mass.: Addison-Wesley.

Table 5–6.4

PROPERTIES OF SEVERAL BRASSES*

Alloy	Composition, %					Tensile strength, lb/in²	Yield strength, lb/in²	Elongation in 2 inches, %	Typical applications
	Cu	Zn	Sn	Pb	Mn				
Wrought alloys (annealed)									
Gilding metal	95	5				30,000	10,000	40	Coins, jewelry base for gold plate
Commercial bronze	90	10				33,000	10,000	45	Grillwork, costume jewelry
Red brass	85	15				36,000	10,000	48	Weatherstrip, plumbing lines
Low brass	80	20				38,000	12,000	52	Musical instruments, pump lines
Cartridge brass	70	30				40,000	11,000	65	Radiator cores, lamp fixtures
Yellow brass	65	35				43,000	16,000	55	Reflectors, fasteners, springs
Muntz metal	60	40				54,000	21,000	40	Large nuts and bolts, valve stems
Low-leaded brass	64.5	35		0.5		43,000	16,000	55	Ammunition primers, plumbing
Medium-leaded brass	64	35		1.0		43,000	16,000	52	Hardware, gears, screws
Free-cutting brass	62	35		3.0		43,000	17,000	51	Automatic screw machine stock
Admiralty metal	71	28	1			53,000	22,000	65	Condenser tubes, heat exchangers
Naval brass	60	39	1			58,000	27,000	45	Marine hardware, valve stems
Manganese bronze	58.5	39.2	1	1 Fe	0.3	65,000	30,000	33	Pump rods, shafting rods
Cast alloys									
Cast red brass	85	5	5	5		34,000	17,000	25	Pipe fittings, small gears
Cast yellow brass	60	38	1	1		40,000	14,000	25	Hardware fittings, ornamental castings
Cast manganese bronze	58	39.7	1 Al	1 Fe	0.3	70,000	28,000	30	Propeller hubs and blades

* From *Physical Metallurgy for Engineers*, by A. G. Guy, Reading, Mass.: Addison-Wesley.

Copper-base alloys. The historical importance of copper-base alloys becomes apparent to us when we read about the Bronze Age, in which Cu-Sn alloys significantly affected the advance of civilization. A "Brass Age" could have followed as a result of the development of Cu-Zn alloys if iron had not been developed concurrently.

Copper-base alloys have two main applications: (1) As corrosion-resistant metals, the bronzes have been important to both modern and ancient man, particularly in marine environments. Copper-nickel alloys have now gained equal importance. (2) As materials for electrical conduction. Here pure copper is best, because, as we saw in Section 4–6, any added alloying elements reduce the electrical conductivity.

Except for beryllium bronze (Cu + 2 to 6% Be), which has a eutectoid similar to the Fe-C eutectoid, the simple copper-base alloys

Table 5–6.5

PROPERTIES OF SEVERAL BRONZES*

Alloy	Condition	Composition				Tensile strength, lb/in²	Yield strength, lb/in²	Elongation in 2 inches, %	Typical applications
		Cu	Sn	Zn	Pb				
Tin bronzes									
5% phosphor bronze	Wrought, annealed	94.8	5		0.2 P	47,000	19,000	64	Diaphragms, springs, switch parts
10% phosphor bronze	Wrought, annealed	89.8	10		0.2 P	66,000	—	68	Bridge bearing plates, special springs
Leaded tin bronze	Sand cast	88	6	4.5	1.5	38,000	15,000	35	Valves, gears, bearings
Gun metal	Sand cast	88	10	2		40,000	15,000	30	Fittings, bolts, pump parts
Aluminum bronzes (eutectoid hardening)									
5% aluminum bronze	Wrought, annealed	95			5 Al	60,000	25,000	66	Corrosion-resistant tubing
10% aluminum bronze	Sand cast	86		3.5 Fe	10.5 Al	75,000	35,000	20	
Same	Sand cast and hardened					100,000	45,000	12	Gears, bearings, bushings
Silicon bronze									
Silicon bronze, type A	Wrought, annealed	95		1	1 Mn 3 Si	56,000	21,000	63	Chemical, equipment, hot water tanks
Nickel bronzes									
30% cupro-nickel	Wrought, annealed	70			30 Ni	60,000	25,000	45	Condenser, distiller tubes
18% nickel silver	Wrought, annealed	65		17	18 Ni	56,000	25,000	42	Table flatware, zippers
Nickel silver	Sand cast	64	4	8	4 20 Ni	40,000	25,000	15	Marine castings, valves
Beryllium bronzes (precipitation hardening)									
Beryllium copper	Wrought, annealed	98			0.3 Co 1.7 Be	70,000	30,000	50	Springs, nonsparking tools
Same	Wrought, hardened					175,000	130,000	5	

* From *Physical Metallurgy for Engineers*, by A. G. Guy, Reading, Mass.: Addison-Wesley.

are seldom heat-treatable, and the hardening mechanism we shall describe in Chapter 6 for steels is not applicable. This means that copper-base alloys are usually strengthened by cold work or by solution-hardening (Section 4–4). Thus most copper-base alloys are single-phase metals with the same fcc structure as copper. Tables 5–6.4 and 5–6.5 set forth the properties of selected copper-base alloys.

Light metals. The light metals are primarily magnesium alloys and aluminum alloys. Magnesium is very attractive for certain applications because of its low density ($\rho = 1.74$ g/cm^3). The hexagonal structure of magnesium, however, precludes extensive plastic deformation because it lacks the greater symmetry that is present in cubic metals. Therefore magnesium is most commonly used as a cast alloy (Fig. 5–6.4).

Fig. 5–6.4 Magnesium casting (lawnmower housing). Although they are not easily rolled or forged, magnesium alloys are widely used for intricate light-weight castings. (Courtesy J. D. Hanawalt and Dow Chemical Co.)

Aluminum, somewhat denser ($\rho = 2.7$ g/cm^3), but still significantly lighter than steels or copper-base alloys, is cubic in structure and thus very amenable to deformation processing. Aluminum sheet and foil illustrate the extensive deformation which is possible. Both aluminum and magnesium are relative latecomers into the materials repertoire of the engineer because they strongly resist separation from their oxide raw materials. It was necessary to develop special technologies before their use became commercially feasible. Their affinity for oxygen is reflected in the oxidation and corrosion tendencies of magnesium and aluminum alloys. Fortunately aluminum oxide (Al_2O_3) produces a protective coating on aluminum. A similar coating over the surface of magnesium is not as protective. In salt water, even the protective coating on aluminum may become unstable. The engineer obviously must take these facts into account as he specifies materials for his designs.

Aluminum can be strengthened by cold work, but is subject to annealing at relatively low temperatures because of its low melting temperature (Section 4–9). In general, solid solubility between the light metals and other common metals is limited. Therefore solution-hardening is not a useful procedure. In Chapter 6 we shall encounter another strengthening process (precipitation hardening) which is particularly adaptable to the light metals. Tables 5–6.6 and 5–6.7 list various light metal alloys and their properties, and Table 5–6.8 gives the alloy designations.

Table 5–6.6

COMPOSITION AND PROPERTIES OF TYPICAL MAGNESIUM ALLOYS*

Use	Am. Soc. for Test. and Mat'ls designation	Composition, %				Condition	Tensile strength, lb/in^2	Yield strength, lb/in^2	Elongation in 2 inches, %
		Al	Zn	Mn	Si				
Sand and permanent mold casting	AZ92	9.0	2.0	0.1	0.2	As-cast	24,000	14,000	2
						Solution treated	39,000	14,000	10
						Aged	39,000	21,000	3
Die casting	AZ91	9.0	0.7	0.2	0.2	As-cast	33,000	21,000	3
Sheet	AZ31X	3.0	1.0	0.2	0.2	Annealed	35,000	20,000	15
						Hard	40,000	31,000	8
Structural shapes	AZ80X	8.5	0.5	0.2	0.2	Extruded	48,000	32,000	12
						Extruded and aged	52,000	37,000	8
	ZK60A		6.0		0.6	Extruded	49,000	38,000	12
					Zr	Extruded and aged	51,000	42,000	10

* From *Physical Metallurgy for Engineers*, by A. G. Guy, Reading, Mass.: Addison-Wesley.

Table 5–6.7

SOME CHARACTERISTICS OF SEVERAL TYPES OF ALUMINUM ALLOYS§

Alloy designation	Principal alloying elements	Hardening process*	Range of properties (soft to hard conditions)				Typical applications
			Tensile strength, lb/in^2	Yield strength, lb/in^2	Elongation in 2 inches, %	Endurance limit (5×10^8 cycles), lb/in^2	
Wrought alloys							
1100	Commercial purity	Cold-working	13,000–24,000	5,000–22,000	45–15	5,000–9,000	Cooking utensils
5052	2.5% Mg	Cold-working	28,000–42,000	13,000–37,000	30– 8	16,000–20,000	Bus and truck bodies
Alclad 2024	4.5% Cu, 1.5% Mg (with protective sheet of pure aluminum)	Precipitation	27,000–68,000	11,000–47,000	22–19	13,000–20,000	Aircraft
6061	1.5% Mg$_2$Si	Precipitation	18,000–45,000	8,000–40,000	30–17	9,000–14,000	General structural
7075	5.6% Zn, 2.5% Mg, 1.6% Cu	Precipitation	33,000–83,000	15,000–73,000	16–11	23,000	Aircraft
Cast alloys							
195	4.5% Cu	Precipitation	32,000–41,000	16,000–32,000	8.5–2	7,000–8,000	Sand castings
319	3.5% Cu, 6.3% Si	Precipitation	27,000–36,000	18,000–26,000	2–1.5	10,000–11,000	Sand castings
356	7% Si, 0.3% Mg	Precipitation	38,000	27,000	5	13,000	Permanent mold castings

* In addition to alloy hardening.

§ From *Physical Metallurgy for Engineers*, by A. G. Guy, Reading, Mass.: Addison-Wesley.

Titanium alloys. ◎ A metal which has come to the forefront relatively recently is titanium. Titanium is interesting to the engineer because it is less dense and therefore lighter than steel (4.5 g/cm^3 versus 7.8 g/cm^3), yet is as stiff as steel, and has a high resistance to corrosion (under oxidizing conditions). However, titanium is relatively expensive because it is difficult to remove all the oxygen from its ore.

Pure titanium transforms from hcp (α) to bcc (β) at 885°C (1625°F):

$$\alpha\text{-Ti} \underset{\text{cooling}}{\overset{\text{heating}}{\underset{885°C}{\rightleftharpoons}}} \beta\text{-Ti.} \qquad (5\text{–}6.3)$$

This transformation in titanium alloys may be compared to the ferrite ⇄ austenite transformation in iron. In titanium alloys, how-

Table 5–6.8

ALUMINUM ALLOY IDENTIFICATION CODES

Composition designations

1xyy	Unalloyed aluminum ($>99\%$ Al)
2xxx	Al + Cu as principal alloying element
3xxx	Al + Mn as principal alloying element
4xxx	Al + Si as principal alloying element
5xxx	Al + Mg as principal alloying element
6xxx	Al + Mg + Si as principal alloying elements
7xxx	Al + Zn as principal alloying element
8xxx	Al + other elements

yy = points of purity. Thus 1060 = 99.60% Al;
       1090 = 99.90% Al; etc.

Temper designations (suffixes to composition designations)

—F	As fabricated	
—O	As annealed	
—H	Strain hardened by a cold-working process	
	—H1X	Hardened only, with X representing fraction hardness (8 = fully hard)
	—H2X	Hardened and partially annealed
	—H3X	Hardened and stabilized
—T	Heat treated	
	—T2	Annealed (cast alloys)
	—T3	Solution heat-treated and cold worked
	—T4	Solution heat-treated and aged naturally
	—T5	Artificially aged only
	—T6	Solution heat-treated and artificially aged
	—T7	Solution heat-treated and stabilized
	—T8	Solution heat-treated, cold-worked, and aged
	—T9	Solution heat-treated, aged, and cold-worked
	—T10	Same as —T5, followed by cold-working

ever, unlike the case of steels, there is not a comparably useful eutectoid with carbon.

Titanium alloys fall into two categories: α alloys and α-β alloys. The α alloys include (1) commercially pure titanium, which transforms readily to the hexagonal phase on cooling, and (2) a few aluminum- and tin-containing alloys, for example, 92Ti-5Al-2.5Sn, which favor α formation. Most alloying elements favor the β, or bcc phase. Therefore we find widespread use of α-β alloys, i.e., alloys which retain some β at room temperature. The α-β alloys have yield strengths about twice as high as those of the α alloys (∼170,000 psi versus 80,000 psi for commercially pure titanium). The α alloys are more desirable for situations which require corrosion resistance.

White metals. The metals of this category not only have a white "color," but also have relatively low melting temperatures. Lead-tin *solders* (Fig. 5–2.4) exemplify much of this group, and are specially selected with regard to their eutectic. *Die-casting alloys* are also selected on the basis of melting temperatures. Here zinc-base alloys are common because they have greater strength than alloys with lead or tin bases. Since die-casting alloys do not have to be plastically deformed, the hexagonal structure of zinc is not a major handicap. (Aluminum die-cast alloys are also common.)

Superalloys. ◎ Steels, even those designed to withstand high temperatures, have limited service above 1500°F (800°C), where oxidation and creep rates can become excessive. Since nickel and cobalt have less tendency toward oxidation than iron, many superalloys—i.e., alloys used for very high temperature service—utilize these as the base constituent (Fig. 5–6.5). Alloys of *refractory metals*, such as molybdenum (T_m = 2625°C, 4760°F), tungsten (3410°C, 6170°F), niobium (2415°C, 4380°F), tantalum (2996°C, 5425°F), and zirconium (1750°C, 3180°F) are very useful in high-temperature applications because the bonding between their atoms is exceptionally strong, and the plastic deformation and creep rates are low. Unfortunately their resistance to high temperature also makes them expensive to refine and fabricate. Also molybdenum and tungsten have very poor resistance to oxidation, and so must have protective coatings added to their surfaces.

The problem of the development and application of metals for use at high temperatures is a particularly challenging one, since the engineer needs to know the structure, properties, and behavior of all available materials.

Example 5–6.1. An AISI 1340 steel which has 1.95 w/o Mn is austenitized (i.e., heated until it is transformed to 100% γ). (a) What temperature is required if the processing specification calls for heating 50°F into the γ-range? (b) What fraction of the steel will be pearlite after it has been annealed from the austenitizing temperature?

Solution. From Fig. 5–6.2, the eutectoid temperature and composition are ∼1320°F and 0.65 w/o C, respectively. Therefore we may redraw the pertinent part of Fig. 5–4.2 in Fig. 5–6.6. (a) ∼1430°F + 50°F ≅ 1480°F (∼800°C), (b) ∼60% γ (= ∼60% P).

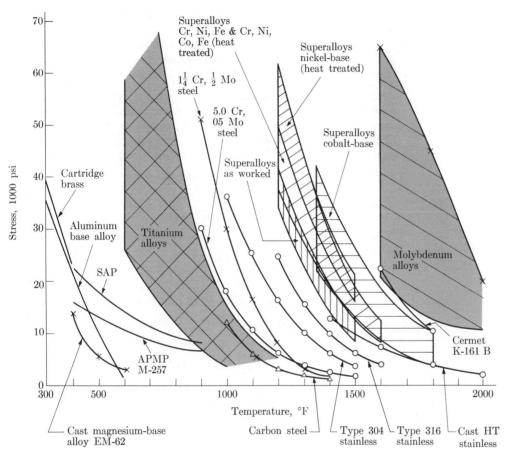

Fig. 5–6.5 Superalloys. Stresses to produce rupture in 1000 hours. (See Section 4–10.) (H. C. Cross and W. F. Simmons, *Utilization of Heat-Resistant Alloys*, Metals Park, O.: American Society for Metals)

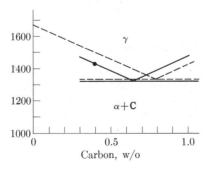

Fig. 5–6.6 Eutectoid in AISI 13xx steels (solid lines). The dashed lines are for Mn-free Fe-C alloys (Fig. 5–4.2). See Example 5–6.1.

Comments. To a first approximation, we can draw the comparable curves of Fig. 5–6.6 parallel to each other.

We shall not be able to determine precisely how much carbide there is in an alloy steel because the carbide, no longer pure Fe_3C, does not contain exactly 6.67% C. ◀

Example 5–6.2. ◎ A malleable cast iron containing 3.0 w/o C was initially solidified as [γ + carbide]; it was then reheated (malleablized) to dissociate the carbide. What is the maximum v/o graphite which may form (a) at 1350°F (730°C)? (b) At 1250°F (680°C)? For both α and γ, $\rho \approx 7.6$ g/cm^3 at these temperatures. The density of graphite is approximately 2.0 g/cm^3.

Solution. Basis: 100 g. Refer to Fig. 5–4.2, but with graphite.

a) At 1350°F:
$$g\ \gamma = \left(\frac{100 - 3.0}{100 - 0.8}\right) 100\ g = 97.8\ g\ \gamma,$$

$$cm^3\ \gamma = 97.8\ g/(7.6\ g/cm^3) = 12.9\ cm^3.$$

$$g\ graphite = \left(\frac{3.0 - 0.8}{100 - 0.8}\right) 100\ g = 2.2\ g\ graphite.$$

$$cm^3\ graphite = 2.2\ g/(2.0\ g/cm^3) = 1.1\ cm^3.$$

$$v/o\ graphite = 100(1.1\ cm^3)/(14.0\ cm^3) = 8.0\ v/o\ graphite.$$

b) At 1250°F:
$$cm^3\ \alpha = \left(\frac{100 - 3.0}{100 - 0.02}\right) \frac{(100\ g)}{(7.6\ g/cm^3)} = 12.8\ cm^3.$$

$$cm^3\ graphite = \left(\frac{3.0 - 0.02}{100 - 0.02}\right) \frac{(100\ g)}{(2\ g/cm^3)} = 1.5\ cm^3.$$

$$v/o\ graphite = 100\ 1.5\ cm^3)/(14.3\ cm^3)$$

$$= 10.5\ v/o\ graphite.$$

Comments. By selecting appropriate processing treatments, one can obtain ferritic malleable irons, pearlitic malleable irons, pearlitic nodular irons, ferritic gray irons, etc. Each has its own set of properties (Table 5–6.3). ◀

REVIEW AND STUDY

5–7 SUMMARY

Many of our highly useful alloys contain more than one phase. Phase diagrams provide us with a means of determining (a) *which* phases are present, (b) the *compositions* of phases, and (c) the *amounts* of phases. The lever rule (Eq. 5–3.2) is used for computing the amounts of phases.

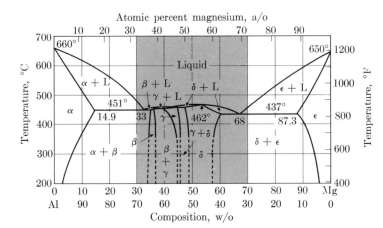

Fig. 5–7.1 The Al-Mg system. (ASM *Handbook of Metals,* Metals Park, O.: American Society for Metals)

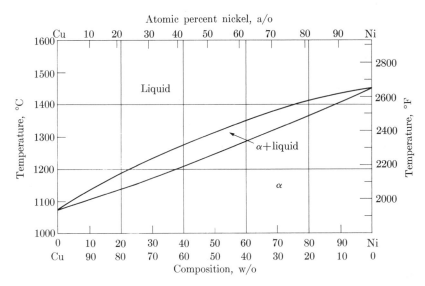

Fig. 5–7.2 The Cu-Ni system. (ASM *Handbook of Metals,* Metals Park, O.: American Society for Metals)

The Fe-C diagram may be used to predict microstructures of plain-carbon and low-alloy steels which are annealed or slowly cooled. Thus the *eutectoid region* of this diagram (Fig. 5–4.2) is worth learning in considerable detail.

The properties of commercial alloys vary considerably from metal to metal; they also vary with the amounts of constituent phases, the phase size, and the phase shape.

The following binary phase diagrams are found in this book.

Terms for Study

AISI-SAE code	Light metals
Austenite, γ	Pearlite
Carbide, **C**	Phase
Cast iron	Phase amount
Cast iron, gray	Phase composition
Cast iron, malleable	Phase diagram
Cast iron, nodular	Proeutectoid ferrite
Cast iron, white	Solder
Die-casting alloys	Solubility limit
Equilibrium diagram	Solution treatment
Eutectic composition	Spheroidite
Eutectic reaction	Steel
Eutectic temperature	Steels, low-alloy
Eutectoid composition	Steels, plain-carbon
Eutectoid ferrite	Steels, stainless
Eutectoid reaction	Sterling silver
Eutectoid temperature	Superalloys
Ferrite, α	Supersaturation
Lever rule	White metals

STUDY PROBLEMS

5–1.1 Lead may dissolve 5 w/o tin in solid solution at 100°C and 18 w/o at 175°C. Scrap (100 lb) containing 95Pb–5Sn is to be melted. How much more tin could one add and still not exceed the solid solubility limit after the material has been frozen and cooled to 175°C?

Answer. 15.8 lb additional tin.

5–2.1 What phases are present at each 100°C interval during the cooling of a 20Ag–80Cu alloy?

Answer. (a) 1000°C, liquid, (b) 900°C, 800°C, (L + β), (c) 700°C, 600°C, 500°C, etc., (α + β).

5–2.2 What are the compositions of α in the previous problem?

Answer. (a) 700°C: 94Ag–6Cu, (b) 600°C: 96.5Ag–3.5Cu, (c) 500°C: 98Ag–2Cu, (d) 400°C: 99Ag–1Cu, (e) <400°C: >99Ag–<1Cu.

5–2.3 Cite the phases and their compositions for a 50Pb–50Sn alloy (a) at 184°C (363°F), (b) at 182°C (360°F).

Answer. (a) α: 80.8Pb–19.2Sn, Liq: 38.1Pb–61.9Sn, (b) α: 80.8Pb–19.2Sn, β: 2.5Pb–97.5Sn.

5–2.4 Pure copper and pure zinc are bonded together and heated to 750°F (400°C) for 60 days. In that period of time at that temperature, zinc diffuses into copper and copper into zinc. What phases develop?

Answer. Numbers refer to % Zn. Phases in sequence: $0 \xrightarrow{\alpha} 39, 46 \xrightarrow{\beta'} 50,$ $59 \xrightarrow{\gamma} 69, 78 \xrightarrow{\varepsilon} 87, 97.5 \xrightarrow{\eta} 100.$

5–3.1 Determine the amount of α and β in 73 lb of a 62Cu–38Zn alloy (a) at 875°C (1607°F); (b) at 700°C (1292°F); (c) at 450°C (842°F).

Answer. (a) 73 lb β; (b) 53 lb α–20 lb β; (c) 73 lb α

5–3.2 Refer to Example 5–2.2. How much γ and ε are there in 100 lb of a 26Cu–74Zn alloy at (a) 600°C (1112°F); (b) 525°C (977°F)?

Answer. (a) No γ, all δ; (b) 53 lb γ–47 lb ε

5–3.3 One hundred grams of liquid of a Pb-Sn eutectic are solidified to α + β at 183°C. How many grams of each are present at 182°C?

Answer. 45.5 g α–54.5 g β

5–3.4 Plot the fractions of α, β, and liquid versus temperature for a 50Pb–50Sn alloy.

Answer. See Fig. 5–3.3

5–4.1 Seventy grams of ferrite are saturated with carbon at 1333°F (723°C). How many additional grams of carbon will be required to saturate that iron with carbon when it transforms to austenite?

Answer. 0.54 g

5–4.2 One hundred grams of austenite containing 0.8 w/o C transform to pearlite. How many grams of ferrite and carbide form?

Answer. 88 g α, 12 g C

5–4.3 Show the phase fractions of ferrite versus austenite and carbide in a 0.6C–99.4Fe alloy as they vary with temperature.

Answer. >1400°F (760°C), all γ; 1335°F (724°C) 0.26 α–0.74 γ; 1331°F (722°F) 0.91 α–0.09 C

5–4.4 What carbon content is required for an Fe-C alloy which at 1300°F contains (a) 0.85 pearlite; (b) 0.03 carbide?

Answer. (a) 0.68 w/o C, or 1.69 w/o C; (b) 0.22 w/o C

5–4.5 Calculate the weight fractions of ferrite, carbide, and pearlite at room temperature in an iron-carbon alloy containing 0.46 w/o carbon.

Answer. 0.93 α, 0.07 C, ~0.56 P

5–4.6 At what temperature will a 0.2 w/o C steel contain (a) $\frac{1}{3}\gamma$–$\frac{2}{3}\alpha$; (b) $\frac{2}{3}\gamma$–$\frac{1}{3}\alpha$?

Answer. (a) ~1420°F (~770°C); (b) ~1525°F (~830°C)

5–4.7 Compare the amount of proeutectoid and eutectoid ferrite in 100 lb of 0.40C–99.6Fe steel.

Answer. 51.6 lb proeutectoid α, 42.7 lb eutectoid α

5–5.1 Tungsten wire is used to reinforce copper to make a composite material for use at elevated temperatures. What is the maximum density possible for the composite? [*Hint:* Since ρ_W is 19.3 g/cm^3 and ρ_{Cu} is 8.9 g/cm^3, maximum density is obtained when the wires are "closed-packed" into a hexagonal pattern.]

Answer. 18.3 g/cm^3

5–5.2 A bronze (95Cu–5Sn) bearing made by powder metallurgy has an overall density of 7.5 g/cm^3. By considering the pores to be another phase with $\rho = 0$ g/cm^3, calculate the porosity of the bearing.

Answer. 15 v/o

5–6.1 Approximately what is the lowest temperature of 100% γ in an AISI 2550 steel containing 5.05 w/o Ni?

Answer. ~1290°F (~700°C)

5–6.2 Approximately how much carbon should be present in an SAE 40xx steel to provide 50% pearlite?

Answer. ~0.33 w/o C

5–6.3 Suppose that you are asked to design a metal support which will be of minimum weight and which will be capable of holding a 5000-pound load without plastic deformation. Three metals (steel, YS = 150,000 psi; copper, YS = 57,000 psi; and aluminum, YS = 50,000 psi) are available. Specify your choice.

Answer. Steel (by a nose)

5–6.4 Estimate the carbon content of the nodular cast iron in Fig. 5–6.3 which contains predominantly ferrite and graphite.

Answer. ~10 v/o graphite $\cong$ 3 w/o C

5–6.5 The maximum solubility of carbon in austenite is 1.6 w/o at the eutectic temperature of silicon-containing cast iron (2 w/o Si). Assuming full carbide dissociation en route, how much graphite separates as the temperature drops *from* the eutectic temperature *to* the eutectoid temperature? Basis: 10.0-lb casting, 1.6 w/o C.

Answer. 0.105 lb graphite. [*Note:* The eutectoid in a 2 w/o Si alloy is at 0.55 w/o C (Fig. 5–6.2).]

PROCESSING
OF
METALS AND ALLOYS

6–1 METAL PRODUCTION ⒠ ⓜ ◎

The major operation in the processing of metals and alloys is removing them from the original ores and refining them into the compositions required by the user. The engineering and scientific accomplishments required in order to achieve this are considerable. We cannot even begin to outline them all here. However, we should cite the principles that underlie the production of various metals because this information often relates to the selection and use of metals.

Extraction from ores. The ores of metals—except for gold, and a small fraction of our copper and silver supply—are not metallic phases. Most commonly they are oxides. However, they may be sulfides: Copper, silver, lead, and zinc sulfides are significant sources of those metals. In either case, the metal must be extracted from its ore. Even when the ore is a sulfide (MS, where M is the metal ion and S is the sulfide ion), it is common to first oxidize the ore to MO, the comparable oxide. Therefore the prime extraction step is one of *reduction*, in which the positive metal ions, M^{2+}, are changed to the metal atoms, M:

$$M^{2+} + 2e^- \rightarrow M^0. \tag{6–1.1}$$

The ease with which reduction occurs varies from metal to metal. Lead oxide reduces readily to metallic lead, whereas it requires an input of considerable energy to effect the reduction of alumina (Al_2O_3). A measure of the stability of the oxides, or of the amount of energy required for extraction, is available from the *"free" energy of formation* of the various oxides (Table 6–1.1). These are the reaction energies which are *released* as the metal is burned to form the oxide (therefore minus quantities). The same amount of energy is *required* to separate the oxygen from the metal (therefore positive quantities).

The metallurgist has several choices of ways to force oxide reduction. As observed in Table 6–1.1, the energy required for reduction decreases

Table 6–1.1

ENERGY‡ OF FORMATION OF OXIDES

| | Energy released on formation, cal/mole oxide,* 25°C | Energy required to reduce oxide to metal | | | | | |
| | | cal/mole oxide | | cal/g metal | | Btu/lb metal | |
		25°C	T_m§	25°C	T_m§	77°F	T_m§
Al_2O_3	− 378,100	+ 378,100	+ 330,800	+ 7,000	+ 6,125	+ 12,600	+ 11,020
MgO	− 136,080	+ 136,080	+ 119,890	+ 5,600	+ 4,930	+ 10,080	+ 8,875
Fe_2O_3	− 176,800	+ 76,800	+ 87,000	+ 1,580	+ 780	+ 2,845	+ 1,404
FeO	− 58,150	+ 58,150	+ 35,900	+ 1,040	+ 643	+ 1,875	+ 1,160
SnO	− 61,400	+ 61,400	+ 56,600	+ 517	+ 475	+ 930	+ 820
CuO	− 31,700	+ 31,700	+ 8,700	+ 500	+ 137	+ 900	+ 247
PbO	− 45,050	+ 45,050	+ 38,150	+ 217	+ 185	+ 390	+ 333
Ag_2O	− 2,500	+ 2,500	− 11,100	+ 12	− 51	+ 22	− 92
CO_2	− 94,260	+ 94,260	—	+ 7,850	—	+ 14,130	—
CO_2 (327°C, 620°F)	− 94,440	+ 94,440	—	+ 7,865	—	+ 14,160	—
CO_2 (1000°C, 1832°F)	− 94,675	+ 94,675	—	+ 7,890	—	+ 14,200	—

‡   Free energy. ($\Delta F_f^o \equiv \Delta H_f^o - T\Delta S_f^o$; however, it will not be necessary for us to use this equation.)
§   At melting temperature of metal.
*   1 calorie = 4.18 joules. Calories are used in this optional subsection, since they are the metric units of thermal energy most widely used by chemists. A mole is 6×10^{23} atoms, or 6×10^{23} formula units of oxide.

as the melting temperature of the metal is approached. In fact, for silver, the sign of the energy requirement reverses, and reduction of the oxide to the metal could proceed naturally. However, this is not a very efficient way to remove the oxygen, because the consumption of energy by the furnace and the losses to the surrounding atmosphere would run high even if elaborate designs were used for the equipment.

A more practical choice of a method of reduction involves the use of some *reducing agent*, such as carbon. In effect, carbon, which is readily available, attracts to itself the oxygen that is in the oxide. Using PbO as an example,

$$C + 2PbO \rightarrow 2Pb + CO_{2(gas)}. \qquad (6–1.2)$$

Observe from Table 6–1.1 that the energy released by forming CO_2 is more than is required to separate oxygen from the lead. [These figures are − 94,440 cal/mole* for the CO_2 versus (2) (38,150 cal/mole*)

*   See footnote to Table 6–1.1.

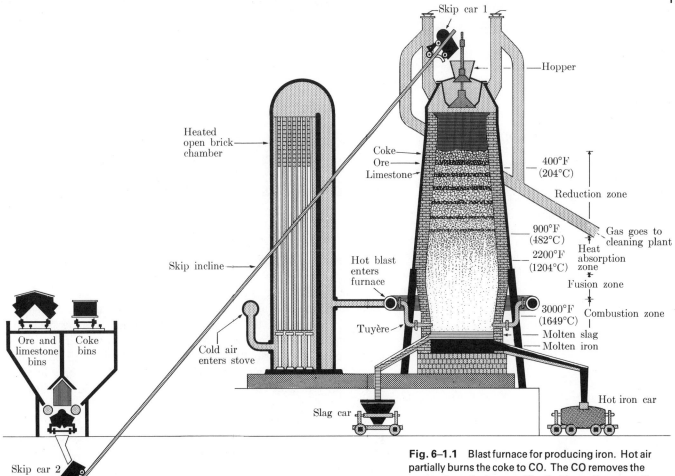

Heated
open brick
chamber

Skip car 1

Hopper

Coke
Ore
Limestone

400°F
(204°C)

Reduction zone

Skip incline

900°F
(482°C)

Gas goes to
cleaning plant

Heat
absorption
zone

Hot blast
enters
furnace

2200°F
(1204°C)

Fusion zone

Tuyère

3000°F
(1649°C)

Combustion zone

Cold air
enters stove

Molten slag
Molten iron

Ore and
limestone
bins

Coke
bins

Hot iron car

Slag car

Skip car 2

Fig. 6–1.1 Blast furnace for producing iron. Hot air partially burns the coke to CO. The CO removes the oxygen from the iron ore (Fe $_2$O $_3$) to produce metallic iron (Eq. 6–1.3). This is done above the melting temperature of the molten iron product, so that the metal may be separated from the residue, i.e., the slag.
(A. G. Guy, *Physical Metallurgy for Engineers*, Reading, Mass.: Addison-Wesley)

for the two PbO's at the melting temperature of lead (327°C). Furthermore, when the resulting CO_2 gas is removed, there is no tendency for the reaction to reverse.

Reduction by carbon may also occur after the initial formation of CO:

$$3CO_{(gas)} + Fe_2O_3 \xrightarrow[\sim 2900°F]{\sim 1600°C} 2Fe_{(liquid)} + 3CO_{2(gas)}. \qquad (6–1.3)$$

Although there are a number of variants, these types of reactions are the ones most widely encountered in metal production. Note that elevated temperatures favor this reaction, since the energy required for the reduction of ore decreases, while the energy released by the formation of CO_2 increases. Figure 6–1.1 shows an iron blast furnace which

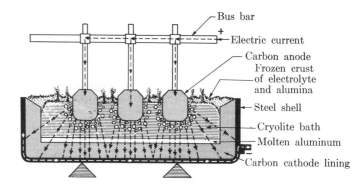

Bus bar
\+ Electric current
Carbon anode
Frozen crust of electrolyte and alumina
Steel shell
Cryolite bath
Molten aluminum
Carbon cathode lining

Fig. 6–1.2 Electric cell for producing aluminum. Since the energy required to remove aluminum from its raw material (Al_2O_3) is great, a high-temperature electroplating process is used. Molten aluminum collects at the cathode, which is at the base of the cell, and CO gas is evolved at the anode (Eq. 6–1.4 and 6–1.5). The critical feature of Hall's (and Heroult's) revolutionary process was the use of fluxes to make a water-free electrolyte. Water cannot be present or else hydrogen gas (H_2) will "plate" at the electrode ahead of the aluminum. (Courtesy Aluminum Company of America)

uses the above reaction to produce several thousand tons of molten iron per day.

The reduction reaction just described is not always feasible. Sometimes the oxidation of carbon does not supply enough free energy. Sometimes the necessary temperature is excessively high. Sometimes the molten metal reacts with the container. Electrical energy offers an alternative, provided that the oxide can be dissolved in a flux. For example, although aluminum-bearing ores had been known for years, the commercial production of aluminum had to await the development of a suitable flux to dissolve the $Al_2^{3+}O_3^{2-}$ for *electrolytic extraction* (Fig. 6–1.2). Here the reaction at 1800°F (980°C) is the same as in Eq. (6–1.1), except that the metal, M, is trivalent aluminum. Carbon and an electric current are used to release the electrons from the oxygen ions and transfer them to the aluminum:

$$3O^{2-} + 3C \rightarrow CO_{(gas)} + 6e^- \qquad\qquad (6\text{–}1.4)$$

$$6e^- + 2Al^{3+} \rightarrow 6Al_{(liquid)}. \qquad (6\text{–}1.5)$$

A few metals, such as zinc, lend themselves to *vaporization* as a means of extraction. Vaporization also may be aided by carbon as a reducing agent because carbon removes the oxygen in the form of CO or CO_2 (Example 6–1.1).

Refining. Most metals, even when they are reduced to remove the oxygen, are not pure enough for commercial use; they must be refined. This is illustrated by an aluminum ore which contains some iron. Since iron reduces more readily than aluminum, any iron in the ore will also appear in the metal. Likewise, a certain fraction of the silicon or phosphorus in an iron ore appears in extracted metallic iron, because these elements are partially reduced along with the iron. As a rule, these impurities are not desired; they must therefore be removed.

Fig. 6–1.3 "Basic oxygen" refining furnace (iron). Oxygen which is blown into the molten metal oxidizes the dissolved carbon (to CO) and dissolved silicon (to SiO_2). The CO leaves the vessel as a gas and completes its combustion to CO_2 in the air. The SiO_2 is dissolved into the basic, CaO-rich slag. (A. G. Guy, *Physical Metallurgy for Engineers,* Reading, Mass.: Addison-Wesley)

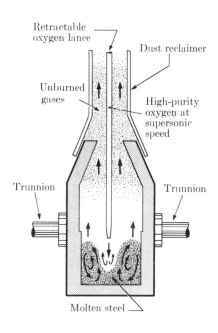

Again there are a variety of metallurgical processes which are available. We shall describe only one, the *basic oxygen process,* which is relatively new (Fig. 6–1.3). Let's assume that we have molten iron containing two weight percent (2 w/o) silicon and 3.5 w/o carbon. This composition is all right provided that we want to use it for certain cast irons (Section 5–6), but we must lower the silicon and carbon contents if we are to use the iron in low-alloy steels (Tables 5–6.1 and 5–6.3). We invariably use oxygen to selectively remove these two impurities. Note, however, that we can't just have a straight reversal of the initial reduction process we described earlier, or the iron would be oxidized along with the silicon and carbon. In simplest terms, we use a CaO-rich slag because it has a great affinity for SiO_2. Thus, as the dissolved silicon and oxygen react, the SiO_2 which forms is deactivated by the CaO in the slag. The oxidation of the carbon produces CO and CO_2, which leave the furnace in the form of gases. Although some of the remaining iron will be oxidized and dissolve in the slag, the iron losses can be minimized by appropriate control of temperature and time.

Casting. The final step in metal production is to solidify the metal into an ingot, or mold it directly into the desired shape. An ingot is simply a large solidified mass of metal which can subsequently be mechanically deformed by rolling or forging.

One of the prime technical considerations in making a casting is that the volume changes when the metal changes from liquid to solid (Fig. 4–7.2). With rare exceptions—e.g., "type" metal, used in printing—shrinkage accompanies solidification. Therefore a reservoir, usually called a *riser,* must be available which will feed additional molten metal into the casting while freezing (i.e., solidification) progresses (Fig. 6–1.4). Also care must be taken that constrictions in the feed channels (*gates*) do not freeze or solidify completely and choke

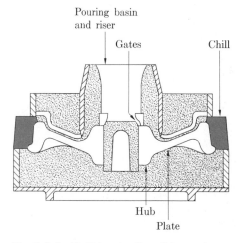

Fig. 6–1.4 Mold (made of sand) for casting metal (railroad car wheel). The pouring basin serves as a riser, or reservoir, to supply metal to compensate for the shrinkage during solidification. The chills direct the solidification from the rim inward. (R. A. Flinn, *Fundamentals of Metal Casting,* Reading, Mass.: Addison-Wesley)

off the inflow of molten metal from the riser to larger sections which are not frozen. The metallurgist can partially compensate for this problem by using *chills*, or *heat sinks*, adjacent to the thicker sections of the casting, thereby increasing the freezing rate in those zones. A casting system, if it is to be satisfactory, must be designed by someone with considerable technical knowledge and accumulated experience.

Example 6–1.1. Zinc oxide (ZnO) and carbon are mixed and placed in a retort, where they are heated so that a reaction similar to Eq. (6–1.2) can occur. Both the zinc and the CO which result are vapors initially; however, the metallic zinc condenses in the cooler part of the retort. (a) How much carbon is required per 100 lb of zinc oxide? (b) How much zinc is reduced?

Solution

$$ZnO + C \rightarrow CO + Zn. \tag{6–1.6}$$

Using atomic weights:

$$(65.37 + 16) + 12 \rightarrow (12 + 16) + 65.37.$$

a) $\dfrac{C}{100 \text{ lb ZnO}} = \dfrac{12 \text{ amu}}{81.37 \text{ amu}},$ b) $\dfrac{Zn}{100 \text{ lb ZnO}} = \dfrac{65.37 \text{ amu}}{81.37 \text{ amu}},$

$$C = 14.75 \text{ lb}. \qquad\qquad Zn = 80.2 \text{ lb}.$$

Comments. If the reaction temperature can be kept low, some of the CO will react with more ZnO to give CO_2 and additional zinc. The lower the temperature, however, the slower the reaction. If the temperature is raised, zinc vapors may escape. The metallurgist must optimize and control the temperature for maximum production efficiency. ◀

Example 6–1.2. According to Eq. (6–1.5), six electrons are required for every two aluminum atoms. How many coulombs are required to extract 27 g of aluminum?

Solution. The atomic weight of aluminum is 27 amu (Fig. 1–2.2); thus 27 g have 6×10^{23} atoms.

$$\text{Electrons required} = (6 \times 10^{23} \text{ atoms})(3 \text{ electrons/atom})$$
$$= 18 \times 10^{23} \text{ electrons}.$$

$$\text{Coulombs required} = (18 \times 10^{23} \text{ electrons})(1.6 \times 10^{-19} \text{ coul/electron})$$
$$= 2.88 \times 10^5 \text{ coul/27 g Al}.$$

Comments. Since a coulomb is 1 amp · sec, a current of one ampere would be required for 288,000 sec (80 hr), or 80 amp for 1 hr (3600 sec).

The term *faraday* ($\mathscr{F}$) is used for 96,000 coulombs, the charge for 6×10^{23} electrons. The above answer is equal to three faradays, since each aluminum atom requires three electrons. ◀

MECHANICAL PROCESSING

6–2 RESHAPING PROCESSES (MECHANICAL FORMING) ⒠ ◎

Although deformation is to be avoided in most service conditions, it is one of the main mechanisms of metal processing. In effect, mechanical forming processes involve "controlled failure," since the metal is deformed well beyond the elastic limit. *Rolling*, *forging*, *extrusion*, *drawing*, and *spinning* (Fig. 6–2.1) are among the more common mechanical forming processes. We saw in Chapter 4 that mechanical deformation leads to strain-hardening. This feature is important for two reasons: First, it affects the properties of the product. Second, it affects the behavior of the material during processing, since a strain-hardened material requires more energy for additional deformation, and because of its reduced ductility, is more subject to fracture.

Hot versus cold working. If a metal is deformed above its recrystallization temperature, annealing occurs almost concurrently with deformation (Section 4–9). As a result, strain-hardening is not encountered in the process. Recall that the recrystallization temperature varies from material to material, and in general is between one-third

Fig. 6–2.1   Common shaping processes. (a) Rolling. (b) Forging (adapted from ASM *Metals Handbook,* Vol. 5). (c) Extrusion (A. Guy, *Physical Metallurgy for Engineers*). (d) Wire drawing. (e) Spinning (adapted from ASM *Metals Handbook,* Vol. 4). Often two or more of these processes are used. Hot working—i.e., shaping above the recrystallization temperature—is necessary for shaping products with large cross sections, or for shaping without the introduction of strain-hardening.

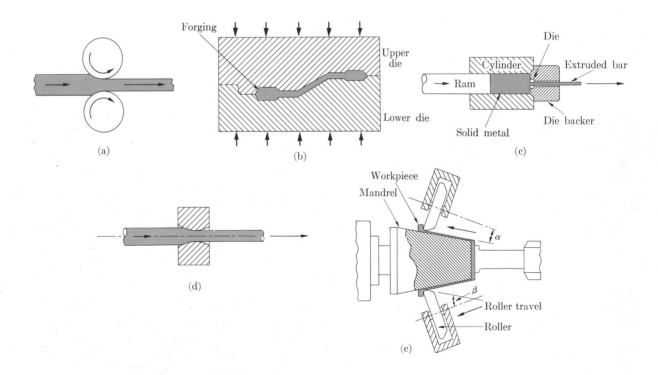

and one-half of the *absolute* melting temperature. Thus, since absolute zero is $-273°C$ ($-460°F$) and the recrystallization temperature of tin is below room temperature ($T_m = 232°C$), mechanical deformation of tin at normal temperature must be classified as hot working.

In contrast, the absolute melting temperature of iron is $273 + 1539°C$ (or $460 + 2802°F$); therefore, depending on the amount and the rate of deformation, iron must be heated about $500°C$ ($930°F$) for hot work without strain-hardening. Tungsten presents special problems in hot working because, throughout the deformation process, it is necessary to maintain the temperature of tungsten above $1200°C$ ($2200°F$) or strain-hardening will result.

It is usually possible to use rolling, extrusion, and forging processes to make large products, providing there is no strain-hardening to demand excessively high energies. Furthermore, large masses of metal cool slowly. Together these factors favor hot working for the *primary working* processes, i.e., for the initial reshaping of the metal from ingots to beams, bars, slabs, etc. But the final shaping (*secondary working*) of lighter-gage materials such as sheet and wire is more commonly done by deep drawing (sheet), wire drawing (Fig. 6–2.1d), and spinning (Fig. 6–2.1e)—all *cold-working* operations. Cold working is desired here because surface oxidation is avoided, close control of dimensions is possible, and the final properties may be adjusted to meet specifications.

Deformation patterns. The strain pattern in a rolling operation is simple in concept. The rolls apply compression in one direction; there is comparable elongation in the longitudinal rolling direction, and only slight (but not zero) strain in the third dimension. The strain in wire drawing or extrusion is only slightly less simple in concept. There is a radial compression and an axial elongation. This axial elongation is equal to the reduction in cross-sectional area.

The details of strain in these rolling and drawing processes are appreciably more complex than the simplified pictures just described, because the grains do not deform uniformly. Those grains that are oriented favorably are subject to extensive shear (Section 4–3). Others deform much less. Furthermore, all strains must adapt to each other to maintain grain-to-grain contacts without discontinuities.

The strains in other commercial processes have more obvious variations due to deformation. This is illustrated for deep drawing in Fig. 6–2.2. Locally the deformation values may vary from zero strain to values several times the average. This introduces significant variation in the resulting properties of the finished product, such as hardness concentrations, local embrittlement, and corrosion differences.

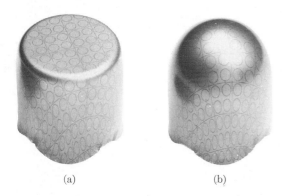

(a)   (b)

Fig. 6–2.2 Deformation during shaping (deep-drawing of sheet). (a) Flat-nosed punch. (b) Round-nosed punch. (Both have a 2 in. diameter.) The 0.5 in. grid and 0.2 in. circles show the variations of deformation with position and punch shape. (Courtesy Bethlehem Steel Co.)

6–3   METAL-REMOVAL PROCESSES Ⓔ Ⓞ

Machining is by far the most widely used metal-removal process. It may be performed on a lathe, in a milling machine, by drilling, or by several other variations. It is a combination of cutting and mechanical deformation, as shown in Fig. 6–3.1, in which various terms associated with cutting tools are listed, and in Fig. 6–3.2, which shows several types of chip formation and distorted steel.

    In practice, the process engineer prefers a chip which breaks as it is cut (Fig. 6–3.2a). This means that it is often desirable to specify some cold work prior to machining, to reduce the ductility of the metal. It also means that certain impurity phases—for example, MnS—may be specified as additions to the steel, to aid chip formation, producing a "free-machining" steel. Chips rather than long turnings not only facilitate the handling of the removed metal, but also provide less friction and therefore less heating at the tool tip. This is important because excessive heating of the tool tip decreases its hardness, which leads to faster dulling and then to more friction and heat, leading to catastrophic failure (Fig. 6–3.3).

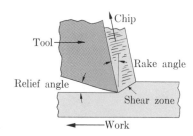

Fig. 6–3.1 Machining. This metal-removal process, which involves both cutting and deformation, depends on the properties of the metal and the geometry of the cutting tool.

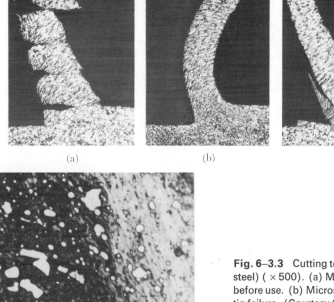

(a)   (b)   (c)

Fig. 6–3.2 Chip formation. (a) Discontinuous. (b) Continuous. (c) Tool-edge buildup. (Cincinnati Milling Machine Co.)

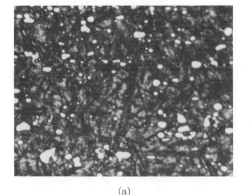
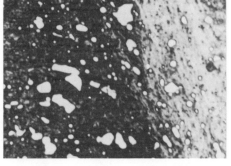

(a)   (b)

Fig. 6–3.3 Cutting tool (high-speed steel) (× 500). (a) Microstructure before use. (b) Microstructure after tip failure. (Courtesy Crucible Materials Laboratory, Colt Industries)

Cutting tools. High-carbon steels can be made very hard by quenching and tempering. That is the reason why they are widely used for tools. We shall observe in Section 6–6 that a quenched and tempered steel contains a very fine dispersion of carbides in a ferrite matrix. At higher temperatures the fine carbides grow (and become fewer in number); in the process the steel softens markedly until it takes on the properties of spheroidite (Fig. 5–5.3), an undesirable microstructure in this case. It is possible, however, to slow down the carbide growth and the softening process considerably by introducing carbide stabilizers such as Cr, W, Mo, and V in steel. If the carbide particles containing these elements are to grow, not only the small carbon atoms, but also the larger atoms just listed must diffuse through the ferrite matrix to the growing particle. This takes time or higher temperature. Metallurgists intentionally include these elements in tool steels so that the cutting speeds may be increased with only minimum softening; hence the term *high-speed steel.*

When one is cutting extremely hard materials, even a high-speed steel will fail. *Cemented carbides*, with WC or TiC as the major phase, may be used. Likewise Al_2O_3 is available as a ceramic material (Chapters 10 and 11). The usefulness of such materials is due not only to their hardness but also to their stability at high temperatures— or their refractoriness—which helps avoid tool-tip melting and the welding of the metal to the tool. Because these carbides are refractory, however, it is impossible to use melting processes when the tools are made. One must use powder-production processes (Section 6–8).

THERMAL PROCESSING

6–4 ANNEALING PROCESSES [e]

The term *annealing* is used generally to denote either a softening or a toughening process. In Section 4–9, the word *annealing* was used to describe the softening which accompanies recrystallization of strain-hardened metals. We shall see in Chapter 11 that the glass technologist uses *annealing* to denote a heat treatment for removing residual stresses and thereby reduce the probability of the development of cracks in a brittle glass. The word annealing may have special connotations when it describes a heat treatment of steel. (We shall discuss these connotations shortly.) In the meantime, we shall simply note that *annealing* entails heating in each case to a temperature at which the individual atoms have added freedom for movement and rearrangement into a more stable structure, i.e., a structure with less energy.

Stress relief. The simplest annealing process is one of stress relief. Nonuniform cooling often introduces residual stresses because part of the metal is still shrinking after the remaining metal has become rigid. Thus some of the material develops tensile stresses, while counteracting compressive stresses are introduced into other portions of the metal. Now consider what would happen if the part of the metal which is in the compression zone were removed by machining. The tensile stresses would be unbalanced and distortion would result. Obviously it is desirable to relieve these stresses before the final, close-dimensional machining is performed.

Residual stresses may also be introduced when there is a change in density which accompanies a change of phase. For example, as steel transforms from a face-centered-cubic structure to a body-centered-cubic structure, the resulting changes in volume may generate stresses of many thousands of psi.

Stress relief proceeds at temperatures lower than those necessary for recrystallization because it is not necessary to grow new crystals. A relatively few dislocation movements will accommodate the stresses.

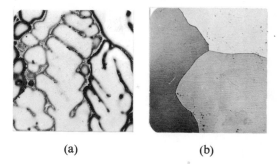

(a)   (b)

Fig. 6–4.1 Solidification segregation (96Al-4Cu) (×100). (a) Cast alloy with copper-rich residual areas between crystals. (b) The same alloy after homogenization. The copper has diffused until it is uniform throughout each grain.
(Courtesy Alcoa Research Laboratories)

Homogenization treatments. During solidification of any material, the initial solid phase is not the same as the overall composition. This is illustrated in Fig. 6–4.1(a) for a 96Al-4Cu alloy. The appropriate phase diagram (Fig. 6–4.2) reveals that κ, the very first solid to form, contains only about 1 w/o copper. While the copper content of the solid gradually increases as the temperature falls, the residual liquid also progressively accumulates a larger fraction of copper and may even form a θ-κ eutectic. Homogenization, sometimes called *soaking*, is required to produce uniformity in the metal (Fig. 6–4.1b). During soaking, some atoms must diffuse long distances compared to the atomic scale ($\sim 10^{-2}$ cm, or $\sim 10^6$ unit cell distances). It is therefore necessary to heat the metal considerably above the recrystallization temperature so that homogenization may occur in a reasonable length of time.

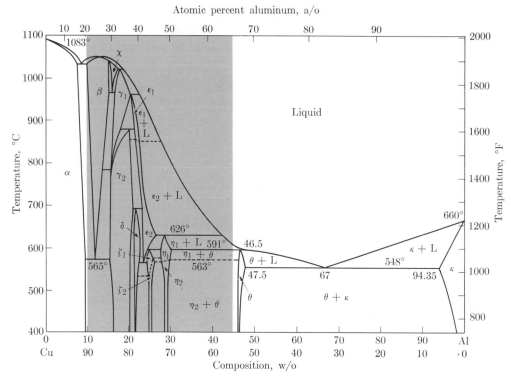

Fig. 6–4.2 Al-Cu system. The solubility limit of copper in the fcc aluminum-rich phase (κ) varies from <1 w/o at room temperature and at the melting temperature (660°C) to more than 5.5 w/o at the eutectic temperature (548°C). The solubility limit of aluminum in the fcc copper-rich phase (α) is essentially independent of temperature at 7–9 w/o (15–20 a/o Al). (ASM *Handbook of Metals,* Metals Park, O.: American Society for Metals)

Fig. 6–4.3 Austenitizing processes. Normalizing (broken line) is commonly performed at about 100°F (55°C) above the A_{c_3} line, the line of no ferrite. Air cooling follows. Annealing (dashed line) is commonly performed at about 50°F (25–30°C) above the A_{c_3} temperature. Slow cooling follows, to give coarser, softer pearlite.

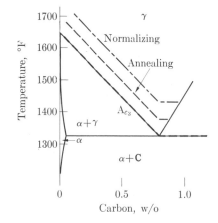

In steel, homogenization is accomplished on a routine basis. The process is called *normalizing*, and involves heating the steel to about 100°F (55°C) into the fully austenitic range for an hour. The austenite is required so that the carbon and the alloying elements are all dissolved to one phase, γ, and not partitioned between two, for example, $[\alpha + \gamma]$ or $[\alpha + C]$. The temperature required to obtain a single γ phase for a eutectoid steel is obviously less than for a low-carbon steel (Fig. 6–4.3).

Full anneal (steel). We observed earlier (Section 5–5) that slow cooling produced a coarser pearlite and a softer steel (Fig. 5–5.2). Engineers often want steels to be like this prior to machining so that cutting them is easier. The commercial procedure requires heating the steel to about 50°F into the fully austenitic range and furnace-cooling, i.e., letting the steel cool at the same rate as the furnace cools. Again the required temperature varies with the carbon content (Fig. 6–4.3). The 50°F margin is necessary to ensure the formation of austenite in the centers of thick sections. At the same time, the 50°F margin is generally sufficient because, in full annealing, unlike the case of normalizing, it is necessary only to form austenite; there is no need for homogenization.

Process anneal. In some steels, particularly steels used in sheet and wire products that have been cold-worked to small dimensions, it is sometimes desirable to soften the metal without forming austenite again.* This must be accomplished at a temperature just below the eutectoid temperature. Since only the cold-work is to be removed, and no major diffusion of atoms is required, the time at the heat-treating temperature is brief.

Spheroidization. This process is accomplished by an extended anneal at the upper end of the $[\alpha + C]$ temperature range so that the former carbide lamellae may coalesce to a spheroidal, i.e., sphere-like, shape with less phase-boundary area (and therefore with lower energy). In this manner a microstructure such as that shown in Fig. 5–5.3(b) will be attained from pearlite after some 12–15 hours just below the eutectoid temperature, say 1275°F. (We shall observe in Section 6–6 that the same structure may be obtained by a much shorter quench and temper treatment.)

The various annealing processes used in heat treatment of steel are summarized in Fig. 6–4.4.

---

* The small cross-sectional dimensions of wire and sheet products allows extremely fast cooling rates, producing brittle martensite if austenite is present (Section 6–6).

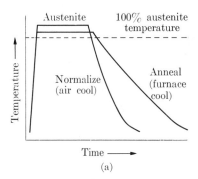

(a)

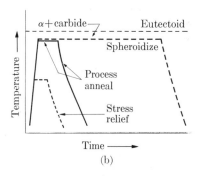

(b)

Fig. 6–4.4 Softening and toughening processes (eutectoid steel). (a) Normalizing and annealing. (b) Process annealing, spheroidizing, and stress relief.

Example 6–4.1. When fcc iron is cooled to 70°F, a small grain of body-centered-cubic iron forms at the fcc grain boundary, but the large mass of surrounding metal keeps it from expanding the necessary 0.3 l/o. What stresses are developed? [Note that the compressibility, β, of the material is

$$\beta = (\Delta V/V)/\sigma_{hyd} \tag{6–4.1}$$

$$= 3(1 - 2\nu)/Y, \tag{6–4.2}$$

where V is volume and ν and Y are Poisson's ratio and Young's modulus, respectively (Section 4–2); σ_{hyd} is the hydrostatic pressure, which is equal in the three orthogonal directions.]

Solution. Recall from Eq. (2–6.3) that volume expansion is approximately triple the linear expansion. Therefore $-0.3\ l/o \cong -0.9$ v/o $= -0.009 = \Delta V/V$. (Minus because expansion is counteracted.)

$$\sigma_{hyd} = \frac{30,000,000 \text{ psi } (-0.009)}{3[1 - 2(0.28)]}$$

$$= -204,000 \text{ psi.}$$

Comments. The stress will not become this high because the fcc matrix will deform plastically and allow some expansion of the body-centered phase. However, stresses may still be high enough to crack the metal or to stop the transformation, either of which is of concern to the process metallurgist. ◀

Example 6–4.2. Let 100 g of a 96Al-4Cu alloy equilibrate at 1150°F (621°C), forming the κ and liquid, as shown by the phase diagram. The alloy is then cooled rapidly to 1020°F (549°C) with no chance for the initial solid to react. A liquid phase will still be present. (a) What will its composition be? (b) What will be its amount?

Solution. At 1150°F,

Liq: 88Al-12Cu, κ: 98Al-2Cu.

Grams liquid $= 100$ g $\left(\dfrac{98 - 96}{98 - 88}\right) = 20$ g liquid.

At 1020°F, and with only the 20 g of liquid reacting,

κ: 94.3Al-5.6Cu

a) Liq: 67Al-33Cu.

b) Grams liquid $= 20$ g $\left(\dfrac{94.3 - 88}{94.3 - 67}\right) = 4.6$ g liquid.

Comments. In the final 4.6 g of liquid, there will be local areas with high copper segregation (33%). There will also be nearly 80 g of metal with only 2% Cu.

Cooling may be considered to take place in a series of small steps, in which the solute in the residual liquid is progressively concentrated. ◀

6–5 PRECIPITATION PROCESSES ⓒ

When we discussed alloys of the light metals in Section 5–6, we indicated that they may be strengthened by *precipitation hardening*
(commonly called *age hardening*). This is an important mechanism
and can be explained on the basis of phase diagrams.

Consider again the 96Al-4Cu alloy of Section 6–4, which was
homogenized for Fig. 6–4.1(b). According to the phase diagram of
Fig. 6–4.2, and also to be seen in Fig. 6–5.1(a), this alloy can be a single-
phase solid solution, κ, between approximately 1080°F ($\sim$ 580°C)
and 920°F ($\sim$ 495°C). The diagrams indicate correctly that, while the
alloy is being cooled after solution treatment, a solubility limit is
reached, and under equilibrium conditions a second phase should
form. This second phase, θ, is an intermetallic compound, $CuAl_2$. It
is hard and brittle.

Now consider various cooling and reheating procedures (Fig.
6–5.1b). An anneal, or slow cooling from the solution treatment
(XD), provides a material that is soft, weak, and not especially ductile—
much less so than pure aluminum. This metal, which is generally
unsatisfactory for engineering purposes, contains large areas of soft
κ that is almost pure aluminum. The brittle, intermetallic, $CuAl_2$
phase, θ, precipitates along grain boundaries and is the locus of crack
propagation when fracture occurs.

Fig. 6–5.1 Age-hardening process (96Al-4Cu).
Maximum hardness occurs when the copper begins to
cluster prior to precipitation as $CuAl_2$; that is, θ
(cf. Fig. 6–4.2).

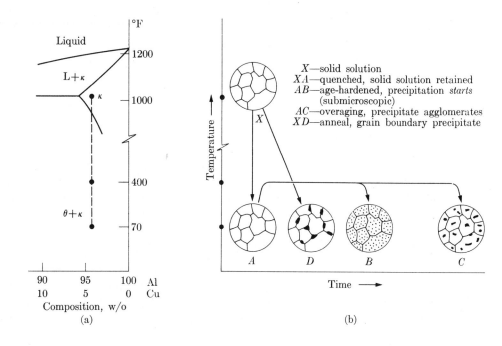

Fast cooling (*XA* in Fig. 6–5.1) provides a stronger alloy which is very ductile (40–50% elongation in 2 in.). This material is attractive to the engineer because it has the advantage of both strength and ductility. A micro-examination reveals only one phase; x-ray diffraction indicates that all the copper is in solid solution, which indicates (Fig. 6–4.2) that equilibrium has not been reached.

Fast cooling, followed by reheating to an intermediate temperature, *XAB*, just about doubles the strength of the alloy (i.e., compared with an alloy processed with the *XA* treatment). This whets the interest of the engineer still further. In explaining this, the materials scientist concludes that the supercooled solid solutions *start* a phase separation, or a precipitation of the $CuAl_2$, θ, during the *AB* reheating. With the available time and temperature, it is not possible for the copper atoms to diffuse into the grain boundaries and combine there with aluminum to give $CuAl_2$. Rather at innumerable locations throughout the grains, there is a *clustering* of the copper atoms, still in solid solution and not yet separated from the κ into the θ phase. At this stage, the κ lattice is greatly distorted around the copper-enriched clusters, so that dislocation movements, and therefore slip, are severely restricted. The increased strength is shown in Table 6–5.1 for a 96Al-4Cu alloy. (Admittedly there is a decrease in ductility compared with an alloy treated by *XA*, but improved from one treated by *XD* in Fig. 6–5.1, particularly when the strength level is considered.)

Table 6–5.1

PROPERTIES OF AN AGE-HARDENABLE ALLOY
(96Al-4Cu)

Treatment (See Fig. 6–5.1)		Tensile strength, psi	Yield strength, psi	Ductility, % in 2 in.
XA	Solution-treated and quenched	35,000	15,000	40
XAB	Age-hardened	60,000	45,000	20
XAC	Overaged	~25,000	~10,000	~20
XD	Annealed	25,000	10,000	15

Extended reheating, *XAC*, of the alloy in Fig. 6–5.1 softens this alloy and also reveals the appearance of distinct $CuAl_2$, or θ, particles that continue to grow with time. This is referred to as *overaging*. Figure 6–5.2 shows the aging and overaging process for one aluminum alloy (2014). Note that both aging and overaging occur in less time at

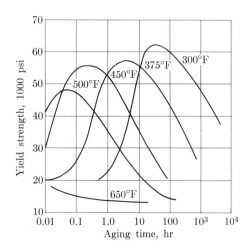

Fig. 6–5.2 Overaging (2014-T4 aluminum). Softening occurs as the precipitated particles grow. This proceeds more rapidly at elevated temperatures. (*Aluminum*, page 116, Metals Park, O.: American Society for Metals)

higher temperatures. Also note that the maximum strength is greatest with low aging temperatures. Here, however, the engineer must choose between a slightly greater strength of the product and higher processing costs for the extended heat treatments.

Table 6–5.2 summarizes an additional alternative available to the metallurgist for developing strength in certain alloys. The alloy, 98Cu-2Be, is age-hardenable because all the beryllium can be dissolved in the fcc structure at 870°C (1600°F). That solubility drops to near zero, however, at room temperature. The data in Table 6–5.2 show the effect of solution treatment, of age-hardening, and also of combinations of age-hardening and cold work. Thus when strain-hardening follows age-hardening, the combined effect is to give a tensile strength of about six times that of the annealed metal (200,000 psi versus 35,000 psi).

Two problems are encountered, however. First, a high-powered production mill is required to cold work a precipitation-hardened metal which already has 175,000 psi tensile strength. Much less energy and power would be required to cold work the solution-treated alloy (72,000 psi TS). Second, the previously age-hardened alloy has lost considerable ductility, so that there can be cracking during cold working. The reverse sequence—cold working of a solution-treated alloy—is possible, and reduces the severity of the above two problems. Admittedly the strength value we realize is only 195,000 psi rather than 200,000 psi, because there was a slight loss of strain-hardening during aging. Fortunately for this purpose, the atom rearrangements during aging are less extensive than during recrystallization. Therefore aging occurs slightly ahead of annealing. Close production control is required.

The prime requirement for precipitation-hardening is a decreasing solubility curve for a second alloying component within the principal metal, i.e., a sloping curve under a single-phase field (Fig. 6–5.3). In addition to the Al (+Cu) and the Be-bronze—that is, Cu (+Be)—alloys described above, this is a common hardening mechanism for magnesium alloys and certain of the stainless steels. In principle, lead-tin alloys could be hardened in this manner; however, because of their low melting temperatures, aging and overaging occur too rapidly at room temperature and the envisioned strengths are not retained.

Precipitation-hardened alloys (other than the above-mentioned Pb-Sn alloys) do not overage at room temperature. However, they all have upper limits of service temperature; for example, aluminum alloys cannot be used in supersonic aircraft, in which aerodynamic heating becomes significant.

Table 6–5.2

TENSILE STRENGTHS OF A STRAIN- AND AGE-HARDENED ALLOY (98Cu-2Be)

Annealed (870°C)	35,000 psi
Solution-treated (870°C) and rapidly cooled	72,000
Age-hardened only	175,000
Cold worked only (37%)	107,000
Age-hardened, then cold worked	200,000 (cracked)
Cold worked, then age-hardened	195,000

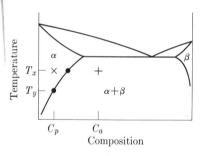

Fig. 6–5.3 Precipitation hardening (age-hardening) and the solubility limit. If the composition C_p is solution treated at T_x, a single phase results. When this is cooled below the solubility limit at T_y, the precipitate starts to form. Since only part of an alloy, C_a, forms α at T_x, only that part can be precipitation-hardened. The remaining part was β at the solution-treatment temperature.

Example 6–5.1. A 90Al–10Mg alloy is precipitation-hardenable. Indicate plausible changes in temperature, composition, and structure which would produce precipitation-hardening.

Solution. Refer to Fig. 5–7.1.

Solution treatment: ($\sim 450°C$, or $850°F$)

Only one phase, α, 90Al–10Mg.

Quench into water ($\sim 20°C$):

Still only one phase, α, 90Al–10Mg.

Age at an intermediate temperature ($\sim 100°C$):

Mg atoms cluster, but remain as part of the α phase.

Comments. Overaging would produce a matrix of α (low Mg content) containing a dispersion of the β phase, which is approximately Al_5Mg_3 (37 a/o Mg). This latter phase is hard and brittle, as are most intermetallic compounds. ◀

6–6 TRANSFORMATION PROCESSES ⓔ

To date we have not considered one very important way of strengthening steels (and a few other alloys): *quenching* followed by *tempering*.

Martensite. When fcc iron or steel is quenched very rapidly from the austenitic temperature range, there is not time for carbon and other alloying elements to relocate from a homogeneous γ solid solution into two phases, $[\alpha + C]$. Almost all the carbon must diffuse to form carbide. Likewise the carbide formers, Cr, Mo, V, etc., should concentrate in the carbide; whereas nickel and silicon must diffuse into the ferrite. These reactions require time.

If the transformation to $[\alpha + C]$ fails to occur before the temperature drops below approximately $750°F$ ($\sim 400°C$), the opportunity for $[\alpha + C]$ to form is essentially lost. Below this temperature, diffusion rates become excessively slow. A driving force, however, develops to transform the fcc structure to a body-centered crystal structure, and at low enough temperatures this driving force becomes sufficient to force a transformation by *shear*. The resulting body-centered structure is not cubic but *tetragonal*, a structure in which the unit cell has only two of the three orthogonal axes equal. The resulting body-centered unit cell traps all the carbon and other alloying elements, giving a highly supersaturated composition. This body-centered tetragonal nonequilibrium phase is called *martensite*.

Martensite is extremely hard for two reasons. (1) Because it is noncubic, it does not have as many slip combinations as a bcc structure.

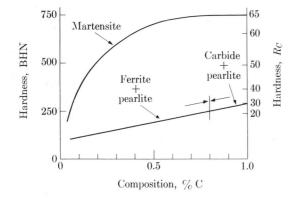

Fig. 6–6.1 Hardness versus carbon content. The [ferrite + pearlite] and the [carbide + pearlite] were formed from austenite by annealing. The martensite formed when austenite was quenched rapidly.

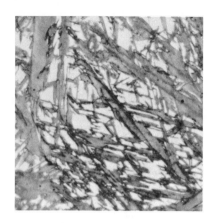

Fig. 6–6.2 Martensite (0.8 w/o C). The individual grains of martensite (gray) are platelike, extremely hard and brittle, and have the same composition as the original austenite (× 1000). (U.S. Steel Corporation)

(2) The carbon which is entrapped interstitially deters slip. All in all, martensite is the hardest iron-rich phase the engineer has available for widespread use. In Fig. 6–6.1, the hardness of martensite is compared with that of ferrite plus carbide. Figure 6–6.2 shows the microstructure of martensite at high magnification.

Martensite is also extremely brittle, so much so that martensite itself has almost negligible use. Fortunately, *tempering* (to be discussed later) changes the martensite and toughens a quenched steel considerably. If done correctly, tempering does not markedly reduce the hardness. Therefore a steel that is both quenched and tempered is ideal. It is both strong and tough. It is this unusual combination of properties which has made steel the prime metal in technology and industry over the past century.

Hardenability. The trick required to form martensite is to cool the steel rapidly enough so that it can*not* form $[\alpha + C]$. The most obvious procedure is to quench the steel in water or an appropriate oil. If you are making a razor blade, for example, you can quench it from 1500°F to 70°F in a fraction of a second without ferrite or carbide being formed. A large shaft, however, is much more massive than a razor blade; you have to remove many more calories, (or joules, or Btu's) through the surface. Even with water quenching, it is impossible to avoid forming some $[\alpha + C]$ in the center of the shaft, where cooling is slowest.

Figure 6–6.3(a) shows a hardness traverse on a 2-inch bar of AISI 1040 steel quenched in water. The surface, which was cooled most rapidly (∼500°F/sec), contains a large percentage of martensite and is

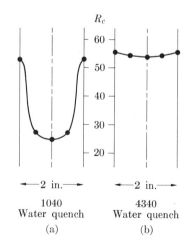

Fig. 6–6.3 Hardness profiles. (a) AISI 1040 steel. (b) AISI 4340 steel. Both steels were cooled at the same rate and have the same carbon content. However, the transformation $[\gamma \rightarrow \alpha + C]$ is slower in the low-alloy steel (4340) than in the plain-carbon steel (1040); therefore 4340 steel forms more martensite. It thus has greater *hardenability*.

therefore very hard. The center of the bar cooled more slowly (~40°F/ sec) and therefore formed considerable $[\alpha + \mathbf{C}]$ which, consistent with Fig. 6–6.1, is much softer than the martensite. Different steels require quenches of differing severities to form martensite in place of $[\alpha + \mathbf{C}]$. Since as a general rule any alloying element slows down the rate of transformation to $[\alpha + \mathbf{C}]$, alloy steels form the harder martensite more readily. We speak of this as *hardenability*; it is demonstrated in Fig. 6–6.3(b), which shows the hardness profile of an AISI 4340 steel which received the same quenching and had the same dimensions as the 1040 steel in part (a) of Fig. 6–6.3. Because the $[\gamma \rightarrow \alpha + \mathbf{C}]$ transformation was retarded due to the alloying elements present, more martensite formed and greater hardness was achieved at the center.

Hardenability curves. Some years ago, in an effort to predict hardness values, a metallurgical engineer by the name of Dr. Walter Jominy devised an end-quench test which is now used as an automobile industry standard. A steel is austenitized, i.e., heated to the γ-region to form 100% austenite, then directionally quenched, as shown in Fig. 6–6.4. The cooling rates indicated at the various locations behind the quenched end are nearly identical for all types of plain carbon and low-alloy steels. Therefore when we plot hardness according to the distance behind the quenched end (Fig. 6–6.5—lower abscissa), we are

Fig. 6–6.4 Jominy or end-quench test. The more rapidly cooled end changes from austenite, γ, to martensite, M, before the $[\gamma \rightarrow \alpha + \mathbf{C}]$ reaction can occur. Therefore it is harder. (After A. G. Guy, *Elements of Physical Metallurgy*, Addison-Wesley)

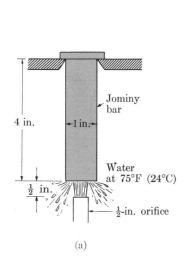

(a)

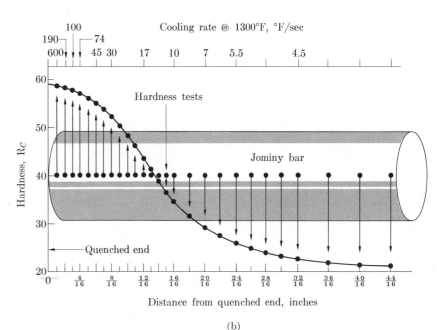

(b)

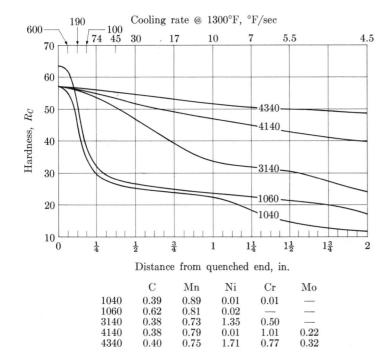

	C	Mn	Ni	Cr	Mo
1040	0.39	0.89	0.01	0.01	—
1060	0.62	0.81	0.02	—	—
3140	0.38	0.73	1.35	0.50	—
4140	0.38	0.79	0.01	1.01	0.22
4340	0.40	0.75	1.71	0.77	0.32

Fig. 6–6.5 Hardenability curves. Since compositions vary slightly, the location of the curves varies somewhat for any given type of steel. The Ni-Cr-Mo steels transform more slowly from austenite, γ, to ferrite plus carbide [α + **C**]; therefore more martensite is formed than in the steels without these alloys. (Adapted from U.S. Steel Corporation data)

also plotting hardness versus cooling rate (Fig. 6–6.5—upper abscissa). These are called *hardenability curves*, and are very useful to the engineer, because if he knows the hardness in a gear tooth, for example, when he is using one steel, he can readily predict the hardness at a comparable location in a substitute steel which is given the same heat treatment (and therefore has the same cooling rate). For example, any point in a 1060 gear which has a hardness of $30R_C$ would have a hardness of $54R_C$ in a gear made of 4140 steel. The latter would have more martensite because of its slower transformation to [α + **C**] with a given cooling rate.

Hardenability curves are also useful because they enable the engineer to predict hardness traverses in the cross sections of common steel shapes, such as round bars. Figure 6–6.6 compares cooling rates for bars of different diameters—at their surface S, mid-radius M-R, and center C—when they are quenched in (a) agitated* water and (b) agitated* oil. From (a) we observe that the center of a 1-inch-diameter

---

*   Agitation is vital; otherwise hot liquid and even vapor which is next to the steel would reduce the quenching rate.

Equivalent distance from
quenched end, in.

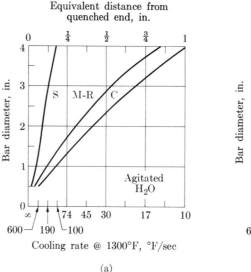

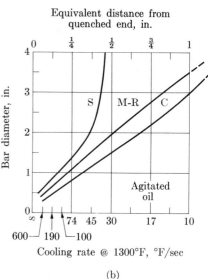

Fig. 6–6.6 Cooling rates in steel bars
(S, surface; M-R, midradius; C, center).
(a) Quenched in agitated water.
(b) Quenched in agitated oil. Bottom scale,
cooling rate at 1300°F (700°C); top scale,
equivalent positions on an end-quenched
test bar (Fig. 6–6.4).

(a)   (b)

bar cools 100°F/sec when quenched in water. This is the same cooling
rate found $\frac{3}{16}$ of an inch behind the quenched end of a Jominy end-
quench bar (upper abscissa of Fig. 6–6.6). For any given steel, these
two locations therefore develop identical microstructures and consequently have the same hardness value. Based on Fig. 6–6.5, this
hardness value would be $35R_C$ in a 1040 steel, $55R_C$ in a 3140 steel, etc.
Specific examples of calculations will be given at the end of this section.

Tempered martensite. As noted before, martensite is much too brittle
to be used just as it is. Therefore we *temper* it. Martensite does not
appear in the phase diagram because it is unstable, i.e., it has more
energy than $[\alpha + C]$. It is also supersaturated with carbon; given an
opportunity at somewhat elevated temperatures, martensite, M,
transforms to $[\alpha + C]$,

$$ \underset{\text{martensite}}{M} \xrightarrow{\text{tempering}} \underset{\text{tempered martensite}}{[\alpha + C]} . \qquad (6\text{--}6.1)$$

The tetragonal body-centered martensite changes to bcc ferrite without
the excess carbon. The carbon collects in extremely small carbide
particles.

Tempered martensite is a two-phase microstructure of $[\alpha + C]$,
but one which differs in properties from pearlite and spheroidite
because of the size, shape, and distribution of the phases. In tempered
martensite, unlike the situation in pearlite, the carbide which forms
is not lamellar, but particulate; it is made up of many extremely fine

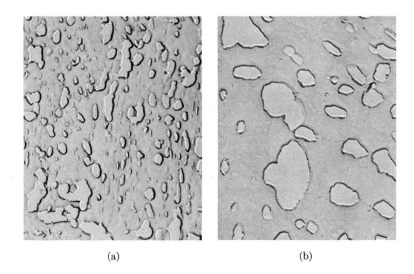

(a)   (b)

Fig. 6–6.7 Growth of carbide particles in tempered martensite ($\times$ 11,000). The two compositions are the same. (a) One hour at 1100°F (590°C). (b) Twelve hours at 1250°F (675°C). (Courtesy General Motors)

particles which are uniformly distributed on a submicroscopic scale, and which greatly interfere with dislocation movements. Thus tempered martensite remains nearly as hard as martensite just after quenching. In addition any crack which might be initiated must progress through a tough ferrite matrix as it advances through the steel. This means that tempered martensite does not have the brittleness of martensite and pearlite.

At higher temperatures the carbide particles grow (and decrease in number) just as the precipitates do during overaging of precipitation-hardened alloys (Fig. 6–6.7), with the result that the hardness also decreases. Figure 6–6.8 relates the hardness to both time and temperature. This explains the softening which occurs in many tool steels as they cut at high speeds. A few minutes at 600–800°F could decrease the hardness far below the values of $65R_C$ available in martensite. Time requirements are increased considerably if the Cr, Mn, Mo, and/or V contents of the carbide are high, because their large atoms must diffuse, as well as the smaller carbon atoms. That is why tool steels usually contain these alloying elements.

Although there is a slight variation from steel to steel depending on the composition, martensite does not temper below 300°F in any reasonable length of time. Therefore martensite remains indefinitely, for all intents and purposes, at normal temperature (70°F). Likewise an extrapolation of the data in Fig. 6–6.8 suggests correctly that over-tempering does not occur in reasonable times at normal temperatures.

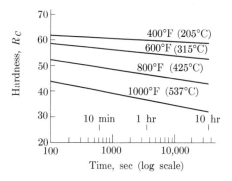

Fig. 6–6.8 Hardness versus tempering time (1080 steel previously quenched to maximum hardness). Softening starts with the martensite, *M*, transforming to ferrite plus carbide (α + **C**), and continues with the growth of the carbide particles (Fig. 6–6).

Atoms rearrange themselves too slowly in steel at room temperature to break up the original structures and form new phases. (See Study problem 6–6.8.)

Example 6–6.1. Martensite with 3 a/o carbon ($\sim$0.66 w/o C) has a body-centered tetragonal unit cell (u.c.) which is 2.83 Å $\times$ 2.83 Å $\times$ 2.91 Å at room temperature. What is its density ρ?

Solution

Basis: 200 atoms = 194 Fe atoms + 6 carbon atoms

$\qquad\qquad$ = 97 unit cells of 2 Fe atoms each, plus 6 interstitial carbon atoms.

$$\rho = \frac{\text{g/97 u.c.}}{\text{vol./97 u.c.}} = \frac{194(55.85 \text{ g/6} \times 10^{23} \text{ atoms}) + 6(12 \text{ g/6} \times 10^{23} \text{ atoms})}{97(2.83 \text{ Å})^2(2.91 \text{ Å})(10^{-24} \text{ cm}^3/\text{Å}^3)}$$

$$= 8.05 \text{ g/cm}^3.$$

Comments. This density represents an expansion in volume from the original austenite, which is slightly denser (8.15 g/cm^3) at room temperature. Therefore the martensite which forms places the remaining austenite under pressure, and in fact may force the last of the austenite to be retained at room temperature without change. ◀

Example 6–6.2. Figure 6–6.3(a) provides a hardness traverse for a 2.0-inch-diameter bar of AISI 1040 steel. Based on the hardenability data of Fig. 6–6.5, sketch a traverse for an AISI 3140 steel bar of the same dimensions and quenched in the same manner.

Solution

	AISI 1040	Cooling rate	Equivalent end-quench position, in.	AISI 3140
Surface	53R_C	400 °F/sec	$\frac{1.5}{16}$	57R_C
Mid-radius	28	60	$\frac{5}{16}$	53
Center	26	37	$\frac{6.5}{16}$	49

Comments. The quenching method affects the cooling rate, and therefore the hardness of the steel. A water quench is more severe than an oil quench because steam forms, requiring a very high heat of vaporization. If the quenching bath is not agitated, the steam can produce insulated spots which slow the heat removal and produce softer spots in the steel. ◀

Example 6–6.3. Requirements for a 2.5-inch-diameter bar call for a surface hardness of more than $50R_C$ and a center hardness of less than $32R_C$. Alternatives include the five steels of Fig. 6–6.5 and two methods of quench: agitated oil and agitated water. Make your recommendations.

Solution

For 2.5-in. bar (Fig. 6–6.6)	Equivalent distance from quenched end, in.	
	Oil quench	Water quench
Surface	$\dfrac{6.5}{16}$	$\dfrac{1.5}{16}$
Mid-radius	$\dfrac{10.5}{16}$	$\dfrac{6.5}{16}$
Center	$\dfrac{13.5}{16}$	$\dfrac{9}{16}$

From Fig. 6–6.5:

3140, 4140, 4340 would all have center hardnesses too high, with either an oil quench or a water quench.

1040 and 1060 would have surfaces too soft with an oil quench.

1040 W.Q.   S: $50R_C$ C: $25R_C$

1060 W.Q.   S: $\sim 55R_C$ C: $27R_C$

Comments. The 1060 W.Q. is the better choice, since a slight variation in the carbon content of the 1040 steel, say to 0.38% C, could cause the surface hardness to fall below $50R_C$.

Engineers often specify high surface hardness for resistance to wear, and require the center to remain soft for greater toughness. ◀

6–7 FABRICATION PROCESSES ⒠ ◎

Whenever two metal surfaces of any type have to be joined, this obviously represents a major problem of metal processing. Welding naturally calls for special consideration of the metals, because in welding both melting and solidification must take place. Metal properties are also important in riveting. Steel rivets are of course generally driven into place while they are hot. And, with no strain-hardening, the rivet conforms closely to the contours of the hole, so that a tight join results. Furthermore, thermal contraction of the rivet compresses the structural members even more tightly. The rivets naturally gain considerable strength as they cool to normal temperatures. However, one must avoid steels which form martensite in the short cooling time, because in a large structure it would be impossible to temper the rivets. In general, steel rivets are made out of plain carbon steels which transform rapidly to $[\alpha + \mathbf{C}]$.

Aluminum rivets, like steel ones, should also be as soft as possible so that they conform closely to the dimensions of the holes they are to be driven into. But aluminum rivets are not inserted while they are hot. Instead, aluminum rivets are solution treated, with no advance strain- or age-hardening. Commonly, engineers choose age-hardening aluminum alloys which harden at normal temperatures but do not overage in reasonable lengths of time (Fig. 6–5.2). Thus an aluminum rivet gains strength in use. However, storing it before you are ready to use it presents a problem. To avoid premature hardening after solution treatment, you must refrigerate age-hardenable aluminum so that the precipitation process is arrested before use.

Welding. In welding steel, we encounter three thermal effects (Fig. 6–7.1): (a) solidification of the weld metal, (b) austenization of immediately adjacent metal, and (c) residual stresses because of severe temperature gradients. If we tried to weld a heat-treated steel, there would be overtempering of the zone which is heated just short of the eutectoid temperature ($\sim 1300°F$ or $700°C$).

The weld metal generally freezes rapidly and is thus fine-grained; in low-alloy steels, it may even form martensite. This could be undesirable since many welded structures cannot be tempered readily because of their size. However, an alternative is available: Weld metal can be added in several passes, so that the later passes temper any martensite formed earlier.

In the majority of weldments, stress relief is necessary. To relieve stresses, one usually reheats the weld zone by means of a welding torch, or some similar procedure, and then lets the weld zone cool slowly.

Brazing and soldering. These two joining processes differ from each other in the melting temperature of the join metal (Table 6–7.1). Otherwise they are similar, in that the adjacent metal is not melted, and bonding occurs when the product is wetted by the joining alloy. Complete contact depends on capillary movement of liquid metal between adjacent surfaces. Thus it is imperative that the metal surface be either a clean, machined surface, or that it have any surface contamination removed by a flux. Although the base metal does not melt (except possibly superficially), a good bond depends on the development of phase and grain boundaries across the joint, similar to those encountered in the interior of any alloy.

Solders are usually alloys of lead and tin, but sometimes other low-melting materials are added for increased strength or wetting

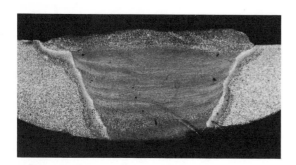

Fig. 6–7.1 Microstructure of a weld. The center of the weld was molten, but solidified rapidly. This heated the adjacent metal, altering its microstructure. (J. W. Freeman)

Table 6–7.1

CHARACTERISTICS OF TYPICAL SOLDERS AND BRAZING FILLER METALS†

Material		Composition	Melting range		Typical uses
			°F	°C	
Solders:	35-65 solder	35Sn 65Pb	361–473	183–245	Wiping solder
	45–55 solder	45Sn 55Pb	361–439	183–226	For automobile radiator cores; general-purpose solder
	60-40 solder	60Sn 40Pb	361–368	183–187	For use where temperature requirements are critical
	Cd-Ag solder	95Cd 5Ag	639–734	337–390	High-temperature solder
Brazing filler metals:	B-CuZn-2	57Cu 42Zn 1Sn	1630–1650	887–900	General-purpose alloy for steel, copper alloys, and nickel alloys
	B-Cu	99+ Cu	1980	1080	Furnace brazing alloy
	B-CuP-2	93Cu 7P	1305–1485	725–807	For self-fluxing brazing of copper alloys
	B-Ag-1*	45Ag 24Cd 16Zn 15Cu	1125–1145	608–618	Low-melting, free-flowing braze for general-purpose work
	B-Ag-8*	72Ag 28Cu	1435	780	For applications, such as vacuum tubes, in which volatile elements are harmful
	B-AlSi-1	95Al 5Si	1070–1165	575–630	General-purpose alloy for brazing aluminum
	B-Mg	89Mg 9Al 2Zn	770–1110	410–600	General-purpose alloy for brazing magnesium
	B-NiCr	70Ni 16.5Cr bal. mostly Fe	1850–1950	1010–1065	Heat-resistant alloy used for fabricating stainless steels or high-nickel alloys for jet engines

* Silver "solders."
† From *Elements of Physical Metallurgy*, by A. G. Guy, Reading, Mass.: Addison-Wesley.

qualities. Table 6–7.1 gives a list of typical solders. The filler metal for *brazing* may be pure metal, such as copper or silver, or copper alloys, such as brass and the various bronzes.

6–8 POWDER PROCESSES ⓔ ◎

Sintering is a thermal process which has been used throughout history for producing ceramic products from clays and/or other solid particles (Section 11–3). In recent years, it has been more and more used in metallurgy, so that now *P/M* (*powdered metal*) products constitute a significant fraction of design products (Fig. 6–8.1).

In principle, P/M products are made by compacting metal powders into a mold. Those metals which are ductile deform at points of contact; however, it is impossible to bring about full density—i.e.,

Fig. 6–8.1 Powdered-metal gear. Processing steps include: (1) compaction into a mold, (2) solid sintering in an oxygen-free atmosphere.

remove all porosity—by this procedure. Compaction is followed by sintering, which (1) binds the powder particles together, and (2) densifies the *compact*, thus reducing the porosity. Here the engineer has some choice. If he is designing products such as filters (e.g., in the gas-feedline of a car), or "oilless" bearings (which actually maintain a lifetime supply of oil in the pores), he may specify porosity. Alternatively, he may specify near-zero porosity where strength is important (such a zero-porosity material requires higher-temperature heating for densification).

The sintering process requires extensive diffusion of atoms, as illustrated in Fig. 6–8.2, which shows nickel powder in the sintering

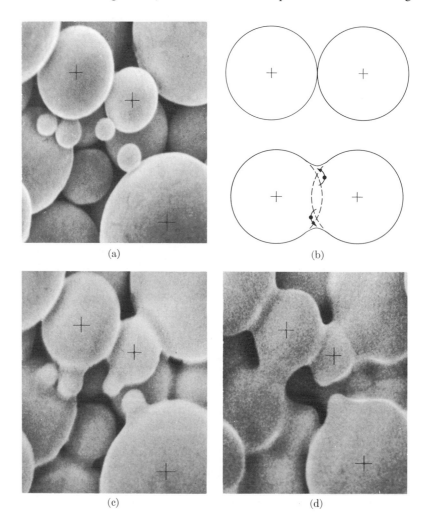

(a)

(b)

(c)

(d)

Fig. 6–8.2 The sintering process (nickel powder). The initial points of contact between metal particles in (a) became areas of contact in (c) and (d) after the powder has been heated at 1100–1200°C for approximately an hour. In order to attain this solid-to-solid bond, atoms must diffuse as sketched in (b), with the resulting shrinkage of dimensions. (Approximately × 1500) (Courtesy R. M. Fulrath, Inorganic Materials Research Division of the Lawrence Radiation Laboratory, University of California, Berkeley)

process. Initially (Fig. 6–8.2a), the particles are simply in contact (in this case not compacted together). The points of contact become areas of contact (Figs. 6–8.2c and d). This requires a diffusion of nickel, as sketched in Fig. 6–8.2(b). We shall not examine the details of this diffusion, which involves the countermovement of vacancies (Fig. 4–8.1) and nickel atoms, except to note that (1) diffusion, and therefore sintering, occurs more readily at high temperatures, and (2) shrinkage accompanies sintering by these mechanisms. The shrinkage becomes evident when one observes the distance between the centers (+) of the nickel particles in Fig. 6–8.2. In Section 11–3, it will be pointed out that the driving force which promotes sintering is the elimination of surface area (Fig. 11–3.1).

The process of sintering metals must be performed in an atmosphere which is devoid of oxygen. Any oxygen present would react with the metal and preclude metal-to-metal bonding of adjacent particles, as developed in Fig. 6–8.2(c). Generally a hydrogen atmosphere is used (with appropriate precautions). The temperature of sintering is normally close to—but short of—the melting point of the metal. This accelerates the required diffusion.

A recent development of major importance involves the use of P/M preforms for hot pressing. A compact is made from powdered metal, followed by a heating step which initiates sintering. The hot preform is then pressed in a closed-die mold to full density. The advantages of this process are twofold. First, nearly 100% of the metal goes into the product. Second, the composition can be more uniform throughout the part than is possible with metal which originates in a casting. As illustrated in Fig. 6–4.1(a), solidification commonly produces segregation.

Example 6–8.1. The powdered metal gear (brass) of Fig. 6–8.1 must be 1.61 in. (4.09 cm) in diameter with near-zero porosity as a final sintered product. What diameter of die is required if the compacted porosity is 20 v/o?

Solution. Basis: A cube, 1 cm × 1 cm × 1 cm, containing 20 v/o porosity. Actual metal volume = 0.80 cm^3. A cube with 0.80 cm^3 and zero porosity = $(0.928 \text{ cm})^3$. Compact diameter:

$$\frac{1.00 \text{ cm}}{0.928 \text{ cm}} = \frac{d \text{ in.}}{1.61 \text{ in.}},$$

$$d = 1.73 \text{ inch.}$$

Comments. If specifications had called for 5 v/o porosity in the final product, the 0.80 cm^3 of actual metal volume would have occupied $[0.80 \text{ cm}^3/0.95]$, or 0.842 cm^3. This is $(0.944 \text{ cm})^3$. Following the above proportions, the initial dimension d would have to be 1.71 inches. ◀

REVIEW AND STUDY

6–9 SUMMARY

The major operations involved in metal production include extraction, refining, casting, mechanical working, heat treatment, joining, and sintering. Extraction and refining are generally chemical in nature, involving reduction and selective oxidation. The technical problems which accompany casting center around solidification shrinkage.

The processes of mechanical working are predominantly reshaping processes—rolling, forging, extrusion, drawing, and spinning—but may also involve considerable removal of metal by machining in the final processing operations. Since a metal does not strain-harden above its recrystallization temperature, hot working is preferred for large metal products. Cold working can be used to adjust final strength and ductility. Machining involves deformation of the metal in the chip as well as cutting.

Thermal processing includes annealing for homogenization, softening, and stress relief. Heat treatments may be designed to bring about precipitation-hardening and also phase transformations which produce considerable strengthening and even toughening. Steel's response to quenching and tempering makes it one of the most versatile engineering materials available to the engineer.

Joining by welding involves all the various phase changes encountered in solidification and heat treatments. Therefore it is necessary to understand the way these affect properties of the final product.

The sintering of powder compacts requires extensive diffusion of atoms to bond the particles together and to eliminate porosity. Allowances must be made for shrinkage.

Terms for Study

Age hardening	Cemented carbides
Aging	Clustering
Annealing	Cold working
Annealing, full	Cooling rate
Annealing, process	Drawing
Basic oxygen furnace, BOF	End-quench test (Jominy bar)
Blast furnace	Extraction
Brazing	Extrusion
Casting	Faraday, $\mathscr{F}$

Forging

Gates

Hardenability

Hardenability curves

Hardness

High-speed steel

Hot working

Ingot

Machining

Martensite

Mechanical forming

Normalizing

Overaging

Precipitation-hardening

Reduction

Refining

Residual stress

Riser

Rolling

Quench

Sintering

Solder

Solidification shrinkage

Spheroidization

Spinning

Strain-hardening

Stress relief

Tempered martensite

Tempering

Tetragonal

Welding

STUDY PROBLEMS

6–1.1 One process for extracting magnesium from MgO involves the use of silicon as a reducing agent. The final products are Mg and SiO_2. How many pounds of silicon are required per pound of magnesium formed?

Answer. 0.58 lb Si/lb Mg

6–1.2 An electrolytic process may also be used to extract magnesium from MgO. How many hours are required to extract one pound (454 g) of magnesium when 50 amp of current are used?

Answer. 20 hr

6–2.1 Copper is rolled at a temperature near its recrystallization temperature. It appears very ductile in the trial rolling at slow speeds, but cracks at the higher speeds of full operation. Give a plausible explanation.

6–2.2 Compare and contrast the stress pattern for wire drawing and for bar extrusion.

6–3.1 Cite two reasons for using a cutting oil for machining.

6–3.2 Detail the movements of atoms required for an increase in the average carbide size in a vanadium-containing tool steel.

6–4.1 Example 6–4.1 calculated stresses at 70°F. Transformation starts at 1000°F (538°C), where the linear expansion is 0.25% and Poisson's ratio is 0.33. What stresses would be encountered if there were no plastic deformation?

Answer. $-169,000$ psi, that is, in compression

6–4.2 Assume that the 98Al–2Cu solid of Example 6–4.2 is separated, remelted, and cooled slowly. (a) What is the composition of the first solid this time? (b) At what temperature?

Answer. (a) 99+ Al and < 1Cu; (b) 1210+ °F (655°C)

6–4.3 Explain why stress relief can be accomplished at a temperature below the recrystallization temperature, while homogenization requires a temperature higher than the recrystallization temperature.

6–4.4 In commercial practice, a normalized steel is heated one hour at 100°F into the full austenitic temperature range. Indicate the normalizing temperature for a 1040, a 1080, and a 9240 steel.

Answer. 1600°F, 1435°F, 1590°F (870°C, 780°C, 865°C)

6–5.1 A 90Mg–10Al alloy is considered for age-hardening. Comment on its possibilities.

Answer. See Example 6–5.1

6–5.2 An alloy of 90Al–10Cu is heated to 1000°F (538°C), then given the treatment *XAB* of Fig. 6–5.1. What fraction of the alloy will age-harden?

Answer. 90%

6–5.3 A very low-carbon steel (~0.05 w/o C) is cold rolled into sheet which is to be used to make metal cans (so-called "tin" cans because they may be coated with tin for protection against corrosion). Its strength arises from strain-hardening. However, it acquires additional hardness and brittleness when food is processed in the cans at 100°C in a food-packing plant. Give a plausible explanation.

6–5.4 Which of the following alloys can't undergo age-hardening? (a) 95Cu–5Al, (b) 95Al–5Cu, (c) 95Cu–5Ni, (d) 95Ni–5Cu, (e) 95Cu–5Zn, (f) 95Zn–5Cu, (g) 95Al–5Mg, (h) 95Mg–5Al, (i) 95Ag–5Cu, (j) 95Cu–5Ag.

Answer. Check the phase diagram for decreasing solubilities.

6–5.5 While they are being stored at an aircraft manufacturing plant, 1000 lb of aluminum-alloy rivets are exposed to 90°F heat long enough to cause them to start to harden. They must, however, be as soft as possible when they are driven into the rivet holes. Can they be salvaged for use? Explain.

6–6.1 A certain steel contains 0.5 w/o carbon. How many carbon atoms are there per 100 unit cells of body-centered tetragonal martensite?

Answer. 4.7

6–6.2 A bar of 1060 steel has a center hardness of $25R_C$ and a surface hardness of $40R_C$. How fast were the center and surface cooled?

Answer. Center, 20°F/sec; surface, 120°F/sec

6–6.3 Plot the hardness traverse for a bar made of 3140 steel, 3 inches in diameter, quenched in water.

Answer. Surface, $56R_C$, mid-radius, $46R_C$, center, $41R_C$

6–6.4 Exact identification was lost concerning a round bar of steel (50 ft long × 3 in. diameter). It is known only that it may be either AISI 3140 or AISI 4340. (a) It is proposed that the bar be heated, quenched in oil, and identified by a hardness traverse. Indicate below the hardness values you would expect.

	3140	4340
Surface	_____ R_C	_____ R_C
$\frac{3}{4}$ in. below surface	_____ R_C	_____ R_C
$1\frac{1}{2}$ in. below surface	_____ R_C	_____ R_C

(b) A furnace is not available to heat the 50-ft bar. However, since the bar is 3 in. longer than required, it is proposed that 3 in. be cut off for the above identification test. Comment. (c) Suggest—and discuss the merits of—alternative possibilities for identification.

Answer. (a) $C_{3140} - \sim 33R_C$, $C_{4340} - \sim 52R_C$

6–6.5 Two round bars, 2 in. and 3 in. in diameter and made of the same steel, yield the following data derived from their hardness traverses.

Diameter	2 in.	3 in.
Quench	Water	Oil
Surface hardness	$59R_C$	$55R_C$
Mid-rad. hardness	$57R_C$	$41R_C$
Center hardness	$55R_C$	$35R_C$

Plot a hardenability curve for this steel. [*Note:* Your answer should consist of one curve *only*!]

6–6.6 Surface hardnesses of 6 bars of the same steel are as follows.

1 in. dia. W.Q. $63R_C$	1 in. dia. O.Q. $55R_C$
2 in. dia. W.Q. $62R_C$	2 in. dia. O.Q. $34R_C$
4 in. dia. W.Q. $50R_C$	4 in. dia. O.Q. $31R_C$

Plot a hardenability curve for this steel. [*Note:* Your answer should consist of one curve *only*!]

6–6.7 Examine Fig. 6–6.7. Suggest a method—other than a 12-hr process anneal (Section 6–4)—for producing spheroidite.

6–6.8 Estimate the time necessary at 600°F to temper an AISI 1080 steel (a) to $50R_C$, (b) to $40R_C$.

Answer. (a) $\sim 10^{5.7}$ sec ($= 140$ hr); (b) $\sim 10^{10}$ sec ($= 300$ yr)

6–7.1 Sketch a plausible hardness profile across the weldment shown in Fig. 6–7.1.

6–7.2 Only the outer $\frac{1}{8}$ inch of a bar 1 in. in diameter and made of 1020 steel contains significant amounts of martensite after quenching. The full 0.5-in. width of weld metal in a weldment of 1040 steel plate (1 in. thick) is high in martensite. Explain.

6–7.3 Beryllium bronze (Table 6–5.2) is welded after various heat treatments. Comment on the effect of the heat treatment on the properties of this bronze.

6–8.1 During sintering, a powder compact shrinks from a cross section 0.57 in. × 1.05 in. to 0.51 × 0.94 in. Assuming uniform shrinkage, what is the volume shrinkage?

Answer. − 28 v/o (original basis)

6–8.2 Work study problem 11–3.2.

6–8.3 Refer to study problem 6–8.1. The P/M product was removed from the furnace when the dimensions were 0.521 in. × 0.96 in. What percent porosity remains? The final product in study problem 6–8.1 was at full density, i.e., 0% porosity.

Answer. 6.2 v/o porosity

6–8.4 Estimate the volume shrinkage during the sintering steps depicted in Fig. 6–8.2.

Answer. (a) → (c) 9 v/o; (a) → (d) 30 v/o

CHAPTER 7

PLASTICS
AND
THEIR STRUCTURES

7–1 GIANT MOLECULES

We shall consider a *molecule* as being a group of atoms which have
strong bonds among themselves, but relatively weak bonds to adjacent
molecules. Some familiar small molecules include water (H_2O),
carbon dioxide (CO_2), methane (CH_4), ethyl alcohol (C_2H_5OH), etc.
Plastics are made up of large molecules which may contain thousands
to millions of atoms. Sometimes these are referred to as *macro-*
molecules, or as *polymers*.

A *micro*molecule, such as ethyl alcohol, illustrates the nature of a
molecule:

$$H : \overset{\displaystyle H}{\underset{\displaystyle H}{\overset{..}{\underset{..}{C}}}} : \overset{\displaystyle H}{\underset{\displaystyle H}{\overset{..}{\underset{..}{C}}}} : \overset{..}{\underset{..}{O}} : H \qquad (7\text{--}1.1)$$

There are strong bonds between the carbon and hydrogen atoms (also
C:C, C:O, and O:H) because two atoms *share* a pair of electrons.
(Recall from Section 1–2 that this is characteristic of nonmetallic
elements.) We commonly call this sharing a *covalent bond*, and use
the presentation:

$$H - \overset{\displaystyle H}{\underset{\displaystyle H}{\overset{|}{\underset{|}{C}}}} - \overset{\displaystyle H}{\underset{\displaystyle H}{\overset{|}{\underset{|}{C}}}} - O - H \qquad (7\text{--}1.2)$$

A carbon atom may have a maximum of four pairs of shared electrons,
or four bonds. Table 7–1.1 shows the maximum numbers of shared
electrons for other nonmetallic elements.

Each atom of the above ethyl alcohol molecule is satisfied with its
maximum number of bonds. Therefore none of the atoms in the
molecule develop strong attractions to adjacent molecules. Weak
attractions do exist between molecules, however; ethyl alcohol and
water are liquids at room temperature because there is enough attraction

Table 7–1.1

SHARED BONDS FOR NONMETALLIC ELEMENTS

Element	Shared bonds
H	1
F	1
Cl	1
O	2
S	2
N	3
C	4

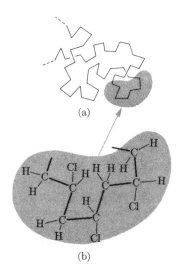

(a)

(b)

Fig. 7–1.1 Linear macromolecules (polyvinyl chloride in liquid solution). Thermal agitation twists the backbone of carbon atoms into a kinked conformation.

Table 7–1.2

BOND LENGTHS BETWEEN NONMETALLIC ATOM PAIRS

Bond	Length, Å	Bond	Length, Å
C—C	1.54	O—H	1.0
C=C	1.3	O—O	1.5
C≡C	1.2	O—Si	1.8
C—H	1.1		
C—N	1.5	N—H	1.0
C—O	1.4	N—O	1.2
C=O	1.2		
C—F	1.4		
C—Cl	1.8	H—H	0.74

at 20°C (68°F) to condense the molecules into close proximity. We know that the attractions are weak because adjacent molecules can flow past each other.

A macromolecule has a long sequence of strong covalent bonds in which adjacent atoms share electrons (Fig. 7–1.1). As we encounter longer and longer molecules, however, liquids of the materials become more and more viscous because of entanglements. In fact, even though some plastics retain the structure of liquids—i.e., they do not crystallize—they become so rigid as a result of their large molecules that we categorize them among the solids.

Molecular weights. Based on the data in Fig. 1–2.2, we know that water has a molecular weight of 18 amu, or 18 g per 6×10^{23} molecules. The molecular weight of ethyl alcohol, C_2H_5OH, is

$$2(12) + 5(1) + 16 + 1 = 46 \text{ g}/6 \times 10^{23} \text{ molecules.} \qquad (7\text{–}1.3)$$

We usually refer simply to this as 46 g/*mole*.

The molecular weights of polymers are invariably more than a thousand grams per mole, and may run into the millions. Consider Fig. 7–1.1, and assume a chain of 700 carbon atoms (1050 hydrogen atoms and 350 chlorine atoms). The molecular weight is nearly 22,000 grams per mole. The 700 carbon count is not a limiting number.

Mers. If we look at Fig. 7–1.1, we observe a unit of C_2H_3Cl which repeats itself along the chain. We call this unit a *mer*, and it is the basic unit of structure in *polymers* (literally, *many units*). If we describe the mer, we can describe the whole molecule. Thus

$$\begin{bmatrix} & \text{H} & \text{H} & \\ & | & | & \\ -& \text{C} - \text{C} & - \\ & | & | & \\ & \text{H} & \text{Cl} & \end{bmatrix}_n \qquad (7\text{–}1.4)$$

describes the molecule of Fig. 7–1.1, and the weight of each mer is $2(12) + 3(1) + 35.5$, or 62.5 amu. If we let n equal 350, the molecular weight is 350×62.5 amu or nearly 22,000 g/mole. The polymer with 350 mers linked together has a *degree of polymerization* of 350.

Molecular dimensions. Adjacent carbon atoms have an interatomic distance of 1.54 Å. Table 7–1.2 lists comparable distances, called *bond lengths*, for other atom pairs. Double bonds involve the covalent sharing of two pairs of electrons. Carbon-to-carbon bond lengths are of special interest to us because they enable us to estimate the length of a molecule. A first approximation indicates that a molecule with 700 carbon atoms would be 700×1.54 Å, or about 1080 Å long.

Actually the bonds are not 180° across a carbon atom, but have an average angle of 109°. Furthermore, the bonds can rotate in three dimensions, so we see considerable twisting and kinking along the length of the chain (Fig. 7-1.1). This will be important to us because, with kinking (see Fig. 7-1.1), the average end-to-end distance $\bar{L}$ of such molecules in noncrystalline polymers becomes

$$\bar{L} = \ell\sqrt{m}. \qquad (7\text{-}1.5)$$

Here ℓ is the interatomic distance (1.54 Å for C—C) and m is the number of bonds. Thus instead of 1080 Å, the value $\bar{L}$ is (1.54 Å) $\sqrt{700}$ or slightly more than 40 Å. The reason this is so important is that many large molecules may be stretched out from this kinked *conformation* to give high strains without noticeably changing the interatomic distances. Rubbers possess this characteristic and may develop high strains at low stresses.

Example 7-1.1. Determine the molecular weight of phenol and formaldehyde, two small molecules which are combined to make bakelite.

a) Phenol:

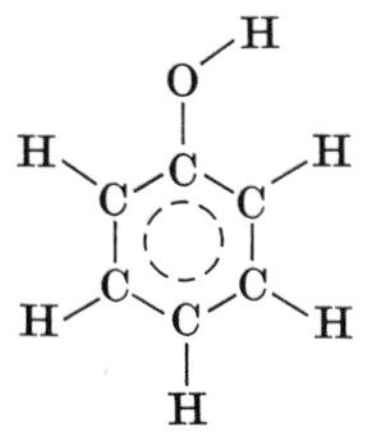

$$(7\text{-}1.6)$$

b) Formaldehyde:

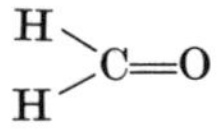

$$(7\text{-}1.7)$$

Solution

a) $[6 \times 12 + 16 + 6 \times 1] = 94$ g/mole
b) $[12 + 16 + 2 \times 1] = 30$ g/mole

Comments. We will not need to know very many molecular compounds. Figure 7-1.2 lists those which will enter our discussions without specific definition. You are probably already familiar with many of them; if not, you will find it helpful to become aware of them. ◀

Example 7-1.2. A polyvinyl chloride plastic (Fig. 7-1.1) has an average molecular weight of 72,130 amu (that is, 72,130 g/mole). What is the degree of polymerization, n?

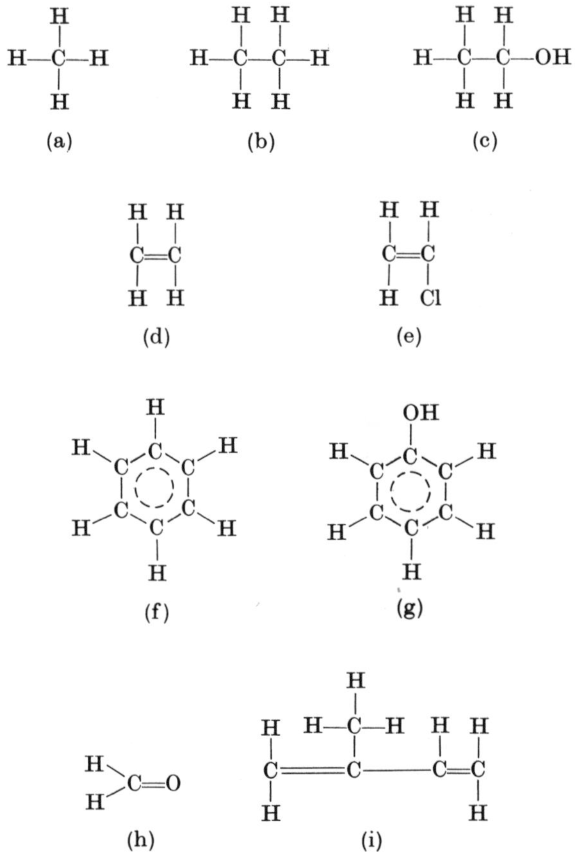

Fig. 7-1.2 Common micromolecules. (a) Methane, CH_4. (b) Ethane, C_2H_6. (c) Ethyl alcohol, C_2H_5OH. (d) Ethylene, C_2H_4. (e) Vinyl chloride, C_2H_3Cl. (f) Benzene, C_6H_6. (g) Phenol, C_6H_5OH. (h) Formaldehyde, CH_2O. (i) Isoprene, C_5H_8.

Table 7-1.3

VINYL COMPOUNDS

$$\begin{pmatrix} \begin{array}{c} H \quad H \\ | \quad\quad | \\ C = C \\ | \quad\quad | \\ H \quad R \end{array} \end{pmatrix}$$

Compound	R
Ethylene	—H
Vinyl alcohol	—OH
Vinyl chloride	—Cl
Propylene	—CH$_3$
Vinyl acetate	—OCOCH$_3$
Acrylonitrile	—C≡N
Styrene	⬡ *

---

* The symbol ⬡ is used to denote the benzene ring,

which in its more complete form is

as a molecule, or

as a radical.

Solution

C_2H_3Cl mer weight $= [2(12) + 3(1) + 35.5]$

$$= 62.5 \text{ amu} = 62.5 \text{ g/mer wt.}$$

Therefore

$$n = 72{,}130 \text{ amu}/62.5 \text{ amu}$$

$$= 1154 \text{ mers/polymer.}$$

Comments. In a plastic, the molecules are not all the same size. Therefore it is necessary for us to speak of the *average molecular weight*, $\overline{MW}$. A larger average molecular weight generally means a higher melting point, more viscosity, and a more stable plastic. ◀

Example 7-1.3. A plastic contains polystyrene with styrene mers as follows:

$$(7\text{-}1.8)$$

The degree of polymerization, n, is 525. How many molecules are there per gram of polystyrene?

Solution. Each molecule of polystyrene has:

$$\{525\,[(8 \text{ C/molecule})(12 \text{ amu/C}) + (8 \text{ H/molecule})(1 \text{ amu/H})]\}$$

$$= 54{,}600 \text{ amu/molecule.}$$

Since there are 6×10^{23} amu per gram,

$$\text{molecules/g} = \frac{6 \times 10^{23} \text{ amu/g}}{5.46 \times 10^4 \text{ amu/molecule}}$$

$$= 1.1 \times 10^{19} \text{ molecules/g.}$$

Comments. Polyvinyl compounds, which include styrene, have the composition

$$(7\text{-}1.9)$$

where R is one of a number of possible *radicals*. Table 7–1.3 lists some commonly encountered ones. ◀

7–2 MOLECULAR CONFIGURATIONS

Just as iron alloys may be either bcc or fcc and retain the same composition, many macromolecules have more than one structure. Consider *natural rubber*, which has a composition [called *isoprene*, $(C_5H_8)_n$] that may be sketched as follows:

$$
\begin{bmatrix}
& & \text{H} & & & \\
\text{H} & \text{H}-\!\!\text{C}-\!\!\text{H} & & \text{H} & \text{H} \\
| & | & & | & | \\
-\text{C}-\!\!\!\!-\!\!\!\!-\text{C}=\!\!=\text{C}-\text{C}- \\
| & & & & | \\
\text{H} & & & \text{H}
\end{bmatrix}_n
\qquad (7\text{–}2.1)
$$

Actually this molecule is highly kinked and exhibits the large strains typical of rubber. An *isomer* of C_5H_8 is

$$
\begin{bmatrix}
& & \text{H} & & \\
\text{H} & \text{H}-\!\!\text{C}-\!\!\text{H} & & \text{H} \\
| & | & & | \\
-\text{C}-\!\!\!\!-\!\!\!\!-\text{C}=\!\!=\text{C}-\text{C}- \\
| & & & | \quad | \\
\text{H} & & & \text{H} \quad \text{H}
\end{bmatrix}_n
\qquad (7\text{–}2.2)
$$

The composition is the same, $(C_5H_8)_n$, but the structure is slightly different. Because the structure is different, its properties are different. As a result, it is not even called isoprene, but *gutta percha*. It does not have the characteristics of rubber, but is more like the polyvinyl chloride tile we use for floors. In polymers we shall encounter many different isomers because there are many different ways in which a large number of atoms can be combined.

Stereoisomers. The simplest polymer of everyday usage is polyethylene

$$
\begin{bmatrix}
\text{H} & \text{H} \\
| & | \\
-\text{C}-\text{C}- \\
| & | \\
\text{H} & \text{H}
\end{bmatrix}_n
\qquad (7\text{–}2.3)
$$

The mer is considered to be C_2H_4 (although a case could be made for calling CH_2 the repeating unit) because this polymer arises from additions of ethylene molecules:

$$\cdots + \overset{\overset{\displaystyle H}{|}}{\underset{\underset{\displaystyle H}{|}}{C}} = \overset{\overset{\displaystyle H}{|}}{\underset{\underset{\displaystyle H}{|}}{C}} + \overset{\overset{\displaystyle H}{|}}{\underset{\underset{\displaystyle H}{|}}{C}} = \overset{\overset{\displaystyle H}{|}}{\underset{\underset{\displaystyle H}{|}}{C}} + \overset{\overset{\displaystyle H}{|}}{\underset{\underset{\displaystyle H}{|}}{C}} = \overset{\overset{\displaystyle H}{|}}{\underset{\underset{\displaystyle H}{|}}{C}} + \cdots$$

$$\longrightarrow \cdots -\overset{\overset{\displaystyle H}{|}}{\underset{\underset{\displaystyle H}{|}}{C}} - \overset{\overset{\displaystyle H}{|}}{\underset{\underset{\displaystyle H}{|}}{C}} - \overset{\overset{\displaystyle H}{|}}{\underset{\underset{\displaystyle H}{|}}{C}} - \overset{\overset{\displaystyle H}{|}}{\underset{\underset{\displaystyle H}{|}}{C}} - \overset{\overset{\displaystyle H}{|}}{\underset{\underset{\displaystyle H}{|}}{C}} - \overset{\overset{\displaystyle H}{|}}{\underset{\underset{\displaystyle H}{|}}{C}} - \cdots \qquad (7\text{-}2.4)$$

The individual C_2H_4 mer is symmetric with no "front" or "back." In contrast, the majority of vinyl compounds are not symmetrical. Consider propylene:

$$\overset{\overset{\displaystyle H}{|}}{\underset{\underset{\displaystyle H}{|}}{C}} = \overset{\overset{\displaystyle H}{|}}{\underset{\underset{\displaystyle H-\overset{\overset{\displaystyle |}{C}}{\underset{\underset{\displaystyle H}{|}}{}}-H}{|}}{C}} \qquad (7\text{-}2.5)$$

The left and right double-bonded carbons are not identical. One has two hydrogen neighbors; the other has a hydrogen and a CH_3 side radical as neighbors. When these are polymerized into chains, there are variations that can occur as shown in Fig. 7-2.1. These variations are called *stereoisomers*. In a later section we shall see that *configuration* differences such as these have an effect on the crystallizability of a polymer and therefore on the properties. The *isotactic* stereo-isomer (Fig. 7-2.1a) crystallizes more readily, is somewhat denser, and is appreciably stronger than the *atactic* isomer (Fig. 7-2.1c).

Branching. Ideally, polymer molecules such as we have studied to date are *linear*, i.e., they are a two-ended chain. There are cases, however, in which a polymer chain branches. We can indicate this schematically as in Fig. 7-2.2. Although a branch is unusual, once formed it is stable because each carbon atom has its complement of four bonds and each hydrogen atom has one bond. The significance of branching lies in the entanglements which can interfere with plastic deformation. Think of a pile of tree branches compared with a bundle of sticks; it is more difficult to move the branches than to move the individual sticks.

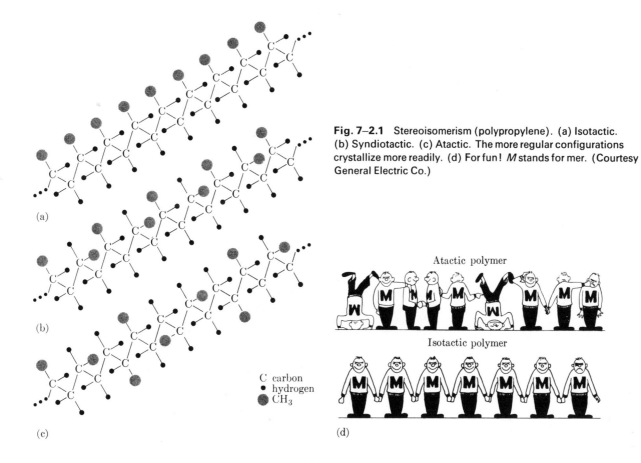

Fig. 7–2.1 Stereoisomerism (polypropylene). (a) Isotactic. (b) Syndiotactic. (c) Atactic. The more regular configurations crystallize more readily. (d) For fun! *M* stands for mer. (Courtesy General Electric Co.)

C carbon
• hydrogen
● CH₃

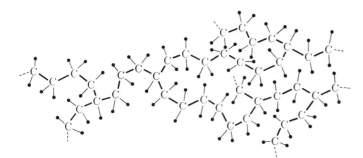

Fig. 7–2.2 Branching of polyethylene (with dots representing hydrogen). Branched molecules do not crystallize readily.

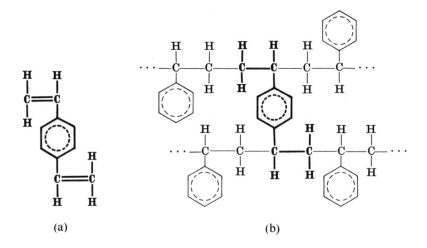

(a)   (b)

Fig. 7–2.3 Cross-linking. (a) Divinyl benzene. (Compare with styrene.) (b) Two chains cross-linked with divinyl benzene. (The symbol ⬡ is used for the benzene ring. Hydrogen atoms are implied at each corner *not* bonded to other atoms.)

Cross-linking. Some linear molecules, by virtue of their structure, can be tied together. Consider the molecule of Fig. 7–2.3(a) and its polymerized combination in Fig. 7–2.3(b). Intentional additives (divinyl benzene in this case, but we do not have to remember the name) tie together two chains of polystyrene. This causes restrictions with respect to plastic deformation.

The *vulcanization* of rubber is a result of cross-linking by sulfur, as shown in Fig. 7–2.4. The effect is pronounced. Without sulfur, rubber is a soft, even sticky material which, when it is near room temperature, flows by viscous deformation. It could not be used in automobile tires because the service temperature would make it possible for molecules to slide by their neighbors, particularly at the pressures encountered. However, cross-linking by sulfur at about 5% of the possible sites gives the rubber mechanical stability under the above conditions, but still enables it to retain the flexibility which is obviously required. Hard rubber has a much larger percentage of sulfur and appreciably more cross links. You can appreciate the effect of the addition of greater amounts of sulfur on the properties of rubber when you examine a hard rubber product such as a pocket comb.

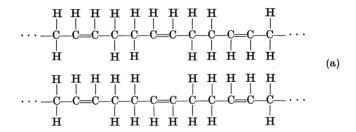

(a)

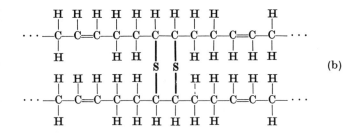

(b)

Fig. 7–2.4   Vulcanization (butadiene). Sulfur atoms serve as anchor points between adjacent molecules of rubber. A possible mechanism is a pair of sulfur atoms which react with the double bonds in a pair of adjacent molecules.

Network polymers. The simpler polymers are the linear ones just described. These are *bifunctional*, i.e., each mer can connect with two—and only two—other mers. Other types of mers can connect with several adjacent molecules. We saw one such mer in Fig. 7–2.3, in which the divinyl benzene bonds to four styrene mers.

A widely used plastic which has polyfunctional units is phenol formaldehyde. The early trade name, *bakelite*, is one name for this combination. The phenol molecules, C_6H_5OH, provide the plastic with *trifunctional* units (Fig. 7–2.5a) and CH_2 groups serve as bridges between adjacent phenol rings (Fig. 7–2.5b). The sketches in Fig. 7–2.5 are intended to reveal that a three-dimensional network structure is formed. We shall see in Chapter 8 that the thermal behavior of network polymers is markedly different from that of linear polymers.

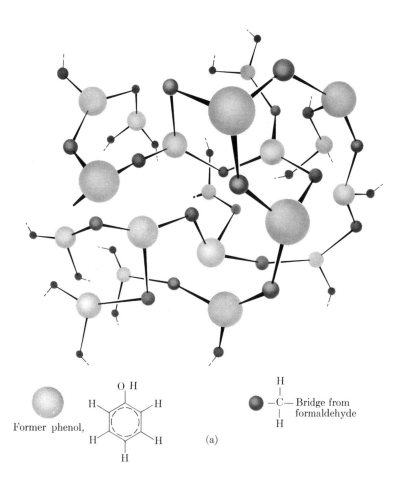

Former phenol,

O H

H   H

H   H

H

(a)

H
|
—C— Bridge from
| formaldehyde
H

Fig. 7–2.5 Network polymer (phenol formaldehyde). (a) A three-dimensional structure forms because each phenol molecule is trifunctional, i.e., it provides three reaction positions. (b) The original phenol and formaldehyde structures are shown in Fig. 7–1.2. Water forms as a by-product.

Originally phenol

Originally CH_2O

H   OH   **H**

H—**C**   **C**   **C**—**H**

H   **C**   **C**   **O**

H   C   C   H   **O**—**H**

     H   **H**

H—**C**—**H**

(b)

In fact, as you may already have guessed, network polymers are more rigid at high temperatures than linear polymers are.

Example 7–2.1. A polyisoprene rubber,

```
        ⎛        H                      ⎞
        ⎜        |                      ⎜
        ⎜  H  H—C—H  H   H              ⎜
        ⎜  |  |  |  |    |              ⎜
     ——— C ———— C ═══ C — C ———         
        ⎜  |           |               ⎜
        ⎝  H           H               ⎠ₙ
```

is vulcanized with 5.0 lb of sulfur per 100 lb of isoprene. If all the sulfur atoms serve as cross links, what fraction of the potential cross links are bridged?

Solution

Basis: 100 amu isoprene + 5 amu sulfur

$(100\ \text{amu})/[5(12) + 8(1)\ \text{amu/mer}\ C_5H_8] = 1.47$ mers C_5H_8

$(5.0\ \text{amu})/(32\ \text{amu/S}) = 0.156$ sulfurs

Sulfurs/mer $C_5H_8 = 0.156/1.47 = 0.106$.

According to Fig. 7–2.4, up to two sulfur atoms could join every *pair* of isoprene mers. This would mean that there could be 1.47 sulfur atoms per 1.47 isoprene mers.

Fraction = 0.106 of possible sulfur cross links.

Comments. This would give a harder rubber than that found in a tire, in which the sulfur/isoprene ratio is commonly about 0.02.

Oxygen will also cross-link with rubber. We see evidence of this when a rubber article hardens with age. ◀

7–3 CRYSTALLINITY IN POLYMERS Ⓜ

Many polymers, unlike metals—which invariably crystallize when they solidify under normal conditions—solidify without complete crystallization. In this section we shall concentrate on those polymers which do crystallize. In the following section, we shall consider polymers which solidify without crystallizing, and which form solids that may be viewed as having the characteristics of a very viscous liquid.

Molecular crystals. The atoms in metals are approximately spherical in shape. In fact, we spoke of the "hard-ball" atom (Section 3–1). Even in those cases in which atoms may be slightly distorted (cf. comments in Example 3–3.1), the atoms are sufficiently equiaxed so

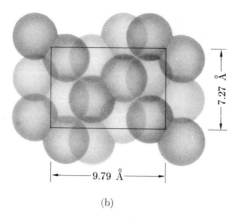

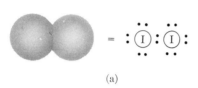

(a)

(b)

Fig. 7–3.1 Molecular crystal (iodine). (a) The iodine molecule, I_2. (b) The unit cell. The c-dimension (not shown) is 4.79Å. The corner angles of the unit cell are all 90°; however, $a \neq b \neq c$. Therefore the cell is not cubic, but orthorhombic.

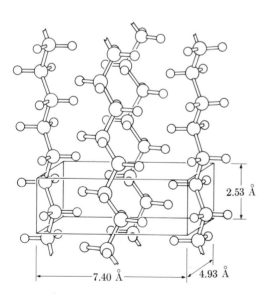

Fig. 7–3.2 Polyethylene unit cell. The molecules extend from one unit cell through the next. (M. Gordon, *High Polymers*, London: Iliffe Books, and Reading, Mass.: Addison-Wesley. After C. W. Bunn, *Chemical Crystallography*, London: Oxford University Press)

that we can view a crystal as a collection of individual balls arranged in a repeating long-range pattern.

Very few molecules—and specifically no polymer molecules—have equiaxed shapes with dimensions that are approximately equal in the three directions of space. In fact, the linear molecules of the previous sections are characterized by their extreme elongation in one dimension.

First let us consider the crystallization of the very simple molecule of iodine, I_2. The atoms of this biatomic molecule share a pair of electrons. Consistent with the first statement of this chapter, in I_2 there is a strong bond between the two atoms of the molecule, but much weaker attraction to other molecules. Thus the I_2 molecule has the shape of a dumbbell (Fig. 7–3.1a), and as such cannot form cubic crystals. The unit cell which results is shown in Fig. 7–3.1(b). Note that the three dimensions are not equal ($a = 7.27$ Å, $b = 9.79$ Å, $c = 4.79$ Å). In this case, however, the unit cell has three 90° angles.

Polyethylene (Table 7–1.3 and Eq. 7–2.3) has long linear molecules. These molecules have strong covalent bonds along the molecular chain. The bonds between chains are much weaker, but are still present, so that the chains tend to line up as shown in Fig. 7–3.2. Observe that a single molecule extends through many unit cells. Because the molecules are long, it is difficult to obtain perfect crystallization. For example, all molecules do not end at the same position because each of the many molecules has a different length. Also the tendency to twist and turn, as shown in Fig. 7–1.1, makes perfect alignment difficult to achieve.

Crystallization shrinkage. When iodine is heated, the individual molecules shown in Fig. 7–3.1(b) are more and more agitated by the heat. This produces thermal expansion. More important, there comes a point, the *melting temperature*, at which the agitation is too great for the weak bonds *between* neighboring molecules to be maintained. At this temperature the iodine suddenly loses its long-range crystalline pattern. (The bonds *within* the molecules remain intact.) All crystals that are close-packed (fcc and hcp) undergo marked expansion in volume at this point, as indicated for lead in Fig. 4–7.2. In fact, only in exceptional cases does expansion fail to take place when a material melts. Figure 7–3.3 shows the volume change for iodine, I_2, at the melting (or freezing) point.

Starting at 200°C with molten iodine, there is a shrinkage in volume as the temperature decreases. The discontinuous *crystallization shrinkage* at 114°C occurs as the iodine molecules pack themselves more efficiently into the crystal lattice. Below 114°C, shrinkage continues, with thermal agitation reduced, but with less change per degree because the molecules must vibrate at fixed crystal locations.

Example 7–3.1. Calculate the density of crystalline iodine.

Solution. Inspect Fig. 7–3.1(b) and observe that an I_2 molecule is centered at the corner of each unit cell, and also at the center of each face. Therefore, in this unit cell, there are 4 molecules (8 atoms):

$$\text{Density} = \frac{8 \text{ atoms } (126.9 \text{ g}/6 \times 10^{23} \text{ atoms})}{(7.27 \times 10^{-8} \text{ cm})(9.79 \times 10^{-8} \text{ cm})(4.79 \times 10^{-8} \text{ cm})}$$

$$= 4.97 \text{ g/cm}^3.$$

Comments. The experimental measured density is 4.94 g/cm³.

A crystal is *orthorhombic* when $a \neq b \neq c$ and the three principal angles are each 90°. ◀

7–4 AMORPHOUS POLYMERS Ⓜ

With small molecules, such as the two-atom iodine molecule, crystals form easily; but the crystallization process with large molecules such as polyethylene is appreciably more difficult. As stated in the previous section, the long, $+C_2H_4+_n$ molecules tend to become twisted and kinked. Entanglements result. Furthermore, one molecule may end and another may not be immediately available to continue into the next unit cell; this makes for poor formation of crystals. Branching also hinders crystallization.

In view of the above complications, it is not surprising that polyethylene may be cooled past the melting temperature (~135°C, or ~275°F) and on down to room temperature without the molecules aligning themselves into a nice crystalline pattern. Supercooling is common among polymers, especially if there are structural irregularities along the molecular chain. For example, the isotactic chain shown in Fig. 7–2.1(a) crystallizes much more readily than the atactic chain of

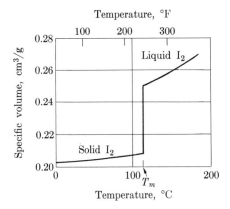

Fig. 7–3.3 Volume versus temperature (iodine). The volume change occurs at the melting temperature ($T_m = 114°C$) because the molecules are packed more tightly in the solid than in the liquid (cf. Fig. 4–7.2).

Fig. 7–2.1(c). The more regular the chains of molecules, the more readily they mesh together into crystals, rather like a zipper closing, to give a familiar analogy.

Glass temperature. Even though the molecules of the polymer are entangled, when there is thermal agitation, there is a continuous rearrangement of the atoms and molecules within a polymer liquid. And because of this, there has to be extra space in the liquid, just as a milling crowd cannot be as close packed as, say, a crowd of stationary people jammed into a football stadium. As the temperature decreases, the thermal agitation lessens, and there is a decrease in volume. This decrease in volume continues below the freezing point into the super-cooled liquid range (Fig. 7–4.1). The liquid structure is retained. As with liquid at a higher temperature, flow can occur; however, naturally flow is more difficult (the viscosity increases) as the temperature drops and the excess space between molecules decreases.

Those polymers which are cooled without crystallizing eventually reach a point at which the thermal agitation is not sufficient to allow for rearrangement of the molecules. Although not crystalline, the polymer becomes markedly more rigid. It also becomes more brittle. When there cannot be continued rearrangement of molecules so that there is better packing, additional decreases in volume mean only that the vibrations of the molecules are of smaller amplitude. Thus the slope of the curve in Fig. 7–4.1 becomes much less steep. This point of change in the slope is called the *glass point*, or the *glass transition temperature*, because this phenomenon is typical of all glasses. In fact, below this temperature, just as with normal silicate glasses, a noncrystalline polymer *is* a glass (although admittedly an organic one). Conversely, we shall observe in Chapter 10 that a normal glass is an inorganic polymer.

The glass transition temperature, or more simply the glass temperature, T_g, is as important to polymers as the melting (or freezing) temperature, T_m, is. The glass temperature of polystyrene (Table 7–1.3) is at $\sim 100°C$ ($212°F$). Therefore it is glassy and brittle at room temperature. In contrast, a rubber whose T_g is at $-100°F$ ($-73°C$) is flexible even in the most severe winter temperatures.

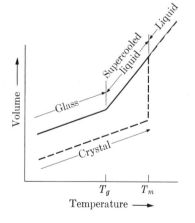

Fig. 7–4.1 Volume versus temperature (glassy material). The liquid curve extends below the melting point, T_m, into the supercooled liquid range. Below the glass temperature, T_g, the molecules cannot rearrange themselves, and the material takes on the rigid, brittle characteristics of glass.

Partially crystalline polymers. Some plastics are partially crystalline. Polyethylene is a good example. As it forms a solid, it is possible for a few unit cells to form relatively good crystals locally. The long molecules, however, extend beyond these crystalline regions into amorphous regions. As a result, various levels of crystallinity are possible. The volume–temperature curves for such polymers lie between the two curves of Fig. 7–4.1.

Example 7–4.1. A polyethylene with no evidence of crystallinity has a density of 0.90 g/cm^3. Commercial grades of polyethylene include "low-density" PE, with a density of about 0.92 g/cm^3, and "high-density" PE, with a density of about 0.96 g/cm^3. Estimate the fraction of crystallinity in each case.

Solution. Referring to Fig. 7–3.2, we may calculate the density of a polyethylene crystal.

Basis: 1 unit cell = $\frac{4}{2}$ mers of $-(C_2H_4)-$.

$$\text{Density} = \frac{(2 \text{ mers})(28 \text{ g}/6 \times 10^{23} \text{ mers})}{(7.40 \times 10^{-8} \text{ cm})(4.93 \times 10^{-8} \text{ cm})(2.53 \times 10^{-8} \text{ cm})}$$

$$= 1.01 \text{ g/cm}^3.$$

Assuming that the change in density is proportional to the crystallization, the crystallization fraction C is:

$C_{\text{HD}} = (0.96 - 0.90)/(1.01 - 0.90) = 0.55$

$C_{\text{LD}} = (0.92 - 0.90)/(1.01 - 0.90) = 0.18$

Comments. The low-density polyethylene softens more between T_g and T_m than the high-density polyethylene does. With less packing, the molecules can rearrange themselves more readily in response to applied stresses.

If polyethylene were 100% crystalline, there would be no glass temperature and the solid would remain essentially rigid up to its melting temperature before suddenly softening into a liquid. ◀

7–5 COPOLYMERS

Each of the vinyl-type mers of Table 7–1.3 has two carbon atoms. Each therefore produces a polymer with a backbone of carbon atoms. It is only logical to ask the question: Can a polymer have more than one type of mer? The answer not only is yes, but also that such combinations are so common that the word *copolymer* has been coined to describe them. Further, the properties of copolymers are often desirable, and are specifically prescribed for engineering components.

Copolymers may be of several types. Figure 7–5.1(a) shows the most general form, in which the chain of molecules contains a *random*

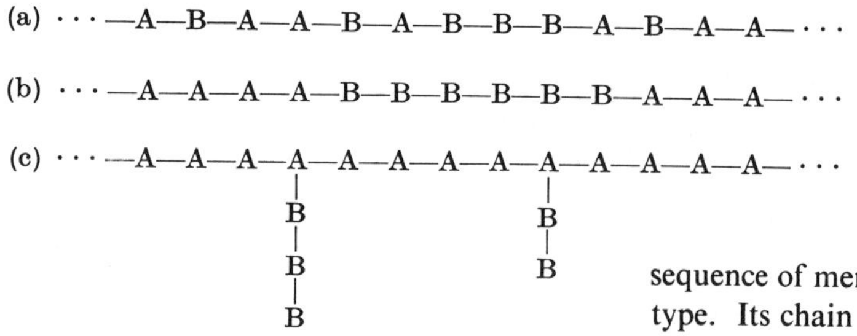

(a) $\cdots$ —A—B—A—A—B—A—B—B—B—A—B—A—A— $\cdots$

(b) $\cdots$ —A—A—A—A—B—B—B—B—B—B—A—A—A— $\cdots$

(c) $\cdots$ —A—A—A—A—A—A—A—A—A—A—A—A—A— $\cdots$
$\quad\quad\quad\quad\quad\quad$ | $\quad\quad\quad\quad\quad\quad\quad\quad$ |
$\quad\quad\quad\quad\quad\quad$ B $\quad\quad\quad\quad\quad\quad\quad\quad$ B
$\quad\quad\quad\quad\quad\quad$ | $\quad\quad\quad\quad\quad\quad\quad\quad$ |
$\quad\quad\quad\quad\quad\quad$ B $\quad\quad\quad\quad\quad\quad\quad\quad$ B
$\quad\quad\quad\quad\quad\quad$ |
$\quad\quad\quad\quad\quad\quad$ B

Fig. 7–5.1 Copolymers (schematic of A and B mers). (a) Random copolymer. (b) Block copolymer. (c) Graft copolymer. The random and graft copolymers seldom crystallize. Local regions of crystal order can occur in a block copolymer.

sequence of mers. One of the early artificial rubbers, Buna-S, is of this type. Its chain contains mers of styrene,

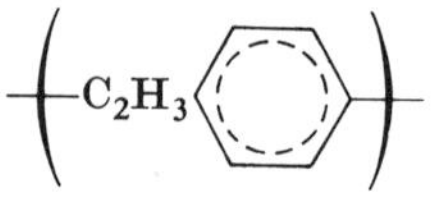

$$\left(\!\!-C_2H_3\!\!\left\langle\bigcirc\right\rangle\!\!-\right)$$

and butadiene, $\left(\!-C_4H_6-\right)$. Styrene by itself is a hard, glassy polymer ($T_g \cong 100°C$); butadiene is a soft rubberlike compound.* Styrene lends hardness and a certain amount of mechanical stability to products made from it. Butadiene lends stretch and a degree of flexibility.

The properties of the copolymer may be modified by varying the ratio of the two component mers. Table 7–5.2 shows data for copolymers of vinyl chloride, $-\left(C_2H_3Cl\right)-$, and vinyl acetate, $\left(C_2H_3Ac\right)-$. In general, copolymers with random sequences of mers in their chains are amorphous, because, when the succeeding mers vary, there is scant likelihood that the meshing required for crystallization will take place.

Block copolymers are sketched in Fig. 7–5.1(b). Because there are longer units of a given species of mer in a block copolymer, the crystallization fraction may be higher than it would be in a comparable random copolymer. The resulting plastic may have the characteristic of a two-phase crystalline microstructure, but with molecular chains that tie the adjacent "grains" together. The ABS plastics are triple copolymers of this type; the letters stand for *a*crylonitrile $\left(\!-C_2H_3C\!\equiv\!N\right)-$, *b*utadiene

$\left(\!-C_4H_6-\right)$, and *s*tyrene $\left(\!-C_2H_3\!\left\langle\bigcirc\right\rangle\!-\right)$, respectively.

The third type of copolymer (Fig. 7–5.1c) is obtained by grafting side branches of a second polymer onto the main chain. A *graft copolymer* exhibits very little, if any, crystallization.

Table 7–5.1

BUTADIENE-TYPE MOLECULES

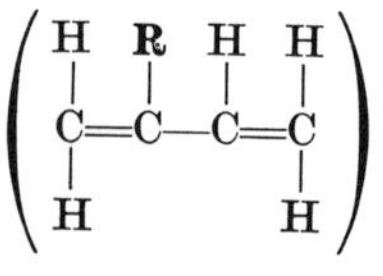

$$\left(\begin{array}{ccc} H & \mathbf{R} & H & H \\ | & | & | & | \\ C\!=\!C\!-\!C\!=\!C \\ | & & | \\ H & & H \end{array}\right)$$

	R
Butadiene	—H
Chloroprene	—Cl
Isoprene	—CH$_3$

* Butadiene has the mer

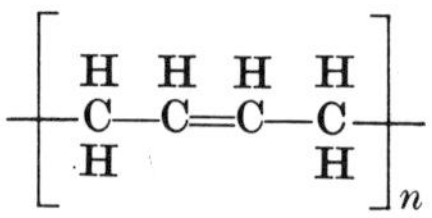

$$\left[\begin{array}{cccc} H & H & H & H \\ | & | & | & | \\ C\!-\!C\!=\!C\!-\!C \\ | & & & | \\ H & & & H \end{array}\right]_n \tag{7–5.1}$$

which may be compared directly with isoprene, Eq. (7–2.1). (See Table 7–5.1.)

Table 7–5.2

VINYL CHLORIDE-ACETATE COPOLYMERS*

Item	w/o of vinyl chloride	m/o of vinyl chloride	Range of average mol. wt.	Typical applications
Straight polyvinyl acetate	0	0	4,800–15,000	Limited chiefly to adhesives
Chloride-acetate copolymers	85–87	90	8,500–8,500	Lacquer for lining food cans; sufficiently soluble in ketone solvents for surface-coating purposes
	85–87	90	9,500–10,500	Plastics of good strength and solvent resistance; molded by injection
	88–90	92	16,000–23,000	Synthetic fibers made by dry spinning; excellent solvent and salt resistance
	95	96	20,000–22,000	Substitute rubber for electrical-wire coating; must be externally plasticized; extrusion-molded
Straight polyvinyl chloride	100	100	—	Limited, if any, commercial applications *per se*; nonflammable substitute for rubber when externally plasticized

*   Adapted from A. Schmidt and C. A. Marlies, *Principles of High Polymer Theory and Practice*. New York: McGraw-Hill.

Example 7-5.1. (M) A copolymer contains 10 m/o vinyl acetate and 90 m/o vinyl chloride (m/o = mer percent). (a) What is the w/o vinyl acetate? (b) What is the w/o chlorine?

Solution. From Table 7-1.3:

Vinyl acetate mer =

$$\left(\begin{array}{cc} H & H \\ | & | \\ -C & -C- \\ | & | \\ H & Ac \end{array}\right)$$

where **Ac** is

$$-O-\overset{\underset{\displaystyle ||}{O}}{C}-\overset{\underset{\displaystyle |}{\overset{\displaystyle |}{H}}}{C}-H$$

Vinyl chloride mer =

$$\left(\begin{array}{cc} H & H \\ | & | \\ -C & -C- \\ | & | \\ H & Cl \end{array}\right)$$

Basis: 100 mers = 10 mers VAc (= 40C + 20O + 60H)

= 90 mers VCl (= 180C + 270H + 90Cl)

10 mers VAc = (12 amu)(40) + (16 amu)(20) + (1 amu)(60)

= 480 amu C + 320 amu O + 60 amu H

= 860 amu.

90 mers VCl = (12 amu)(180) + (1 amu)(270) + (35.5 amu)(90)

= 2160 amu C + 270 amu H + 3195 amu Cl

= 5625 amu.

a) w/o vinyl acetate = 860/(5625 + 860) = 13.3 w/o.
b) w/o chlorine = 3195/(5625 + 860) = 49.3 w/o.

Comments. A copolymer may be viewed as a solid solution of the contributing mers. Just as with the solid solutions in Section 3-4, the overall structural pattern is one which exists with one, two, or several types of components. ◀

REVIEW AND STUDY

7-6 SUMMARY

The reason that the atoms within a molecule are so tightly bonded is that they share pairs of electrons. Bonds *between* molecules are much weaker. The molecules that go to make up plastics, or polymers, are very large, with thousands of atoms in each molecule.

Giant molecules commonly involve chains of carbon atoms. We can understand the structure of these molecules readily if we pay attention to the mer, or repeating unit. The vinyl units of

$$\left(\begin{array}{cc} \text{H} & \text{H} \\ | & | \\ -\text{C}-\text{C}- \\ | & | \\ \text{H} & \text{R} \end{array}\right)$$

are the most widely encountered mers, where R is one of several possible side radicals such as —H, —OH, —Cl, —CH_3, etc. (See Table 7–1.3.)

If the contributing units have more than two connecting points, a network polymer can result. In Chapter 8 we shall see the chain polymers and network polymers have markedly different properties, which presents the engineer with a variety of options.

The configuration of a polymer also affects its properties. Variations such as stereoisomers, branching polymers, cross-linking of molecules, and copolymers influence the crystallinity and therefore the behavior of the polymers in service.

The glass temperature is a basic thermal characteristic of amorphous materials, just as the melting temperature is a basic thermal characteristic of crystalline materials.

Terms for Study

Amorphous	Isotactic
Atactic	Macromolecule
Bifunctional	Mer, $\left(\ \ \right)$
Block copolymer	Molecular crystals
Branching	Molecular weight
Chain polymer	Network polymers
Configuration	Polyfunctional
Conformation	Polymer
Copolymer	Rubber
Crystallinity	Stereoisomers
Crystallization shrinkage	Syndiotactic
Degree of polymerization, n	Thermal agitation
Glass temperature, T_g	Vinyl compounds
Isomer	Vulcanization

STUDY PROBLEMS

7–1.1   What is the mer weight of isoprene?

Answer. 68 amu (or 68 g/6 × 10^{23} mers)

7–1.2   The degree of polymerization of polyvinyl acetate is 117.   What is the molecular weight?

Answer. 10,070 g/mole

7–1.3   A calculation shows that there are 10^{20} polypropylene molecules per gram. (a) What is the average molecular weight? (b) What is the degree of polymerization?

Answer. (a) 6000 g/mole; (b) 143 mers/molecule

7–1.4   The average end-to-end distance of a molecule of polyvinyl alcohol, $\{C_2H_3OH\}_n$, is about 100 Å.   (a) What degree of polymerization is required? (b) What is its average molecular weight?

Answer. (a) 2100; (b) 92,000 g/mole

7–2.1   Refer to Example 7–2.1.   (a) What weight percent sulfur would be present if all the possible cross links of polyisoprene were occupied by sulfur? (b) If 50% of the cross links were occupied by sulfur?

Answer. (a) 32 w/o S; (b) 19 w/o S

7–2.2   A chloroprene rubber has a chlorine atom in place of the $-CH_3$ side radical of isoprene (Eq. 7–2.1). (a) What is the fraction of sulfur present when 25% of the cross links are occupied? (b) What fraction of the cross links are occupied when 8.3 w/o oxygen is present in the rubber?

Answer. (a) 0.083 S; (b) 0.50

7–3.1   The density of polyethylene is 1.01 g/cm³ when it is fully crystalline. (See Fig. 7–3.2 for the dimension of a unit cell.) (a) How many $\{C_2H_4\}$ mers are there per unit cell? (b) Check your answer by examining Fig. 7–3.2.

Answer hint. The $\{C_2H_4\}$ mer along each side is shared by *two* unit cells.

7–3.2   Based on Fig. 7–3.3, determine the percent solidification shrinkage in iodine.

Answer. 17 v/o

7–4.1   Assuming that the density of 20°C of supercooled liquid iodine is 4.30 g/cm³, what is the crystallization fraction in iodine which has a measured density of 4.85 g/cm³?

Answer. 87%

7–5.1   What is the ratio of ethylene $\{C_2H_4\}$ mers to $\{C_2H_3Cl\}$ mers in a copolymer of the two in which there is 30 w/o chlorine?

Answer. m/o PVC = 0.34; m/o PE = 0.66; PE/PVC = 1.95

7–5.2   The mer ratio in a styrene $\{C_2H_3 \langle \rangle \}$, butadiene $\{C_4H_6\}$ copolymer is 1/3. What is the w/o carbon?

Answer. 90 w/o carbon

THE MAKING
AND SHAPING
OF PLASTICS

8–1 RAW MATERIALS Ⓜ ⓜ Ⓞ

The earliest types of polymeric materials to be used—the naturally occurring materials such as wood, leather, cotton, wool, etc.—still have widespread applications. These materials may be used more or less directly except for some type of mechanical shaping process. They too contain large molecules, molecules which in principle are similar to those we encountered in the last chapter.*

There are several advantages in using materials which have naturally occurring molecules. For one thing, they are already formed, and do not have to be manufactured, except for having geometric changes performed on them. They are usually relatively cheap. And they typically have high strength-to-weight ratios. However, these naturally occurring materials also have certain disadvantages: (1) They absorb water, (2) they are combustible, (3) their strength varies with the direction of their structures, (i.e., they have strength anisotropies).

Not surprisingly, therefore, attempts have been made to improve on nature. Plywood, for example, has less anisotropy than plain wood cut directly from the tree because the "grain" is laid in two directions. It is also possible for plywood to be produced as large sheets, an impossibility with naturally occurring wood. Wood may also be impregnated with other materials to reduce its permeability to moisture (e.g., railroad ties are treated with creosote), and resins may be used to bond fabrics into products that are impervious to water.

Another advance in the state of technology came about years ago when means became available to extract cellulose from cotton and wood, to modify it in various ways, and to regenerate it as a polymeric raw material. Rayon (as a fiber) and celluloid (as a plastic) originated

---

* Although the molecules are similar to those of the man-made materials (e.g., Table 7–1.3), they are more complex; in Chapter 7 we intentionally limited our discussions to the simpler macromolecules.

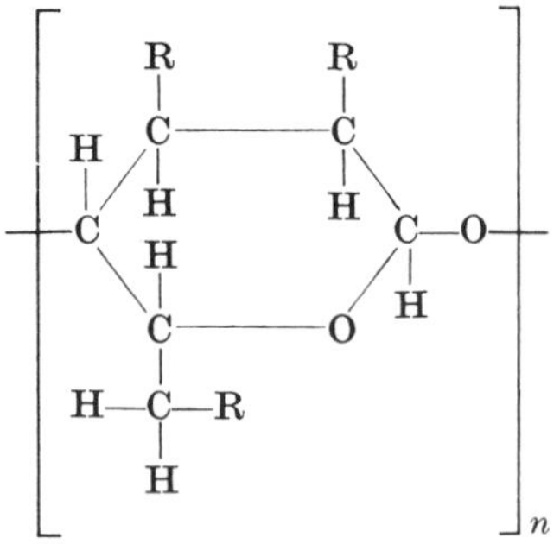

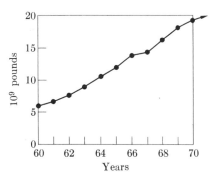

Fig. 8–1.1 Cellulose, a naturally occurring polymer. In nature the radical, R, is —OH. Man has learned how to modify cellulose by replacing the —OH's with —OCOCH₃ to give cellulose acetate, or with —ONO₂ to give cellulose nitrate. The properties depend on the modification. Cellulose acetate is the basis of rayon; cellulose nitrate is guncotton (if all of the —OH's are replaced by —ONO₂). Other cellulose derivatives— i.e., modifications—are possible.

with the invention of the method of extracting cellulose. The basic structure of cellulose, cellulose acetate, and cellulose nitrate are shown in Fig. 8–1.1, where the radicals R are —OH, —OCOCH₃, and —ONO₂, respectively. The three materials have a very similar structure, except for the added radical. This difference in radical, however, greatly affects their properties. The products made from these materials range from lacquers, to plastics, to explosives (nitrocellulose), depending on what fraction of the —OH units of cellulose are replaced by —ONO₂. If —OCOCH₃ (that is, acetate radicals) are used as replacements rather than —ONO₂ radicals, the product is non-flammable, and is widely used as "safety" film in photography; millions of pounds of it are used to make fibers for the manufacturer of textiles, as well.

Raw materials for many of the newer-type plastics are extracted from coal and/or petroleum products. As an example, coke, the carbonaceous part of coal, plus the methane that is the chief component of natural gas together yield acetylene, C_2H_2:

$$3C + CH_4 \rightarrow 2(H—C\equiv C—H). \qquad (8–1.1)$$

Acetylene, in turn, may be converted to various vinyl compounds, as follows:

Ethylene	$C_2H_2 + H_2 \rightarrow C_2H_4$	(8–1.2)
Vinyl chloride	$C_2H_2 + HCl \rightarrow C_2H_3Cl$	(8–1.3)
Styrene	$C_2H_2 + C_6H_6 \rightarrow C_2H_3(C_6H_5)$	(8–1.4)
Vinylidene chloride	$C_2H_2 + Cl_2 \rightarrow C_2H_2Cl_2$	(8–1.5)

Other sources of raw material require chemical reactions that are somewhat more complex, and we shall not consider them here, except to note that a tremendous industry has developed for the production of raw materials from which plastics can be made. The expansion of this production through the 1960's is plotted in Fig. 8–1.2.

Example 8–1.1. One pound of cellulose is two-thirds converted to cellulose acetate, i.e., two of the three radicals of Fig. 8–1.1 are changed from the —OH of cellulose to —OCOCH₃. What is the percent gain in weight?

Solution

$$\underset{\text{cellulose}}{C_6H_7(OH)_3O_2} + \underset{\text{acetic acid}}{2HOCOCH_3} \rightarrow \underset{\text{cellulose diacetate}}{C_6H_7(OH)(OCOCH_3)_2O_2} + \underset{\text{water}}{2H_2O}$$

$$[(6)(12) + (7)(1) + (3)(17) + (2)(16)] + \cdots$$
$$\rightarrow [(6)(12) + (7)(1) + 17 + (2)(59) + (2)(16) + \cdots$$
$$162 + \cdots \rightarrow 246 + \cdots$$
$$\tfrac{246}{162} = 152\%, \text{ or } 52\% \text{ gain in weight.}$$

Fig. 8–1.2 Production of raw materials for plastics in the 1960's. The trend continues upward as new and improved plastics are developed which require these raw materials.

Comments. Rayon normally requires about a two-thirds replacement of hydroxyls by acetate radicals in cellulose, as indicated above. A triacetate, in which all the R's of Fig. 8–1.1 are acetate, $C_6H_7(OCOCH_3)_3O_2$, was harder to develop because the triacetate was not deformable enough to make fiber production feasible. Fortunately these difficulties have been overcome. Now fibers of triacetate can be made, and their resistance to deformation can be used advantageously because they retain their shape in "permanent-press" textiles and other crease-resistant fabrics. ◄

8–2 BIG MOLECULES OUT OF SMALL MOLECULES Ⓜ ⓜ Ⓞ

The real breakthrough in the development of plastic materials came when the chemist and the engineer learned how to make large molecules out of small ones, in a commercially feasible way. This achievement opened the way to vast new developments. People looking for raw materials were no longer limited to natural polymers such as cellulose or isoprene rubber. Today a wide range of polymers is made by *synthesis* of macromolecules from micromolecules. There are two chief procedures for growing large molecules: (1) *chain-reaction polymerization*, sometimes called *addition* polymerization, and (2) *step-reaction polymerization*, sometimes called *condensation* polymerization.

Chain-reaction (addition) polymerization. The reaction of Eq. (7–2.4) for ethylene which forms polyethylene may be rewritten as a general equation for all vinyl polymers:

$$n\begin{pmatrix} H & H \\ | & | \\ C & = C \\ | & | \\ H & R \end{pmatrix} \longrightarrow \begin{pmatrix} H & H \\ | & | \\ C & - C \\ | & | \\ H & R \end{pmatrix}_n \qquad (8\text{–}2.1)$$

As in Table 7–1.3, R stands for any of a variety of radical groups. The characteristic of this reaction is that each small original molecule, called a *monomer* (i.e., a single unit), must be *bifunctional*. That is, it must have two reactive sites at which it may connect with two adjacent molecules, much as a railroad freight car needs two coupling links to become part of a train.

A chain reaction actually involves three steps: (1) initiation, (2) propagation, and (3) termination. We shall focus our attention on the second step, *propagation*. Assume that we have a growing vinyl chain already containing a large number of mers and an additional vinyl monomer:

$$
\cdots -\underset{\underset{\text{H}}{|}}{\overset{\overset{\text{H}}{|}}{\text{C}}} -\underset{\underset{\text{R}}{|}}{\overset{\overset{\text{H}}{|}}{\text{C}}} -\underset{\underset{\text{H}}{|}}{\overset{\overset{\text{H}}{|}}{\text{C}}} -\underset{\underset{\text{R}}{|}}{\overset{\overset{\downarrow}{\overset{\text{H}}{|}}}{\text{C}}}\bullet \quad + \quad \underset{\underset{\text{H}}{|}}{\overset{\overset{\text{H}}{|}}{\text{C}}} =\underset{\underset{\text{R}}{|}}{\overset{\overset{\text{H}}{|}}{\text{C}}}
\qquad (8\text{–}2.2)
$$

Note that the carbon below the arrow has only three bonds rather than the more stable four (Table 7–1.1). It therefore possesses a *reactive site* •, and will combine readily with the neighboring vinyl monomer:

$$
\cdots -\text{C}-\text{C}-\text{C}-\text{C}-\text{C}-\text{C}\bullet
\qquad (8\text{–}2.3)
$$

Although the reacting carbon gains the required fourth bond, such a reaction obviously does not eliminate the reaction tendencies at the end of the chain, because the new end carbon now has only three bonds and it is now reactive. Sequentially another—and yet another—reactive site is produced, with continued growth proceeding as long as the small C_2H_4 molecules are available.

The process of *termination* occurs when the reactive ends of two molecules join:

$$
\cdots -\text{C}-\text{C}-\text{C}-\text{C}\bullet \qquad \bullet\text{C}-\text{C}-\text{C}-\text{C}-\text{C}-\text{C}-\cdots
\qquad (8\text{–}2.4)
$$

$$
\cdots -\text{C}-\text{C}-\text{C}-\text{C}-\text{C}-\text{C}-\text{C}-\text{C}-\text{C}-\text{C}-\cdots
\qquad (8\text{–}2.5)
$$

This termination can occur whenever the active ends of growing molecules of any size happen to encounter one another. Thus few, if any, of the final molecules are the same size. This emphasizes our comments of Example 7–1.2 about average molecular weights. Since he usually wants materials which have high molecular weights, the polymer technologist strives to develop ways of postponing reactions (8–2.4) to (8–2.5) until the molecules have grown to large sizes. However, his ways of achieving that are beyond the scope of this text.

Step-reaction (condensation) polymerization. The production of dacron (called terylene, trevira, and tergol in other countries) requires a somewhat different reaction. In a very simplified form, this reaction is:

$$HO-\overset{\overset{O}{\|}}{C}-X-\overset{\overset{O}{\|}}{C}-O(H \quad HO)-Y-O(H \quad HO)-\overset{\overset{O}{\|}}{C}-X-\overset{\overset{O}{\|}}{C}-(OH \quad H)O-Y-(OH \cdots \longrightarrow$$

$$HO-\overset{\overset{O}{\|}}{C}-X-\overset{\overset{O}{\|}}{C}-O-Y-O-\overset{\overset{O}{\|}}{C}-X-\overset{\overset{O}{\|}}{C}-O-Y-O-\cdots + z\, H_2O \qquad (8\text{–}2.6)$$

The X and Y may be various groups of atoms, such as $+(CH_2)+_n$, in the centers of the initial molecules. Those atoms do not enter directly into the reaction. In each joining step, however, a small *by-product* molecule such as H_2O is released; hence the adjective "condensation." Those large molecules which have

$$\left(\begin{matrix} \overset{O}{\|} \\ -C-O- \end{matrix} \right)$$

units along their chains are called *polyesters*, a term which is often used to identify this class of plastics.

We need not remember the chemistry of Eq. (8–2.6), but let us compare some features of this reaction with the addition or chain reactions of Eq. (8–2.2). The earlier chain reaction required only one type of monomer, and no by-product. The step-reaction polymerization typically involves two types of initial micromolecules and commonly produces a by-product (in this case H_2O). Both reactions involve bifunctional molecules, and therefore produce linear polymers. In Fig. 7–2.5 and Example 8–2.1, we observe that some condensation reactions may involve trifunctional molecules to give a network polymer. [Those addition-type monomers with more than one double bond—e.g., the rubbers (Table 7–5.1) and divinyl benzene (Fig. 7–2.3)—may be polyfunctional and provide multiple connections. Thus a cross-linked linear polymer also has a network-like structure.]

Example 8–2.1. Melamine, $C_3N_6H_6$, and formaldehyde, CH_2O, have the following structures:

Each CH_2O can react with $-NH_2$ radicals, i.e.,

$$-N\diagdown^{H}_{H}$$

joining two melamine molecules and producing a by-product molecule of water (H_2O) as follows:

$$-N\diagdown^{\textbf{H}}_{\textbf{H}} \ + \ {\overset{H}{\underset{H}{C}}}{=}0 \ + \ {\overset{\textbf{H}}{\underset{\textbf{H}}{\diagdown}}}N- \ \longrightarrow \ -N{\overset{\overset{H}{|}}{\underset{\underset{H}{|}}{C}}}N- \ + \ \textbf{H}_2\textbf{O} \qquad (8\text{–}2.7)$$

Show how this condensation (step-reaction) polymerization can produce a framework structure.

Solution

Comments. This melamine-formaldehyde polymer, which goes by various trade names, remains rigid at dishwashing temperatures, and therefore is widely used in home and restaurant dishes; probably the most familiar trade name is Melmac. It is moderately expensive.

The second H of each

$$-N\diagdown^{H}_{H}$$

is much less readily removed for a second $-CH_2-$ bridge because the space available for the new connecting units is limited. As a result, each melamine unit is generally restricted to receiving three connecting $-CH_2-$ bridges to adjacent molecules. ◀

8–3 ADDITIVES ⓔ ⓜ Ⓜ

The major component of the plastics made and/or used by engineers are polymeric molecules. Almost invariably, however, these products have additional components which are added to attain specific desired properties.

The materials added to plastic products may be added for purposes of reinforcing and strengthening the product; or they may be added to introduce more flexibility. Thus polyvinyl chloride (PVC) is often used for floor tile; for this usage, an abrasion-resistant *filler* is added. The same polymer is used for raincoats, but with a *plasticizer* added to attain flexibility and to give it characteristics that are markedly different from those of floor tiles. Additives may also be used as *stabilizers*, to keep the polymer from deteriorating, or as *flame retardants*. Finally, additives may be included as *colorants*, for esthetic purposes. Sometimes certain materials are added to reduce the cost of the product, if at the same time other properties can be improved. For example, adding a filler often makes for a more rigid plastic which also costs less per cubic inch than it would otherwise. Additives frequently serve multiple purposes. For example, carbon black strengthens rubber and also absorbs ultraviolet light; this means that the rubber offers more resistance to deterioration during service.

Fillers. We shall discuss those additives which are used in larger proportions than others. Most of them are added to give strength or toughness to plastics. Thus wood flour (a very fine sawdust) is commonly added to a PF plastic (phenol formaldehyde, Fig. 7–2.5) to increase its strength (Fig. 8–3.1). More important is the fact that 35 v/o wood flour more than doubles the toughness of a PF plastic. As a final fringe benefit, wood flour costs less than half what an equal volume of PF would cost. Thus the service properties of the product are improved at the same time the cost of it is being reduced. This is real engineering! Note from Fig. 8–3.1 that the improvement in strength is not a result of the adding of the wood flour alone, because 100 v/o wood flour has nil strength. Rather it is a result of the mutual interaction of the two components, just as a steel made of ferrite plus carbide is stronger than either one of them alone (Fig. 8–3.2).

Fillers may be of various types. We have already cited wood flour, which in reality is a polymer itself, one containing cellulose (Section 8–1.1). Wood flour has the advantage of being essentially the same density as PF; therefore the product retains a low specific gravity. Silica flour (finely ground SiO_2 made from quartz sand or quartzite rock) is also used (Example 8–3.1). It is appreciably harder than wood flour, and therefore adds abrasion resistance to the plastic product.

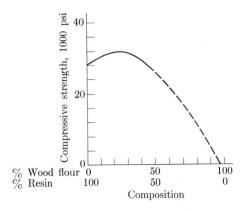

Fig. 8–3.1 Addition of a filler to a plastic (wood flour added to phenol formaldehyde). The mixture of the two is stronger than either alone.

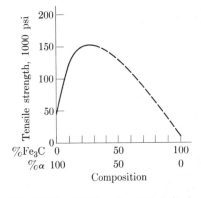

Fig. 8–3.2 Addition of carbide to ferrite. The more rigid carbide prevents slip in the soft ferrite matrix; however, since carbide is brittle, 100% carbide is weak.

Furthermore, it neither burns nor softens at high temperatures; as a result, it adds thermal stability to the product. Of course, it does increase the product's total density because the specific gravity of the SiO_2 is about twice that of common resins. (See Appendix B.)

Fibrous fillers are especially effective when you want to add strength to a product. These fillers may be glass fibers, organic textile fibers, or mineral fibers such as asbestos. They are often chopped into short lengths so that they may be mixed with the polymeric materials which are subsequently molded as reinforced plastics (RP). In Chapter 13 we shall give specific attention to composites in which a high volume fraction of continuous fibers reinforce polymers, often with considerable enhancement of properties.

Plasticizers. At normal temperatures, the small molecules of Section 7–1 and Fig. 7–1.2 are generally liquids or gases. In contrast, the macromolecules which we have been describing here are solids because they involve long chains or networks. When small molecules are intimately mixed with macromolecules, they reduce the rigidity of the macromolecular—or polymeric—product. When small molecules surround large ones, the large molecules move more readily, in response to both thermal agitation and external forces. In brief, the small molecules *plasticize* the larger ones. Plastics that are normally stiff, such as polyvinyl chloride, can be made flexible, an obvious requirement if the product is to be used in film or sheet form. A plasticizer, in effect, lowers the glass transition temperature T_g, which we spoke of in Section 7–4 and Fig. 7–4.1, so that molecular movements and rearrangements can occur at room temperature in polymers that would otherwise be rigid.

A plasticizer must have certain characteristics. It should have a high boiling temperature (low vapor pressure) so that it will not readily evaporate. You can understand that a plasticized raincoat would become useless if the plastic were to become stiff and brittle with time; this, of course, could happen if the plasticizer gradually evaporated, leaving a plastic which would be below its glass temperature.

The plasticizer must not be soluble in the liquids with which it comes in contact. Although some plasticizers are nonvolatile, solvents such as petroleum products can dissolve a large variety of micromolecules. Thus the inner surface layer of a plastic container may be embrittled as the plasticizer is depleted by solvents, which are put into the container.

A plasticizer must be "compatible" with the polymer. Although we leave the details to the polymer scientist, let us note that, in the case of some small molecules, the small molecule may have greater attraction to another of its kind than to the surface of large molecules.

Likewise, strong attraction between two adjacent large molecules may prevent the entry of the small molecules between them for plasticizing purposes. For a plasticizer to be suitable, the small molecules should be attracted to the surfaces of the large molecules; the small molecules should not segregate within the plastic. The choice of the most suitable plasticizer depends on the details of the characteristics of the molecular structure, and usually involves extended testing on the part of the manufacturer.

Stabilizers. Polymers deteriorate in two general ways: by degradation and by oxidation. *Degradation* occurs when the large molecules break up into smaller molecules, thus causing the plastic product to lose strength. Light radiation, particularly by ultraviolet light, can break the molecules into fragments because the light ray, or photon, may supply a powerful "kick" of energy locally onto the bond along the backbone of the molecule. Carbon black is commonly used as a stabilizer because it absorbs the light radiation before the radiation has a chance to degrade the molecule of polymer or rubber. A typical recipe for rubber for a tire to be used on a passenger car includes about 50 lb of carbon black (soot from an oxygen-deficient flame) for every 100 lb of isoprene, for the double purpose of filler and stabilizer.

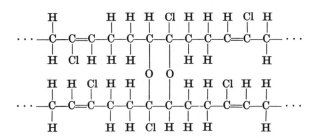

Fig. 8–3.3 Vulcanization of rubber in the presence of oxygen. Like sulfur (Fig. 7–2.4), oxygen from the air can cross-link rubber molecules (chloroprene). This hardens the rubber. Still further oxidation would degrade it to micromolecules.

Deterioration by *oxidation* results when oxygen reacts with the polymeric molecule. In the case of rubber, a change is first noticed when the rubber becomes harder and more brittle. Just as with vulcanization by sulfur (Fig. 7–2.4), oxygen cross-links the rubber (Fig. 8–3.3). As we would expect, therefore, the molecules are no longer independent and their movements are restricted. For cross-linking to occur readily by oxidation, the polymer molecule must contain double bonds. Thus we encounter this type of oxidation much more commonly in rubbers than in other plastics. Ozone (which is O_3 rather than O_2) provides a very reactive source of oxygen because it breaks down to O_2 and a single reactive oxygen:

$$O_3 \rightarrow O_2 + O \bullet. \tag{8–3.1}$$

Thus rubbers and plastics are usually tested in atmospheres which have a high ozone content as an accelerated test to check their resistance to oxidation. Ultraviolet light also speeds the oxidation of rubber (Fig. 8–3.3) by supplying energy locally to start that reaction.

Some plastics are flammable; however, in general those plastics which contain significant amounts of chlorine and/or fluorine have reduced flammability. Polyvinyl chloride (PVC), polytetrafluoroethylene (Teflon), and chloroprene rubber (the C_4H_5Cl of Table 7–5.1) fall in this category. Not only do the chlorine and fluorine not support combustion, but their presence interferes with the access of air to a hot plastic, so that oxidation is retarded or completely precluded. (However, excessive heating may still soften or char these plastics.) *Flame retardants* are added to other plastics which are to be used where flammability must be avoided. The polymer chemist usually includes antimony trioxide (Sb_2O_3) or compounds containing chlorine or bromine as flame retardants.

Colorants. The purposes of these additives are self-evident. The specifications with respect to colorants are stringent because consumers have specific color preferences; also even slight alterations in color are readily apparent to the human eye. The *dyes* and *pigments** used for colorants must be evenly distributed throughout the plastic. For example, if the dye of a textile fiber is only superficial, slight wear or abrasion produces a color change. A colorant must also be extremely stable against deterioration by oxidation or ultraviolet light, because changes in color usually precede other evidences of deterioration. The technical field of colors and dyes is highly specialized, and is beyond the scope of this text.

Example 8–3.1. Forty-five pounds of silica flour (i.e., finely powdered SiO_2) are mixed with each 100 pounds of melamine formaldehyde (MF). The specific gravities of each are 2.65 and 1.5, respectively. What fraction of the volume is filler?

Solution

Basis: 100 g MF and 45 g SiO_2

100 g MF/(1.5 g/cm^3) = 67 cm^3 MF = 0.8 MF

45 g SiO_2/(2.65 g/cm^3) = <u>17</u> cm^3 SiO_2 = 0.2 SiO_2

Total = 84 cm^3

---

* *Dyes* are molecules which *dissolve* into the structure of the plastic. *Pigments* are *fillers* which retain their identity as a separate phase.

Comments. A filler is used not only because it is an inexpensive diluent for a moderately expensive polymer, but also because it gives added thermal and dimensional stability to the plastic. ◀

8–4 THE SHAPING OF PLASTIC PRODUCTS ⓔ

Thermoplasts and thermosets. Materials used in plastic products fall into two general categories: (1) the thermoplasts, which soften when they are heated and become rigid again when they are cooled, and (2) the thermosets, which cure, or "set," when they are heated. This contrasting behavior naturally requires distinctly different processing procedures.

The *thermoplasts* soften at high temperatures because they are *linear polymers* (Eqs. 8–2.3 and 8–2.6) with only limited (if any) cross-linking or branching. As the temperature rises, the molecules can respond to pressure by sliding by one another. Thus it is possible to inject these materials into a mold while they are hot, or to draw them into fibers or sheets. Most of the vinyl compounds of Table 7–1.3 are linear and fall into this category. Of course, for thermoplasticity to be effective, the temperatures must be above the glass transition temperature, T_g (Fig. 7–4.1). Further, the normal temperatures of use must be in a range in which the plastic is rigid, so that dimensions and shapes are retained (Fig. 8–4.1). One economic advantage in the production of thermoplasts (as contrasted to thermosets) is that there is no net chemical or structural change through the production cycle. Therefore, quite often, scrap product may be crushed and recycled into later production batches.

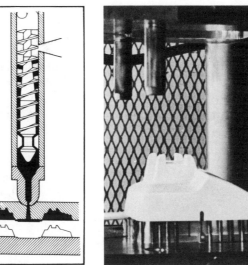

Fig. 8–4.1 Injection-molded telephone receiver. (a) The warm thermoplastic is forced through sprues to the die cavity, where it quickly hardens. (b) The finished product lifted from the base die. (Courtesy Western Electric Co.)

(a)   (b)

The *thermosets* are altered both chemically and structurally during thermal processing. They are characterized by *network* structures (Fig. 7–2.5 and Example 8–2.1). These structures are not completely polymerized before forming, but become one large three-dimensional structure when they are processed in the presence of heat and pressure. As an example, the melamine formaldehyde (MF) of Example 8–2.1 is usually only partially reacted before it is heated and compressed into the mold. Thus the MF plastic is still deformable, since only a few of the melamine units have the maximum of three connecting $-CH_2$ bridges which lend rigidity to the network structure. The "curing" within the heated mold completes the formation of the networks. Therefore this plastic gains rigidity before the added pressure and temperature are removed. The production of thermosets, unlike the production of thermoplasts, is slow because a curing time is necessary; furthermore, scrap material cannot be recycled because the network-forming reaction is irreversible. However, thermosets have obvious advantages in applications in which high temperatures are usual.

Mixing. The plant which manufactures the final plastic product must usually "compound" the polymer and the various additives prior to the final forming operations. If there is to be consistent control of the process, there must be a thorough mixing of the several components. However, mixing them is a major problem. Even if the raw materials are as small as one millimeter in size, there must be provision for still more intimate mixing. Fillers must be entirely surrounded by polymer, and plasticizers, dyes, etc., must be dispersed on an intermolecular basis.

One of the common mixers for thermoplasts is an *open-roll mill*, as sketched in Fig. 8–4.2. This mill is also called a *rubber mill*, since it is widely used for compounding the ingredients in automobile tires. These mills are much like the old washing machine wringer, except that the rolls do not move at the same speed; furthermore, one of the rolls may be internally heated or cooled (by steam or by a circulating fluid). At the entrance to the nip (where the two rolls are closest) there is continuous shearing action on the thermoplastic, which blends the additives into the viscous polymer. When the two rolls are at different temperatures, the plastic coats one preferentially at the exit of the nip, so that the batch may be continually mixed for desired periods of time. (Rubber usually follows the hotter of the two rolls, while other plastics usually follow the colder roll. This, however, is unimportant to us at the moment.) The operator simply cuts the plastic off the roll when it is sufficiently blended.

Other mixing procedures include drums with internal rotors and blades. These processes lend themselves more readily to the use of

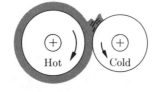

Fig. 8–4.2 Open-roll mill. The components of the plastic, or rubber are blended by a shearing action in the nip region between the rolls. The two rolls rotate at slightly different speeds. They are also maintained at different temperatures.

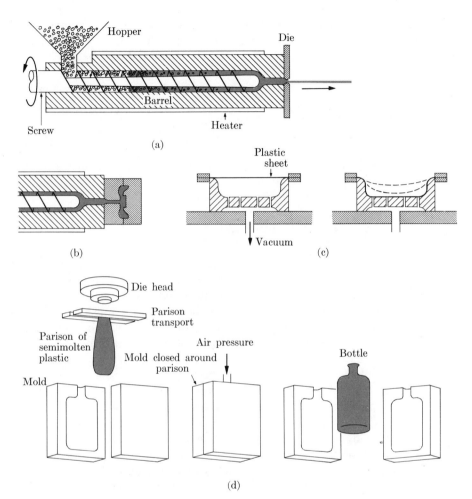

Fig. 8–4.3 Molding operations. (a) Extrusion. (b) Injection molding. These two operations are similar, except that extrusion requires an open-ended die, while injection is into a closed die. Either may be by screw or ram pressure. (c) Vacuum forming. (d) Blow molding. These latter two processes are patterned after glass-molding procedures. (Cf. Fig. 11–4.3.)

inert atmospheres for those polymers which are sensitive to oxidation during the shearing operation.

Molding. The sketches of Fig. 8–4.3 summarize various shaping procedures which combine pressure, heat, and dies. *Extrusion* requires massive equipment, but also lends itself to producing many thousands of tons of products such as plastic pipe, plastic sheet, or any other product which has a constant cross-sectional profile. Extrusion is essentially injection under pressure through an open-ended die. The raw materials—pellets or granules—must be heated into the thermoplastic range. Extrusion pressure is commonly applied by an auger, or screw. Alternatively, a ram or piston may be used, but in general this provides a less uniform operation.

The extruder of Fig. 8–4.3(a) has external heaters surrounding the compression barrel. These are used primarily to heat the equipment and the initial charge. Heat is also obtained from the mechanical operations, because essentially all the power of the auger is dissipated into the material. An equation may be developed which relates the temperature rise, ΔT, in °F, and the horsepower input, P. Since there are 2545 Btu/hr for each horsepower,

$$P = R c_p \, \Delta T / 2545, \tag{8–4.1a}$$

where R is the rate in pounds/hr and c_p is the heat capacity in Btu/lb · °F (or in cal/g · °C). This equation does not include a factor for motor efficiency. The units of the above equation are:

$$\text{hp} = \left(\frac{\text{lb}}{\text{hr}}\right) \left(\frac{\text{Btu}}{\text{lb} \cdot {}°\text{F}}\right) ({}°\text{F}) / \left(2545 \, \frac{\text{Btu/hr}}{\text{hp}}\right). \tag{8–4.1b}$$

Injection molding utilizes the same extrusion principle, but uses closed dies into which the plastic is forced. Here a ram or plunger is more common than a screw or auger because a back pressure is developed (Fig. 8–4.3b). If thermoplasts are being molded, the mold is water-cooled so that the injected plastic becomes rigid immediately and a high rate of processing is possible. When thermosets are injection-molded, however, they must be forced into a heated mold so that polymerization reactions can be completed. It is not necessary to chill a thermoset product before stripping the mold; however, in general the operation is rate-limited by the reaction time.

Stamping and *vacuum forming* offer two variants to the molding process by using starting materials in sheet form. The former uses a mechanical force to mold the sheet. As shown in Fig. 8–4.3(c), the vacuum process provides an effective pressure up to 14.7 lb/in² from the external air pressure. The resulting force becomes significant for products which have large dimensions. More important, however, is the fact that a vacuum provides an easy mechanism for distributing pressure uniformly.

Blow molding provides our final example of molding. (It does not exhaust all the molding procedures, however. See a polymer text or a trade magazine, such as *Modern Plastics*.) Suppose you want to make a plastic container. Just as in the production of glass containers, a parison of soft plastic is extruded (Fig. 8–4.3d). This "hollow drop" is surrounded by a mold, and then air pressure is used to expand the parison until it takes on the shape of the mold. We shall see in Chapter 11 that the glass industry has developed this process to the extent that light bulbs are made by an automated machine at the rate of several thousand per minute. The process is becoming equally important in the production of plastic containers. This highly complex process utilizes our knowledge about the relationship between viscosity and temperature, and about viscoelasticity. Use of the process is highly dependent on the experience and ingenuity of the production specialist.

Films and fibers. These are generally made by variations of the extrusion process, in which the die slit is very narrow, or in which extrusion takes place through a multiorificed die called a *spinnerette*. The characteristic production of films and fibers, however, involves post-die treatments (Fig. 8–4.4). These may be chemical treatments to harden the fiber, dyeing treatments to color it, or mechanical treatments for crystallization. The latter warrants further comment.

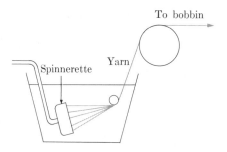

Fig. 8–4.4 Fiber drawing. This is the basic operation for rayon, nylon, and various polyester fibers.

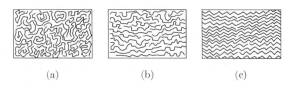

(a)   (b)   (c)

Fig. 8–4.5 Mechanical orientation and crystallization of molecules. (a) Structure of amorphous polymers. (b) Orientation by strain above the glass temperature. (c) Crystallization. When the material is cooled below the glass temperature, the molecules will not rekink into an amorphous polymer. The oriented and crystallized plastic is strong in the longitudinal direction.

Molten polymers are, of course, noncrystalline or amorphous (Fig. 8–4.5a). The rapid cooling of film or fibers precludes crystallization. Therefore, as we shall see in Chapter 9, they are weaker than crystalline polymers. It would be desirable, if possible, to crystallize these plastics. This can be done by deformation. An otherwise amorphous structure can be oriented by elongation (Fig. 8–4.5b). The orientation, in turn, provides greater opportunity for the meshing of the molecules into a crystal by a zipper effect (Fig. 8–4.5c). If the stretched, oriented, and crystallized polymer is cooled rapidly, the crystallization may be locked in, because the kinking described in Section 7–1 cannot recur. When this is done to a fiber, the longitudinal strength is increased several-fold. (It turns out that the transverse strength is reduced by this process. However, that is not important, because who is going to load an individual fiber crosswise?)

A film can also be strengthened by deformation. Here the strength in a second direction *is* important. To gain effective strength advantage, a film must be deformed in two directions at once, so that the molecules have their preferred orientations in the plane of the film. (Here again a film is weak in the thickness direction, a relatively unimportant consequence.) The mechanical engineering procedures by which biaxial orientation—i.e., molecular alignment in two of the three directions—is obtained are interesting. One such procedure is the *bubble method*, which forms a cylindrical film. This film is made when

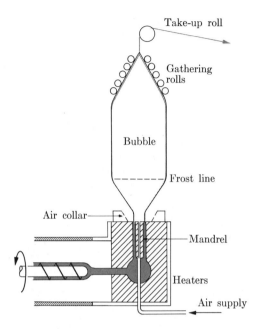

Take-up roll

Gathering rolls

Bubble

Frost line

Air collar

Mandrel

Heaters

Air supply

Fig. 8–4.6 Bubble forming. The cylindrical sheet is expanded simultaneously in two directions as it cools below the glass temperature. Therefore it develops biaxial strength in the two dimensions of the film.

a die containing a mandrel is used. Air is blown into the mandrel, so that the material expands in two directions before the final packaging (Fig. 8–4.6). Another procedure is to use side tension to spread the material in width at the same time as it is stretched longitudinally. This process is descriptively called *tentering*.

Example 8–4.1. Polystyrene, which is to be extruded through a 4-inch auger at a rate of 360 lb/hr, must maintain a temperature of 400°F. Once the production rate is established, how much power is required, given 70% motor efficiency under steady operating conditions? (The heat capacity of polystyrene is 0.3 Btu/lb · °F.)

Solution. From Eq. (8–4.1), and assuming a normal temperature of 70°F,

$$P = \frac{(360 \text{ lb/hr})(0.3 \text{ Btu/lb} \cdot °F)(400°F - 70°F)}{(2545 \text{ Btu/hr} \cdot \text{hp})(0.70)}$$

$$= 20 \text{ hp.}$$

Comments. This would be the power required for continuous operation. Note, however, that the engineer would need to specify a motor with a significant margin of excess power, so that it could operate under start-up or other high power conditions. ◄

REVIEW AND STUDY

8–5 SUMMARY

Polymers, or plastics as they are commonly called, originate from a variety of sources. Before the advent of synthetics, polymers were modified forms of natural organic products, such as the cellulose, which is extracted from wood, or cotton (Fig. 8–1.1). Currently polymers are usually synthesized from micromolecules. There are two main polymerization reactions: (1) Addition (or chain-reaction) polymerization, in which mers are added to a growing chain, rather like the way railroad cars are added to a train. The vinyl compounds belong to this category. (2) Step-reaction polymerization, commonly called condensation polymerization. This involves a chemical reaction in which the ends of two molecules join, and at the same time release a small by-product molecule such as water (H_2O) or methyl alcohol (CH_3OH). Polyesters and polyamides are examples of condensation polymerization.

Additives are widely incorporated into plastics to modify their characteristics. Fillers may be simply diluents of a less expensive material; often, however, fillers serve to strengthen a polymer against mechanical deformation, so there is usually a twofold advantage to adding them. Plasticizers serve as an "internal lubricant" for a polymer, so that molecular rearrangements are easier. Some polymers, such as polyvinyl chloride, can be made flexible where they would otherwise be highly rigid at normal temperatures. In simplest terms, a plasticizer is composed of small or *micro*molecules which are absorbed between larger or *macro*molecules. Stabilizers are additives which serve to reduce the rate of deterioration of the organic product. In general, this involves additives which prevent oxygen from getting into the plastic, or reduce the product's ability to absorb radiant energy, such as ultraviolet light. Colorants are also used as additives, for obvious reasons.

The contrast between thermoplastic and thermosetting materials is of major importance in the fabrication and application of plastic products. Thermoplastics have linear molecules, so that heating them markedly increases their molecular movements, and they flow visco-elastically when subjected to pressure. On the other hand, thermosetting plastics are subject to cross-linking or network growth at high temperatures. As a result they become more rigid when they are given thermal processing, and consequently they are more useful in applications in which they are subjected to heat.

Processing commonly involves (a) mixing of ingredients and (b) shaping. There are many kinds of shaping processes. The choice among these depends on the dimensions required and the quantities of product to be made. Extrusion, injection molding, vacuum forming, and blow molding are all common procedures. Fibers and films present an added processing option because they make possible post-forming treatments. In particular they can be strengthened by orientation of their molecules and by crystallization.

Terms for Study

Addition polymerization	Extrusion
Additives	Fibers
Blow molding	Fillers
Bubble processing	Flame retardants
Cellulose	Injection molding
Chain-reaction polymerization	Linear polymers
Colorants	Molding
Condensation polymerization	Monomer
Deformation crystallization	Network polymers
Degradation	Orientation

Phenol-formaldehyde, PF   Stabilizer

Plasticizer   Stampings

Plywood   Step-reaction polymerization

Polyester   Tentering

Propagation   Termination

Rayon   Thermoplasts

Reactive site, ●   Thermosets

Sheet   Vacuum forming

STUDY PROBLEMS

8–1.1 How many tons of acetylene, C_2H_2, and hydrochloric acid, HCl, would it take to produce 6,000,000 tons of vinyl chloride? (This is the approximate annual production of polyvinyl chloride predicted for 1975–1980.)

Answer. 2,500,000 tons C_2H_2 + 3,500,000 tons HCl

8–1.2 Cellulose has a specific gravity of about 1.35 and is the main component of dry wood. White pine weighs 32 lb/ft^3. Estimate the fraction of dry pine that is pore space. (Water weighs 62.4 lb/ft^3.)

Answer. 62 v/o porosity. [Note: Some of these pores are enclosed within the wood and are not open to the outside.]

8–2.1 Show how the double bonds change during the addition polymerization of isoprene to polyisoprene.

Answer

8–2.2 Show how phenol formaldehyde can form a network polymer.

Comment. There is room for —CH_2— bridges to connect to only three of the six carbons.

Answer. See Example 8–2.1 and make appropriate substitutions of —CH_2— bridges between the benzene (⬡) rings.

8–3.1 A 3.2-g sample of polyvinyl chloride-base plastic floor tile is heated to 1000°F so that only 1.3 g of silica-flour ash remains. What fraction of the volume was filler? The specific gravities of PVC and silica flour (SiO_2) are 1.3 and 2.65, respectively.

Answer. 25 v/o

8–3.2 As a maximum, a pair of isoprene (C_5H_8) mers may be joined by two oxygen atoms (Fig. 8–3.3). By this reaction, one pound of isoprene rubber would increase in weight to _____ pounds.

Answer. 1.23 pounds

8–3.3 Suppose that you manufacture plastic ball bearings, and you want to use 50 v/o of finely chopped glass fibers ($\rho = 2.4$ g/cm^3) in nylon ($\rho = 1.15$ g/cm^3). What weight of glass fibers should be batched with each pound of nylon in making the plastic–glass composite?

Answer. 2.08 lb glass per lb nylon

8–4.1 An outdoor sign 25 inches in diameter is vacuum-formed into its final shape. What force would be required behind a die to provide the same force in a die-stamping operation?

Answer. 7250 lb force

8–4.2 A dynamometer indicates that 8 hp of actual power is used in a 2-inch extrusion auger. How many pounds/hour must be extruded if the temperature rise of methylmethacrylate is to be limited to 360°F? [$c_p = 0.35$ Btu/lb · °F (or 0.35 cal/g · °C)]

Answer. 160 lb/hr

8–4.3 A certain paint weighs 15 lb/gal and contains 30 w/o of a volatile thinner (which weighs 8 lb/gal), the rest being a pigment and a reactive polymer. One gallon of paint covers 400 ft^2 of surface. How thick is the paint film after setting? [One cubic foot = 7.5 gallons (U.S.).]

Answer. ~0.002 in., if one assumes that the only volume change is the evaporation of the thinner.

CHAPTER 9

CHARACTERISTICS OF PLASTICS

9–1 THERMAL BEHAVIOR Ⓜ Ⓞ

The two prime thermal properties we cited in Chapter 2 were thermal expansion and thermal conductivity. In addition a very important characteristic of many plastics is thermal softening.

Thermal expansion. In general, plastics have a high thermal expansion coefficient, α_l. As shown in Appendix B, the values for the vinyl types (polyethylene, polystyrene, etc.) are approximately 50×10^{-6} in./in./°F (90×10^{-6}/°C). This is in contrast to 5 to 40×10^{-6}/°F (10–70×10^{-6}/°C) for the majority of the network polymers containing CH_2 bridges, and for metals, and $<5 \times 10^{-6}$/°F ($<9 \times 10^{-6}$/°C) for ceramics. These contrasts are to be expected, since strongly bonded materials require a greater input of thermal energy (and therefore a greater temperature increase) to expand the structure than do weakly bonded materials. The linear (thermoplastic) polymers of the vinyl type are characterized by weak bonds between molecules. Like metals, the network (thermosetting) polymers have a three-dimensional framework of strong bonds.

Fortunately the high thermal expansion does not make plastics especially sensitive to thermal cracking. This is because they also have very low elastic moduli. As a result their large thermal strains do not induce high stresses. (See Example 9–1.1.)

Thermal conductivity. Plastics generally have low thermal conductivity because they have no free electrons to transport energy across the temperature gradient. Furthermore the majority of polymers are amorphous with irregular internal structures. Thus the atomic mechanism of transferring energy by coordinated atomic vibrations is generally ineffective. In this latter connection, it is worth our attention to compare the thermal conductivity of "high-density" polyethylene and "low-density" polyethylene (Table 9–1.1 and Example 7–4.1).

Table 9–1.1

THERMAL PROPERTIES OF POLYETHYLENE*

	Low-density polyethylene (LDPE)	High-density polyethylene (HDPE)
Density, g/cm^3	0.92	0.96
Thermal expansion, °C^{-1}	18×10^{-5}	12×10^{-5}
°F^{-1}	10×10^{-5}	7×10^{-5}
Thermal conductivity		
(joule · cm)/(sec · cm^2 · °C)	0.0034	0.0052
(Btu · in.)/(sec · ft^2 · °F)	0.00065	0.0010
Continuous use		
heat resistance, °F	130–175	180–250
°C	55–80	80–120

* $T_g < 20°C\ (68°F) < T_m$.

The 50% extra conductivity comes primarily from the greater crystal-linity that accompanies the high-density (HD) polyethylene rather than from the 5% extra density. All crystalline materials have higher conductivities than their comparable glassy-state materials because thermal vibrations move more readily through a well-ordered structure.

Softening. Since many plastics are made with linear polymers and are therefore thermoplastic, we are aware that plastics may soften at higher temperatures. We use the softening to advantage in processing plastics by extrusion, injection molding, etc. This affects their service behavior, however, because it places an upper limit on the tolerable temperatures during use. This can be illustrated with a very familiar example. A plastic spoon (injection molded) for eating an ice cream sundae at a drive-in will soften out of shape if you use it to stir sugar in hot coffee. The plastic was chosen correctly for its intended service; it does not have the properties required for the more severe service. In fact, it would have been overdesigned for the intended use if specifications had also called for stability in hot water.

  Figure 9–1.1(a) shows the shear stress, τ, required to deform a polymethyl methacrylate, PMM,* plastic one linear percent (1 l/o) as a function of temperature. It is obvious that a marked softening

* *Lucite* is a common trade name.

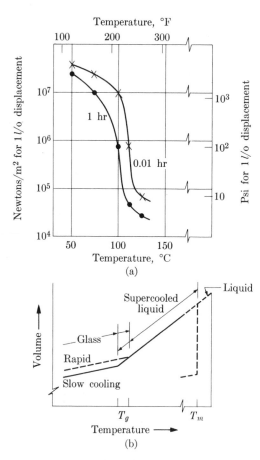

Fig. 9–1.1 (a) Softening (polymethyl methacrylate). The shear stress required for one l/o displacement drops markedly at the glass temperature. The required stresses are also lower when the time is lengthened. (b) Glass transition temperature, T_g, varies with the cooling and heating rates. When slowly cooled, the molecules of the supercooled liquid have opportunities for rearrangements as the temperature continues to drop.

occurs slightly above 100°C.   This temperature corresponds to the glass temperature, T_g, of Fig. 7–4.1. Recall from that earlier discussion that above the glass temperature the molecules have the freedom to kink and turn by thermal agitation.   When they drop below that temperature, there is insufficient thermal agitation to permit rearrangements of molecules into closer packing.   Thus this represents a discontinuity in the thermal behavior of the material.   Return to Fig. 9–1.1(a) and observe that the stress required for a deformation of 1 l/o changes by more than two orders of magnitude at the glass temperature.   Obviously the glass temperature is a temperature important for polymer behavior.

The two curves of Fig. 9–1.1(a) indicate that less stress is required when the time at stress is increased from 36 sec (0.01 hr) to 1 hr. The two curves also indicate that the glass temperature drops 5–10°C as the time is increased from 0.01 to 1.0 hr. This change is reflected in Fig. 9–1.1(b) as a change in the glass transition temperature with slower cooling.   With the slower cooling, the molecules can continue to rearrange themselves within the supercooled glass until somewhat lower temperatures are reached.

Example 9–1.1

a) A polystyrene block, 2 in. × 2 in. × 1 in., is restricted in its short dimension to zero expansion as it is heated from 0°C to 30°C. What stresses develop?

b) Repeat for a copper block of the same dimension.

Solution. As in Example 2–6.1,

$$(\Delta l/l)_{total} = 0 = (\Delta l/l)_{mech.} + (\Delta l/l)_{thermal}$$

$$= \sigma/Y + \alpha_l\,\Delta T.$$

Using data from Appendix B, we have

a) $0 = \sigma/400{,}000 \text{ psi} + (40 \times 10^{-6}/°\text{F})(30°\text{C} \times 1.8°\text{F}/°\text{C})$,

$\sigma = -(400{,}000 \text{ psi})(40 \times 10^{-6}/°\text{F})(54°\text{F}) = -860$ psi.

b) $0 = \sigma/(16 \times 10^6 \text{ psi}) + (9 \times 10^{-6}/°\text{F})(54°\text{F})$,

$\sigma = -7800$ psi.

Comments. The minus value for stress implies compression.

Thermal cracking depends not only on thermal expansion and the elastic modulus, but also on thermal conductivity, because a high conductivity reduces thermal gradients and therefore local strains. ◀

(a)

(b)

9–2 MECHANICAL DEFORMATION Ⓜ Ⓞ

The softening described in the previous section has a dominant role in the mechanical behavior of many plastics. We can understand that the producer aims for controlled softening. We might surmise that the user wants full rigidity; actually this varies with the application. For all intents and purposes, a telephone receiver (Fig. 9–2.1a) must remain rigid. An automobile tire, on the other hand, must support a car with near rigidity, but must itself have some flexibility in order to withstand transient forces (Fig. 9–2.1b).

Viscoelasticity. A polymeric material, when placed under a shear stress, is subject to both elastic (spring-like) and viscous (slow-flow) behavior. This means that total deformation is not instantaneous, and usually not fully recoverable. The viscous or flow characteristics are to be expected for many amorphous polymers, since they are really supercooled liquids which are between the glass temperature and the melting temperature.

We shall consider shear stresses, τ, and shear strains, γ (Fig. 4–2.3) that lead to the shear modulus, G, of Eq. (4–2.3),

$$G = \tau/\gamma, \tag{9–2.1}$$

rather than the axial stresses, σ, and linear strains, ε, that provide Young's modulus, Y, because the deformation of plastics is almost entirely by shear. In effect, viscous flow is the internal shear of a material with respect to itself. Since the two types of deformation occur simultaneously, it normally requires somewhat detailed calculations.

Fig. 9–2.1 Service requirements. (a) The telephone receiver must remain rigid in service. (b) The automobile tire must respond to rapid deformations without failure. (Courtesy Western Electric Co. and Atlas Supply Co.)

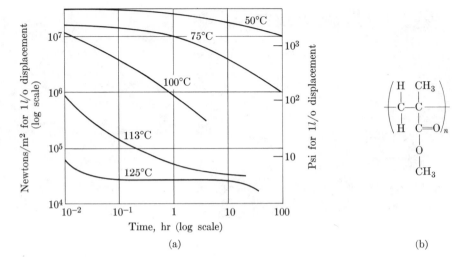

Fig. 9–2.2 Stresses for deformation (1%).
(a) Lower stresses are required when temperatures are higher and when the loading time is extended. [Cf. Fig. 9–1.1. Both figures are for polymethyl methacrylate (b), commonly called lucite.]

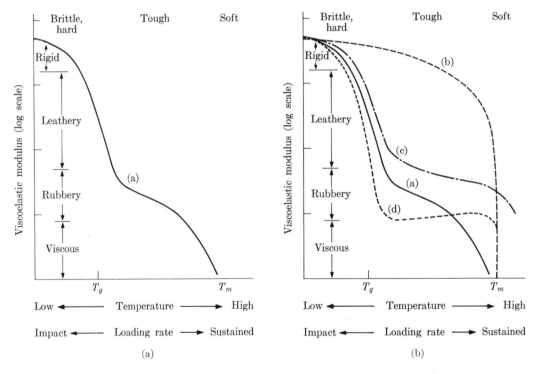

Fig. 9–2.3 Viscoelastic modulus versus structure. (a) Amorphous linear polymer. (b) Crystalline polymer. (c) Cross-linked polymer. (d) Elastomer (rubber).

As a result, we shall consider only special simplified cases. Before doing that, however, let us see how the internal structure of a polymer affects its deformation.

Figure 9–1.1(a) showed how the stress required for 1 l/o displacement is a function of temperature. Figure 9–2.2 draws from the same original data for polymethyl methacrylate as Fig. 9–1.1(a), but shows how the displacement is a function of time at given temperatures. Note that lower stresses are required to produce the same displacements at elevated temperatures.

Conversely, high temperatures have the same effect on a thermoplastic polymer as sustained loading. This is shown schematically in Fig. 9–2.3(a), where rapid softening again makes evident the importance of glass temperature. (Note the log scale.) There is, however, a plateau between T_g and T_m. As the melting temperature, T_m, is approached, the required stress again drops rapidly.

The ordinate of Fig. 9–2.3 is plotted as a viscoelastic modulus, M_{ve};

$$M_{ve} = \tau/(\gamma_e + \gamma_f), \tag{9–2.2}$$

where τ is the shear stress, and $(\gamma_e + \gamma_f)$ is the total of the elastic displacement and the viscous flow (or viscous displacement) after a specified period of time.

We may relate the various regions of the displacement curve in Fig. 9–2.3(a) to the deformation characteristics of polymeric materials. Below the glass temperature, T_g, where only elastic deformation can occur, the material is comparatively *rigid*; a clear plastic triangle used by a draftsman is an example. In the range of the glass temperature, the material is *leathery*; it can be deformed and even folded, but does not spring back quickly to its original shape. In the *rubbery plateau*, the plastic deforms readily but quickly regains its previous shape if the stress is removed. A rubber ball and a polyethylene "squeeze" bottle serve as excellent examples because they are soft and quickly elastic. At still higher temperatures, or under sustained loads, the plastic deforms extensively by *viscous flow*.

The second part of Fig. 9–2.3 compares the deformation behavior for the different structural variants of Chapter 7. A highly *crystalline* polymeric material does not have a glass temperature. Therefore it softens more gradually as the temperature increases until the melting temperature is approached, at which point fluid flow becomes significant. The higher-density polyethylenes (Example 7–4.1) lie between curves (a) and (b) of Fig. 9–2.3(b) because they possess between $\frac{1}{2}$ and $\frac{2}{3}$ crystallinity.

The behavior of *cross-linked* polymers is represented by curve (c) of Fig. 9–2.3. A vulcanized rubber, for example, is harder than a

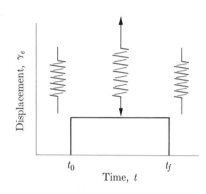

Fig. 9–2.4 Spring model. All deformation is elastic. The displacement is recovered when the load is removed. This model represents metals at low temperatures and plastics below their glass temperature.

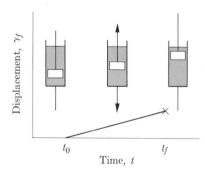

Fig. 9–2.5 Dashpot model. All displacement is by viscous flow (around the edges of the piston). The displacement is not recovered after the load is removed. This model represents the behavior of common fluids such as water and certain liquid fuels.

nonvulcanized one. Curve (c) is raised more and more as a larger fraction of the possible cross links are connected. Note that the effects of cross-linking carry beyond the melting point into the true liquid. In this respect, a network polymer like phenolformaldehyde (Fig. 7–2.5) may be considered as an extreme example of cross-linking, which gains its thermoset characteristics by the fact that the three-dimensional amorphous structure carries well beyond an imaginable melting temperature.

Elastomers. This is a technical term for the rubber family of polymers. They have an extended rubbery plateau. Once the glass temperature is exceeded, the molecules can be deformed by unkinking to produce considerable strain. If the stress is removed, the molecules quickly snap back to their kinked conformations (Fig. 7–1.1). This rekinking tendency increases with the greater thermal agitation at higher temperatures. Therefore the behavior curve increases slightly to the right across the rubbery plateau (Fig. 9–2.3d). Of course, finally the elastomer reaches the temperature at which it becomes a true liquid, and then flow proceeds rapidly.

Viscoelastic models. $\boxed{e}$ On a simplified basis, we may consider viscoelastic deformation as series or parallel combinations of elastic and viscous deformations. To do that, let us make a model for each. Figure 9–2.4 shows that the elastic deformation of a *spring* is immediate when the load is applied at t_0; the strain does not change with time; and both the strain and the energy which was introduced are fully recoverable when the stress is removed at time t_f. This type of deformation can be used as a model for the elasticity of steel and for plastics below their glass temperature.

Viscous flow can be modeled by a leaky pump piston (called a *dashpot*). The flow starts when the load is applied and continues as long as the load forces liquid around the edge of the piston (Fig. 9–2.5). The *rate* of displacement is directly proportional to the applied load and inversely proportional to the viscosity of the leaking fluid. Thus a viscous material requires more stress for shear than is required by a fluid material. Finally, when the stress is removed, there is no recovery of the flow nor of the energy which was introduced. Molten asphalt is an easily visualized example of such a material.

On a simplified basis, some viscoelastic materials are comparable to a spring and dashpot in *series* (Fig. 9–2.6). This model could be used if a constant load were applied and held for varying lengths of time. A rapid or impact load produces only elastic displacement, because there isn't time enough for viscous flow to occur. Extended loading, however, does make possible viscous flow. With enough time, the permanent displacement exceeds the original elastic displacement. Road asphalt is an example of this. When a car drives over it, it is strained elastically; when a car parks on it, the asphalt flows gradually, particularly if the weather is warm, so that the molecules can readily slide by one another.

For our final simplified example, let us consider a viscoelastic model of a spring and dashpot in *parallel* (Fig. 9–2.7). When the load is applied to t_0, the elastic deformation cannot occur immediately because the rate of flow is limited by the viscous component. Displacements continue until the strain equals the elastic deformation of the spring and it resists further movement. When the load is removed, the spring reverses the strain, eventually returning to zero net strain.

This latter model (Fig. 9–2.7) helps one understand the behavior of plastics that are between their glass temperature and their melting temperature. A rubber, for example, cannot deform fully at the moment a stress is applied because it takes some time for the molecules to unkink. The rearrangement and straightening of molecules makes possible considerable deformation, until finally full elastic strain is arrived at, and deformation stops. When the load is removed, it takes a certain amount of time before the molecules rekink back to zero net deformation. The amount of time necessary varies with temperature, from milli-seconds to seconds or minutes, as illustrated by Study Problem 9–2.1.

Stress relaxation. [e] The discussions of the preceding subsection assume constant stresses. In many engineering applications, materials may be subject to *constant strain*. A simple—and nonengineering—example is the following: Stretch a rubber band over a book and let it lie on the shelf for a period of time. An elastic stress has been introduced. This of course first required that the molecules unkink, a process which took relatively few milliseconds. However, they are now in a highly stressed conformation, and unless they have extensive cross links with their neighbors, they can start to slide by one another and eventually rekink. This creep process is slow unless the temperature is high; but while the rubber is in the process of creeping, the stresses gradually relax. The rate at which the stress is relaxed ($-$) is proportional to the amount of stress remaining. This leads to a formula which indicates the relationship of the remaining stress σ to the original stress σ_0 and time t as

$$\sigma = \sigma_0 e^{-t/\lambda}, \tag{9–2.3a}*$$

---

* Equation (9–2.3a) can be derived by using the model of Fig. 9–2.7, and by calculus from the information stated above, which in equation form is

$$\lambda(-d\sigma/dt) = \sigma. \tag{9–2.4}$$

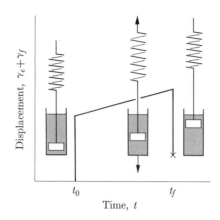

Fig. 9–2.6 Series model. The elastic displacement is immediate; the viscous flow requires time. After the load is removed, the elastic displacement is recovered immediately; the viscous displacement is not recovered. This model (Maxwell) approximates the behavior of asphalt, and metals at high temperatures $> T_m/2$.

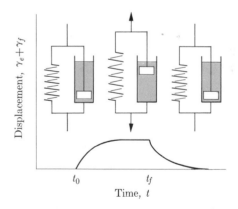

Fig. 9–2.7 Parallel model. The displacement requires concurrent elastic deformation and viscous flow. The maximum displacement occurs when the total load is carried by the spring. The initial rate of displacement is determined by the viscous flow through the dashpot. After the load is removed, the spring recovers the displacement by reversing the flow through the dashpot. This model (Voigt) roughly approximates the behavior of rubber.

or more conveniently,

$$2.3 \log_{10} (\sigma_0/\sigma) = t/\lambda, \qquad\qquad (9\text{–}2.3b)$$

where λ is a characteristic of a material and its temperature, called the *relaxation time*.

Example 9–2.1. (e) A rubber loses half its stress in 30 days. What is its relaxation time?

Solution. From Eq. (9–2.3b), and with $\sigma = 0.5\sigma_0$,

$$2.3 \log_{10} (\sigma_0/0.5\sigma_0) = 30 \text{ days}/\lambda.$$

$$\lambda = 30 \text{ days}/(2.3 \times \log_{10} 2)$$

$$= 30 \text{ days}/(2.3)(0.301) = 43.3 \text{ days}.$$

Comments. If this reduced stress is removed, then only part of the original strain is recovered. ◀

9–3 MECHANICAL FAILURE OF POLYMERS Ⓜ ⓔ Ⓞ

Softening and creep may be the initial steps in terminating the usefulness of a plastic product. However, failure by fracture, fatigue, and/or tearing is much more obvious to the average consumer. These latter responses, called *ultimate failures*, are as much characteristics of the dimensions and shapes of the product or of the geometry of the load as they are of the material itself. As an illustration, a plastic bag used for merchandizing food exhibits considerable strength on a psi basis. But take a knife and make a slit in it; then it is easy to tear. Likewise cracks in plastic toys and other products usually start at reentrant corners or other points at which stress is concentrated. When designing objects to be made of plastic, the engineer uses standardized tests to index the ultimate properties of various plastics he is considering, but also depends heavily on stress analyses and simulated service testing before he makes his design final. This leads to a variety of important specialized tests which have been developed under the auspices of ASTM* or other producer-user groups. Examples, among others, include bag-bursting tests, cloth-tear tests, and—for film— needle-penetration tests. In any of these, however, it is apparent that an increase in the inherent properties of the material leads to improvements in service behavior.

Tests to failure. Tensile loading and impact loading are performed much in accordance with the tests described in Sections 2–2 through

---

*   American Society for Testing and Materials.

2–5. Special variants of the geometry of the test sample are required, however, to replace the machined tensile-test specimens shown in Fig. 2–3.2.   Figure 9–3.1(a) shows a test specimen which can be cut from rubber sheet, or formed by injection molding of a thermoplastic. The large grip areas prevent failure outside the central area that is being tested.   Figure 9–3.1(b) is a ring specimen, made with a molded surface, which prevents the damage to the surface that accompanies cutting a test piece.   In contrast, a tear test (Fig. 9–3.1c) is specifically designed with a cut in it as a stress-raiser.

Test data.   Ultimate failure below the glass temperature is by brittle fracture.   Even in the rubbery range, failure does not involve the plastic deformation that is characteristic of a ductile metal.   As a result failure of a brittle plastic almost always originates at the defect or at the point at which there is the greatest concentration of stress.   This leads to data with a degree of *variance* in the sample, so that successive tests on the same material may produce differences as great as $\pm 20\%$ (Fig. 9–3.2).   Consequently, results of strength tests must be reported as a *mean* (average) value or the *median* (middle) test value.   In Fig. 9–3.2, these are 3001 psi and 3030 psi, respectively.   Although the mean and median values are close to each other, they are usually not identical. This difference arises in this case because the individual results are more widely dispersed at the lower strengths.   Reporting either the mean or the median does not fully satisfy the design engineer, because he does not know how strong or how weak individual samples may be.   Thus he may ask for the *standard deviation*, $\bar{\sigma}$.   The standard deviation is a statistical measure of scatter, and is calculated as follows:

$$\bar{\sigma} = \sqrt{\frac{(X_1 - \overline{X})^2 + (X_2 - \overline{X})^2 + \cdots + (X_n - \overline{X})^2}{n}}, \qquad (9\text{–}3.1)$$

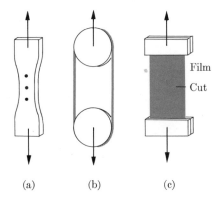

Fig. 9–3.1 Tensile tests for plastics. (a) "Dog bone" sample. (b) Ring sample. (c) Tear tests. The first may be cut from sheet products or injection molded. The second is useful for the manufacturer who molds rubber products. The third is of specific value to producers and users of film products.

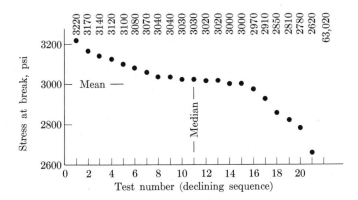

Fig. 9–3.2 Variations in test data. In the absence of ductile failure, a considerable variance occurs in the data relating to strength. The *mean* value is the average value. The *median* value is the middle value. If the data distribution is skewed, the two values are not the same.

where $\overline{X}$ is the mean and X_i are the individual values (up to n). In practice the standard deviation has significance in that approximately two-thirds of the results lie within $\pm 1\sigma$, and approximately 5% of the statistical results lie outside $\pm 2\sigma$.

Fracture.   The actual fracture mechanism in a polymer is complicated because the structure is complex.   As a result we shall mention only those features that affect fracture of polymers and not attempt to assign specific behaviors for various plastic materials.

   Below the glass temperature, a plastic is brittle.   This could lead to a fracture surface which cuts through molecules with very little absorption of energy.   Often, however, the propagating crack can cause local deformation of molecules (Fig. 9–3.3), with considerable consumption of energy.*   This effect tends to toughen a polymer.

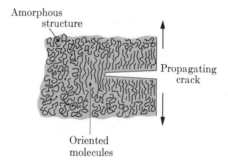

Fig. 9–3.3   Absorption of energy by molecular reorientation. If molecular reorientation accompanies fracture, the crack is slowed, and may even be stopped. This type of behavior is called *crazing*.

   The path of a fracture through a plastic that contains a filler is affected by any second phase. Commonly, the filler is harder, more rigid, and stronger.   As a result, the path of the fracture must either remain in the polymer phase, or follow along the phase boundary.   This becomes particularly important in composite materials (Chapter 13) such as glass-reinforced plastics (GRP), in which special attention is given to the treatment of the surface of the glass fibers to improve bonding.   For a filler to be effective, the phase boundary must be at least as strong as the matrix polymer phase.

   Polyblends of rubbery and glassy polymeric phases or block copolymers (Fig. 7–5.1b), with both amorphous and crystallizable segments, provide the engineer with materials that have built-in toughness on a microscale with little opportunity for interphase fracture.   In

---

*   In some plastics this can also occur at temperatures below the glass temperature because strain energy is dissipated as additional thermal energy to permit rearrangement of molecules. Alternatively, imposed strains lower the glass temperature which was described in Fig. 9–1.1(b).

these materials, individual molecules extend from one of the regions into the other.

Example 9–3.1. Determine the mean, $\bar{X}$, median, and standard deviation, $\bar{\sigma}$, for the following tensile test data.

Solution

Test data

Test no.	Strength, psi	$(X - \bar{X})^2$
1	1,510	400
2	1,730	40,000
3	1,450	6,400
4	1,690	25,600
5	1,370	25,600
6	1,430	10,000
7	1,590	3,600
8	1,520	100
9	1,480	2,500
$n = 9$	$\Sigma X = 13,770$ psi	$\Sigma(X - \bar{X})^2 = 114,200$ psi^2

$$\bar{X} = \Sigma X/n = 1530 \text{ psi} \qquad \bar{\sigma} = \sqrt{\frac{114,200}{9}} = 113 \text{ psi}$$

$$\text{Median} = 1510 \text{ psi}$$

Comments. Note that the median, 1510 psi, is slightly below the mean, 1530 psi, chiefly because the highest value, 1730 psi, diverged more markedly from the majority of the results than did the lowest value, 1370 psi.

In this case, two-thirds of the values lay within 1530 ± 113 psi. However, with this small number of tests (only nine), exceptions to the two-thirds rule are not unusual. ◀

9–4 DETERIORATION OF POLYMERS Ⓜ Ⓞ

The deformation and fracture of a plastic, as described in the preceding two sections, originate from external sources of applied stress. However, the integrity of a polymer may also disappear by reactions which are basically chemical, or bond-breaking. We use the term *degradation* to describe a broad category of undesired changes that range from *scission* to *swelling*.

Scission. Plastics that contain large molecules have greater strength and resistance to deformation than those containing only small molecules. For this reason, the chemist and materials engineer study polymerization reactions (Section 8–2) which increase the degree of polymerization. The reaction may be reversed, and then *depolymerization* occurs. This requires considerable energy input, because 1.5 ×

10^{-19} cal (6.25×10^{-19} joules) are required to break an average
C—C bond. That value may appear to be a small number, but compare it with only 0.75×10^{-22} cal (3.1×10^{-22} joules) needed to raise the temperature of a mer of polyethylene from 5° to 35°C (41° to 95°F). It requires 2000 times as much energy to break the average C—C bond of a mer as to raise the temperature of those atoms 30°C (54°F).

Local sources of high energy which can break bonds within the polymer chain include light photons or other radiation. We shall see in Example 9–4.1 that ultraviolet light ($\lambda = 3000$ Å) has photons of energy of 1.6×10^{-19} cal (6.6×10^{-19} joules), slightly more than enough to break the average C—C bond. Visible light is only about half as energetic as ultraviolet light; however, as it hits a C—C bond, there is still a measurable probability that the bond will be broken because the 1.5×10^{-19} cal cited above is only an *average* energy. At any instant of time, some of the bonds may possess significantly more energy than average, and therefore require less than the average energy per bond for scission (literally *cutting*). Of course, a fraction of a moment later, such a bond may require appreciably more than this average energy per bond for scission.

Other sources of energy which breaks bonds are neutron radiation (Fig. 9–4.1) or gamma radiation (γ-rays) from nuclear reactions. The effects may not be all detrimental. As shown in Fig. 9–4.2, a side atom may be stripped, rather than the main backbone of the polyethylene chain, to open up a reactive site for side branching. Gamma radiation is used commercially to modify the structure of low-density polyethylene; in this case the polyethylene develops a higher temperature of softening because of the molecular entanglements resulting from branching.

Swelling. The scission processes described above break *intra*molecular bonds. A second method of deterioration involves micromolecules which enter as solutes between the macromolecules to make direct polymer-to-polymer contact impossible. In effect this is a breaking of *inter*molecular bonds, although admittedly these bonds are relatively weak.

Micromolecules are intentionally added to polymeric materials to make them more flexible (Section 8–3). This is not always desirable, however. Consider a polyvinyl alcohol, $(C_2H_3R)_n$, with the R being —OH (Table 7–1.3). Water molecules can be absorbed among the vinyl chains. This leads to a weakening and a swelling of the polymer. Likewise, unless the necessary structural adaptations are made, petroleum molecules can be absorbed into the rubber of a

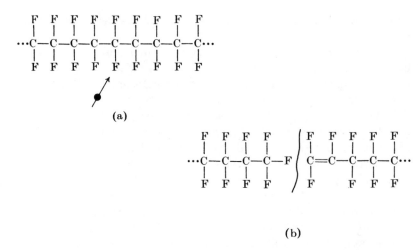

Fig. 9–4.1 Scission by neutron radiation (PTFE). Radiation damage occurs when the average size of molecules decreases.

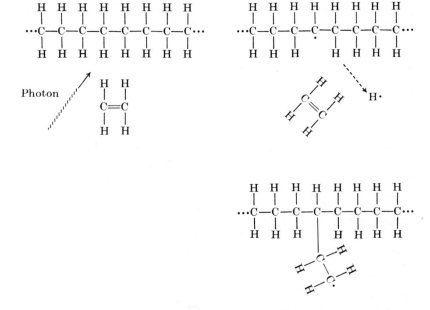

Fig. 9–4.2 Branching by irradiation (PE). Branching can occur where a bond is broken by radiation.

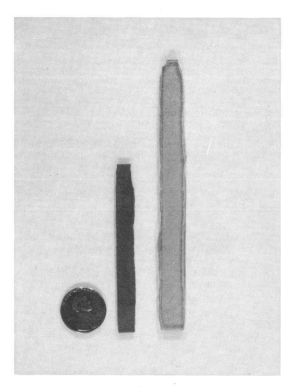

Fig. 9–4.3 Swelling. Micromolecules of benzene are absorbed between large molecules of isoprene rubber in the sample at the right, expanding it 60 l/o. Rubber, if it is not sufficiently cross-linked, is particularly subject to this type of deterioration. (Courtesy G. S. Y. Yeh, The University of Michigan)

gasoline hose, which produces swelling and reduces the usefulness of the hose.

Swelling (Fig. 9–4.3) is not compatible with engineering specifications, of course, so the materials engineer looks for ways to prevent it. Cross-linking reduces swelling, since the molecules are linked together. Also crystallized plastics are less subject to swelling than amorphous polymers because crystallized plastics have more closely intermeshed molecular structures. Finally, the engineer can choose among a number of polymers in order to avoid swelling. Small molecules distribute themselves among macromolecules more readily when the two types are chemically similar. For example, the polyvinyl alcohol, C_2H_3—OH, and water, H—OH, described above are closely related. Therefore water is readily absorbed among these vinyl molecules. Likewise, hydrocarbon petroleum fluids are absorbed into hydrocarbon rubbers. Where swelling is critical, such similarities are to be avoided through judicious selection of polymers.

Example 9–4.1. The energy, E, of a light photon (sometimes described as a small energy "packet" within a light ray) may be calculated as follows:

$$E = h\nu, \qquad\qquad (9\text{–}4.1)$$

where h is a constant with a value of 6.62×10^{-34} joule $\cdot$ sec and ν is the frequency of light. What is the energy of a photon of ultraviolet light with a wavelength of 3000 Å?

Solution. Since the frequency of light ν is the velocity c ($= 3 \times 10^{10}$ cm/sec) divided by the wavelength λ ($= 3000$ Å, or 3×10^{-5} cm),

$$E = (6.62 \times 10^{-34} \text{ joule} \cdot \text{sec})(3 \times 10^{10} \text{ cm/sec})/(3 \times 10^{-5} \text{ cm})$$

$$= 6.6 \times 10^{-19} \text{ joules} \qquad (= 1.6 \times 10^{-19} \text{ cal}).$$

Comments. The constant h is called *Planck's constant*, and enters widely into calculations which involve the wave characteristics of energy. Usually it is expressed as 6.62×10^{-27} erg $\cdot$ sec, or 1.58×10^{-34} cal $\cdot$ sec. Also, 4.13×10^{-15} eV $\cdot$ sec is used when energies are expressed in electron volts (Example 12–2.3). ◀

Example 9–4.2. A nylon absorbs 1.66 w/o water (dry basis). Because of differences in density, this means that 100 cm³ of dry nylon absorbs 1.9 cm³ of H_2O to produce 101.5 cm³ of saturated nylon. Account for the missing 0.4 cm³.

Solution. In this case, where absorption occurs, the attraction between unlike molecules is greater than the attraction between like molecules, i.e., nylon–H_2O attractions are greater than the average of nylon–nylon and water–water attractions. A contraction in volume results.

Comments. If the attraction between like molecules had been greater than between unlike nylon–water pairs, the water and nylon would have become segregated and absorption would have been limited. ◀

9–5 ELECTRICAL PROPERTIES OF POLYMERS ⓜ Ⓜ Ⓞ

Resistivity. The resistivity data of Appendix B indicate that polymers are insulators, since the resistivity values all exceed 10^{12} ohm · cm. This is in comparison to $\sim 10^{-5}$ ohm · cm for metals and $10^{-2 \text{ to } +4}$ for semiconductors (Chapter 12).

One can make rubbers and other plastics conductive by incorporating conducting fillers (typically graphite) in them. Thus an appropriately compounded plastic or rubber can ground static charges, or even be used as nonmetallic ignition "wires" on cars.

The above resistivity is *volume resistivity*, and the conductivity (the reciprocal of resistivity) is volume conductivity, because the charge is transported through the bulk of the material. Conductivity also occurs in plastics as a consequence of the transport of surface charges, or *surface conductivity*. (Refer to Fig. 9–5.1.) For example, polyesters which we described in the discussion of Eq. (8–2.6) as having

$$\left[\begin{matrix} & O \\ & \| \\ -C & -O- \end{matrix} \right]$$

units provide exposed electrons on oxygen atoms adjacent to the surface. These *polar groups* attract moisture, hydrogen ions, and other surface contaminants, which in turn make the surface a conductive one. In order to avoid this, one may use highly resistive, nonpolar coatings such as selected silicones.

Relative dielectric constant. As discussed in Section 2–9, the dielectric constant is the ratio of charge carried by a condenser with and without intervening material. This is illustrated again in Fig. 9–5.2. Table 9–5.1 lists dielectric values of commercial plastics. We observe that the values of the dielectric constants are higher in those plastics which (1) are above their glass temperature, and which (2) have atoms with polar sites (Fig. 9–5.3). Condition (1) makes it possible for the

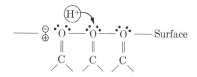

Fig. 9–5.1 Surface conduction. A polar molecule— i.e., a molecule with different centers of positive and negative charges—provides opportunities for ion adsorption. These adsorbed charges can move in an electric field to produce surface conductivity.

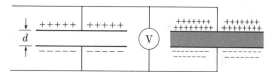

Fig. 9–5.2 Relative dielectric constant, κ. The charge densities on the electrodes are proportional to the relative dielectric constant. This constant for a vacuum is 1.0; values for common plastics are listed in Table 9–5.1.

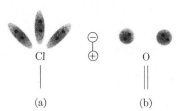

Fig. 9–5.3 Polar sites. (a) —Cl, for example, in polyvinyl chloride. (b) = O, for example, in polyesters. Molecules containing polar sites have electric dipoles because their centers of positive and negative charges are not coincident. Above their glass transition temperatures, plastics with polar molecules have somewhat higher relative dielectric constants than other molecules because the dipole moves with the electric field.

Table 9–5.1

RELATIVE DIELECTRIC CONSTANTS OF PLASTICS
(20°C, UNLESS OTHERWISE STATED)

	At 60 Hz*	At 10⁶ Hz
Nylon 6/6	4.0	3.5
Polyethylene, PE	2.3	2.3
Polytetrafluorethylene, PTFE	2.1	2.1
Polystyrene, PS	2.5	2.5
Polyvinyl chloride		
plasticized ($T_g \approx 0°C$)†	7.0	3.4
rigid ($T_g = 85°C$)	3.4	3.4
Rubber (12 w/o S, $T_g \approx 0°C$)		
−25°C	2.6	2.6
+25°C	4.0	2.7
+50°C	3.8	3.2

* Hz = cycles per sec.
† T_g = glass temperature.

molecules to rearrange their orientations so that the electric dipoles favor a build-up of positive and negative charges onto the electrodes. (See the data for rubber in Table 9–5.1.) A combination of conditions (1) and (2) provides locations in which the centers of positive and negative charges are displaced with respect to each other, to give an electric *dipole*.

The dielectric constant is sensitive to frequency (Fig. 9–5.4a) and to temperature (Fig. 9–5.4b). At higher frequencies, there is less time for the electric dipoles to rearrange themselves so that they align with the fields. With respect to temperature, the relative dielectric constant jumps sharply as the materials rise above the glass temperature. But at still higher temperatures, when there is greater thermal agitation, the relative dielectric constant drops because the excited electric dipoles do not stay as well aligned with the electric field.

Example 9–5.1. The electrical engineer describes *surface resistivity* ρ_s in units of *ohms/square*. Explain these units.

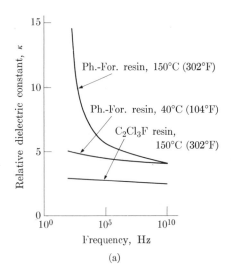

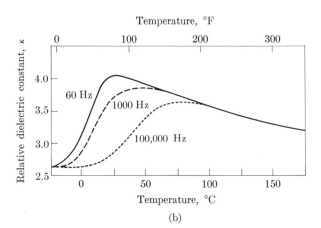

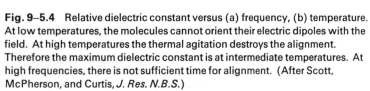

Fig. 9–5.4 Relative dielectric constant versus (a) frequency, (b) temperature. At low temperatures, the molecules cannot orient their electric dipoles with the field. At high temperatures the thermal agitation destroys the alignment. Therefore the maximum dielectric constant is at intermediate temperatures. At high frequencies, there is not sufficient time for alignment. (After Scott, McPherson, and Curtis, *J. Res. N.B.S.*)

Solution. The surface resistance is proportional to the length of the path, ℓ, and inversely proportional to the width of the path, w:

$$R = \rho_s(\ell/w). \tag{9–5.1}$$

Thus a square surface, where $\ell = w$, gives $\rho_s = R$, and the units are simply ohm · cm/cm, or ohm/☐.

Comments. Less commonly, *volume resistivity* ρ_v may be identified as *ohms/cube*, since from Eq. (2–8.4) we obtain

$$R = \rho_v(\ell/A), \tag{9–5.2}$$

and in a *unit* cube $\ell = A$. In practice, it is better to identify ρ as ohm · cm, since, in order to make our dimensions consistent, we would have to use only unit cubes. This restriction is not as confusing in Eq. (9–5.1). ◀

Example 9–5.2. Figure 9–5.5 shows a test cell for measuring the volume resistivity of plastic film. This is of course important, because film is often used as a dielectric spacer in capacitors. According to this design, what range and accuracy are required of an ohmmeter to determine a resistivity of 10^{14} ohm · cm within 5% for 0.001-in. film?

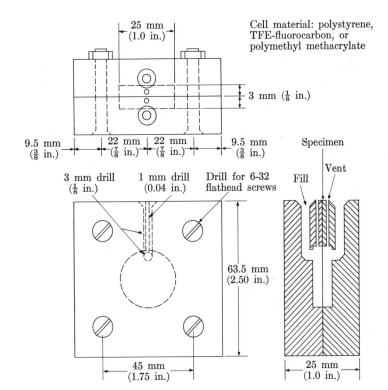

Cell material: polystyrene, TFE-fluorocarbon, or polymethyl methacrylate

Fig. 9–5.5 ASTM film test for volume resistivity (D257). (See Example 9–5.2.) Mercury is filled into the two side wells to provide electrodes. (American Society for Testing and Materials)

Solution. As sketched, the conducting area has a diameter of 2.5 cm. Also 0.001 in. = 0.00254 cm. From Eq. (9–5.2),

$$R = (10^{14} \pm 5\% \text{ ohm} \cdot \text{cm})(0.00254 \text{ cm})/(\pi/4)(2.5 \text{ cm})^2$$

$$= 5 \times 10^{10} \text{ ohm} \pm 2.5 \times 10^9 \text{ ohm}.$$

Comments. Measuring instruments must not only cover the range of reading (in this case $\sim 10^{10}$ ohm), but also be of known accuracy for use in property specifications. ◀

Example 9–5.3. The capacitance C of a parallel-plate capacitor can be calculated as follows:

$$C = \frac{\kappa A}{(4.452 \times 10^{12} \text{ volt} \cdot \text{in./coul})(d)}, \tag{9–5.3}$$

where κ is the relative dielectric constant, A is the area, and d is the thickness of the dielectric in inches. The relative dielectric constants for polyvinyl chloride (PVC) and polytetrafluoroethylene (PTFE) are as follows:

Frequency, Hz	PVC	PTFE	Vacuum
10^2	6.5	2.1	1.0
10^3	5.6	2.1	1.0
10^4	4.7	2.1	1.0
10^5	3.9	2.1	1.0
10^6	3.3	2.1	1.0
10^7	2.9	2.1	1.0
10^8	2.8	2.1	1.0
10^9	2.6	2.1	1.0
10^{10}	2.6	2.1	1.0

a) Plot capacitance versus frequency curves for three capacitors with 1 in. × 50 in. effective area separated by 0.001 in. of (1) vacuum, (2) PVC, and (3) PTFE.

b) Account for the decrease in the relative dielectric constant of PVC (that is, C_2H_3Cl) with increased frequency, and the absence of change for PTFE (that is, C_2F_4).

Solution

a) Sample calculations at 10^2 Hz (that is, 10^2 cycles per second):

$$C_{vac} = \frac{(1.0)(50 \text{ in.})(1 \text{ in.})}{(4.452 \times 10^{12} \text{ volt} \cdot \text{in./coul})(0.001 \text{ in.})} = 11.2 \times 10^{-9} \text{ coul/volt},$$

of, since a farad (f) is a coul/volt, and it is common to express capacitance in microfarads ($1f = 10^6 \mu f$),

$$C_{vac} = 0.0112 \ \mu f.$$

See Fig. 9–5.6(a) for the remainder of the results.

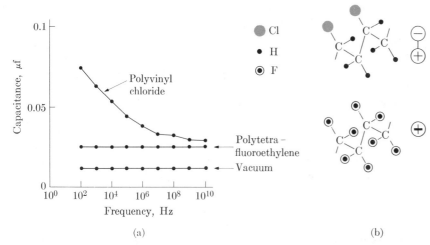

(a)

(b)

b) The relative dielectric constant of PVC is high at low frequencies because the chlorine, with its large complement of electrons, is displaced strongly toward the positive electrode; this gives a significant difference between the centers of positive and negative charges of the mer (Fig. 9–5.6b). At higher frequencies the chlorine atom cannot shift as fast as the frequency changes. The PTFE is a symmetric mer; therefore a large electric dipole does not exist. The relative dielectric constant of PTFE arises from the shifting of the electrons from one side of the atom to the other.

Comments. Above 10^{17} Hz, the electrons cannot shift as fast as the field does. Therefore the relative dielectric constants of both PVC and PTFE drop to 1.0, the relative dielectric constant of a vacuum. ◀

Fig. 9–5.6 Capacitance versus frequency. (a) See Example 9–5.3. (b) Symmetry of polyvinyl chloride (PVC) and polytetrafluoroethylene (PTFE) molecules. At low frequencies the PVC orients with the field to increase the charge density on the capacitor. At all frequencies below 10^{16} Hz, the electrons respond with the field, so that the charge density is greater than when nothing is present (vacuum).

9–6 OPTICAL CHARACTERISTICS OF PLASTICS Ⓜ ⓜ Ⓞ

Light transmission. Figure 9–6.1 shows the use of a plastic, in which light transmission is obviously of paramount concern. A *transparent* material is one which does not scatter light. *Transmittance*, T, is the ratio of the transmitted light intensity, I_x. to the incident light intensity, I_0,

$$T = I_x/I_0. \tag{9–6.1}$$

Many plastics are basically transparent; however, when fillers and other additives are incorporated, they scatter the light. Such materials are said to be *translucent*. The scattering occurs because the light rays are reflected and refracted at the phase boundary between the polymer and the additive (or between crystalline and noncrystalline regions). With sufficient scattering, a plastic will become, for all intents and purposes, opaque.

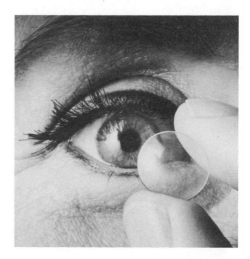

Fig. 9–6.1 Transparent plastics (contact lenses). Specifications for plastics to be used for this purpose include transparency, index of refraction, exact geometry, and inertness to the surroundings. (Courtesy International Optical Co.)

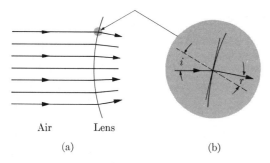

Fig. 9–6.2 Refraction. Light travels more slowly in the denser medium. The index of refraction, *n*, is defined in Eq. (9–6.2). It is also equal to the ratio of the sine of the incident angle, *i*, to the sine of the refracted angle, *r*: $n = \sin i/\sin r$.

Colorants, which were cited in Section 8–3 as either dyes or pigments, are effective because they absorb all that part of the spectrum *except* the color that we see. Thus a red dye transmits only red light; it absorbs all the other colors. In Study Problem 9–6.2, we shall observe that if two colors are not equally absorbed by a transparent material, the color transmitted depends on its thickness. The color chemist, as he controls specifications, must be aware of these factors.

Birefringence. A light ray is refracted—i.e., bent—as it goes from one material to another (Fig. 9–6.2). The lens of Fig. 9–6.1 is contoured to do this very precisely in order to achieve an accurate focus. The amount of refraction *n* depends on the velocity *v* of the light within the material,

$$c = vn, \tag{9–6.2}$$

as contrasted to *c*, the velocity of light in a vacuum (3×10^{10} cm/sec). Many materials transmit light at only one velocity, regardless of the plane of vibration of the light. However, if a material is not isotropic—i.e., if its structure varies with the orientation of the material—the velocity of the transmitted light also varies with the structural orientation. For plastics, two situations are encountered in this respect. The first involves polymers which crystallize in the spherulitic form, as shown in Fig. 9–6.3. The study of transmitted polarized light helps the polymer scientist to interpret the structure and deformation behavior.

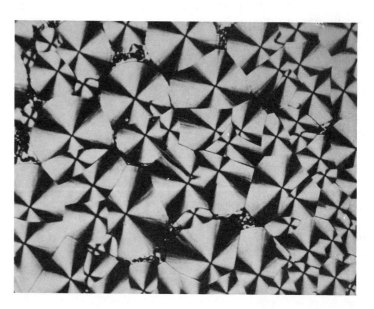

Fig. 9–6.3 Spherulites in polymers, as revealed by polarized light. They originate during crystallization of the polymer. (Courtesy G. S. Y. Yeh, The University of Michigan)

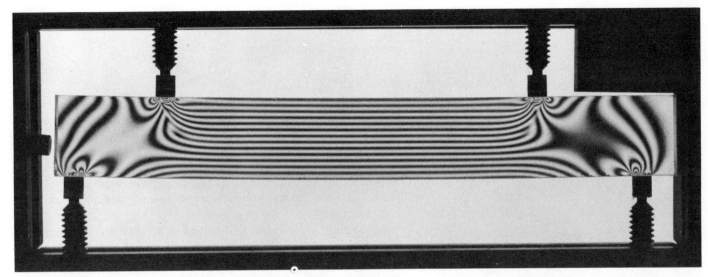

Our second encounter with optical anisotropy has widespread use in experimental stress analysis. Strains modify the index of refraction of any material. Specifically, light vibrating in one plane within a plastic that is strained travels faster than light vibrating in a plane at right angles. The difference in the two velocities shows up as a birefringence, $n_{slow} - n_{fast}$. With appropriate optical analyzers, *photoelasticity* may be used to study the distributions and concentrations of stress (Fig. 9–6.4). It is a very useful tool for the design engineer.

Fig. 9–6.4 Photoelasticity. The stress patterns are revealed because the strain produces different indexes of refraction for light vibrating vertically and horizontally. (Courtesy Vishay Research and Education)

Example 9–6.1. Experiments show that the transmittance, T, is related to thickness as follows:

$$\log T = \log (I_x/I_0) = -\beta x/2.3 \qquad (9\text{–}6.3)$$

where β is *absorptivity* and x is the thickness. What is the absorptivity if 90% of the light is transmitted through 1.0 cm of plastic?

Solution   Since $I_x = 90\% \, I_0$,

$$\log 0.90 = -\beta(1.0 \text{ cm})/2.3 = -0.046,$$

$$\beta = 0.106/\text{cm}.$$

Comments. The incident light is the light that actually enters through the surface of the plastic. Typically, approximately 5% of the external light is reflected at the surface. ◀

REVIEW AND STUDY

9–7 SUMMARY

Polymers and plastic products have properties and service characteristics which are directly related to their internal structures. Polymers have relatively high thermal expansion because some of the bonds between molecules are weak. Likewise, they have low thermal conductivity because they are electrical insulators with no free electrons to transport thermal energy; furthermore, the conductivity is still further reduced because polymers are commonly amorphous, so there is considerable scattering of molecular vibrational energy.

Softening and viscous deformation arise because many polymers are predominantly amorphous, with little crystallinity. A large discontinuity in rigidity exists at the glass temperature, T_g. Sustained loading has the same effect on polymers as higher temperatures, because under either condition molecules can be rearranged in response to stresses. Conversely, plastics are brittle at low temperatures or under impact loading because there is little opportunity for stress redistribution by molecular rearrangements.

Polymers deteriorate by a variety of mechanisms. The most severe is by scission, in which intramolecular bonds are broken. This requires high local energy. However, this energy is available from ultraviolet radiation, and from neutrons. Not surprisingly, the presence of these in the environment has great effects on polymers. A second type of deterioration is by swelling, in which small molecules of solute penetrate among the macromolecules. Fortunately the materials engineer can select polymers that are resistant to swelling by virtue of their cross-linking, composition, or crystallinity.

Plastics are electrical insulators unless conductive additives are introduced or unless surface impurities are adsorbed. They are not inert to electric fields, however. Their relative dielectric constant indicates that electric dipoles form with their negative ends displaced toward the positive electrode; and conversely, the positive toward the negative.

Since they are insulators with no free electrons, plastics transmit light. However, to be transparent, they must be free of internal boundaries due to pores, fillers, or microcrystals. Such boundaries scatter the light, producing translucency. The birefringence of structurally oriented plastics is useful for photoelastic analyses.

Terms for Study

Absorptivity, β	Relative dielectric constant, κ
Birefringence, $n_2 - n_1$	Relaxation time, λ
Dashpot	Rubbery plateau
Degradation	Scission
Depolymerization	Standard deviation, $\bar{\sigma}$
Dielectric constant	Stress relaxation
Dye	Surface resistivity, ρ_s
Elastomers	Swelling
Glass temperature, T_g	Translucency
Mean, $\overline{X}$	Transmittance, T
Median	Transparency
Photoelasticity	Viscoelastic modulus, M_{ve}
Pigment	Viscoelasticity
Polar group	Viscous flow, γ_f
Polyblend	Volume resistivity, ρ
Refraction, n	

STUDY PROBLEMS

9–1.1 Transparent polystyrene "glasses" are used for serving cold drinks on airplane flights. Discuss their use for serving hot coffee.

Answer. Consider T_m and T_g

9–1.2 The two surfaces of a 0.5-cm-thick polyethylene sheet (107 cm × 23 cm) are at 10°C and 27°C. Thus heat is conducted through the plastic. Compare the number of joules conducted per hour if a choice is available between the two polyethylenes of Table 9–1.1.

Answer. HD, 1.5 × 10⁶ joule/hr; LD, 10⁶ joule/hr (or 1.9 × 10⁵ Btu/hr)

9–1.3 A methylmethacrylate (lucite) rod is heated from 0°C to 10°C and allowed to expand. It is then loaded by compression. What force is required to return the rod to the original length? (The original dimensions were 1.0 in. long × 1.0 in. diameter.)

Answer. 350 lb force (compression)

9–2.1 Demonstrate the effect of temperature on the rekinking time of rubber molecules. Stretch a 2-inch rubber band over the length of a 7.5-inch unsharpened pencil. (a) Snap it off the pencil at room temperature. (b) Place it in a freezing compartment of a refrigerator (or take it outdoors on a sub-zero day) until it is cooled. Snap it off the pencil while it is still cold.

9–2.2 Draw a schematic curve of deformation for the spring of Fig. 9–2.4 in series with the spring-dashpot of Fig. 9–2.7.

9–2.3 Repeat 9–2.2, but for the dashpot of Fig. 9–2.5 in series with the spring-dashpot model of Fig. 9–2.7.

9–2.4 What fraction of the stress remains in a plastic (relaxation time = 7 hr) at the end of (a) 1 hr, (b) 7 hr, (c) 14 hr, (d) 28 hr?

Answer. (a) $\sigma = 0.866\sigma_0$ (b) $0.368\sigma_0$ (c) $0.135\sigma_0$ (d) $0.018\sigma_0$

9–2.5 The relaxation time of a rubber is 70 days. (a) How long will it take to relax 50% of the original stress? (b) 75%?

Answer. (a) 48.5 days (b) 97 days

9–3.1 What is the standard deviation of the data in Fig. 9–3.2 for "stress at break."

Answer. 137 psi

9–3.2 Determine the (a) mean, (b) median, and (c) standard deviation of the following set of impact data for an ABS plastic: 2.1, 2.2, 2.4, 1.9, 2.2, 2.0, 2.3, 1.0, 2.5, 2.2, 2.5, 2.9, 2.1, 1.5, 1.9, 1.4, 2.4, 2.5, 2.1, 2.7, 2.7, 1.8, 3.1, 2.2, 2.3, 3.0, 1.7, 2.3, 2.9, 2.1, 2.4, 1.6, 1.9, 2.2, 2.7, 2.3, 1.5, 2.0, 2.1, 1.7, 2.2, 2.8, 1.8.

Answer. (a) 2.18 (b) 2.2 (c) 0.44

9–3.3 Other factors equal, which will have the greatest *toughness*? (a) Rubber subject to low temperatures and a high rate of impact. (b) Rubber subjected to high temperatures and a high rate of impact. (c) Rubber subjected to high temperatures and a slow rate of impact. (d) Rubber subjected to low temperatures and a slow rate of impact. Why?

9–4.1 Green light has a wavelength of about 5250 Å. (a) What is its frequency? (b) What is the energy of its photons?

Answer. (a) 5.7×10^{14}/sec (b) 0.9×10^{-19} cal (3.8×10^{-19} joules)

9–4.2 The average energy of a C—Cl bond in polyvinyl chloride is 81,000 cal/mole, i.e., 81,000 cal (or 340,000 joules) per 6×10^{23} bonds. Will visible light have enough energy to break the average C—Cl bond? [Visible light lies between 4000 Å (violet) and 7000 Å (red).]

Answer. 1.35×10^{-19} cal/bond (or 5.7×10^{-19} joule/bond) requires 3500 Å.

9–4.3 From the data in Example 9–4.2, calculate the density of (a) dry nylon and (b) water-containing nylon.

Answer. (a) 1.144 g/cm³ dry (b) 1.146 g/cm³ wet

9–5.1 The surface resistivity of a plastic is 200 ohms/square (Example 9–5.1). (a) What is the surface resistivity along the length of a 3.1 cm × 2.3 cm film when measured on only one side? (b) 2.3 cm × 3.1 cm film? (c) 3.1 ft × 2.3 ft? [*Note*: By convention, lengths are cited first for the purposes of this problem.]

Answer. (a) 270 ohm (b) 148 ohm (c) 270 ohm

9–5.2 A 0.01-cm plastic film separates two electrodes which have a potential difference of 110 volts. An area of 6500 cm² of film is to be considered. What wattage can move through this film if the resistivity is 10^{10} ohm · cm?

Answer. 0.8 watt

9–5.3 The film in the previous problem serves as a capacitor. Its relative dielectric constant is 7. What is its capacitance? (1 in. = 2.54 cm.)

Answer. 0.4×10^{-6} farad, or 0.4 μf

9–6.1 Refer to Example 9–6.1. Suppose that 90% of the light is transmitted through a plastic that is 1.0 cm thick. How thick must the plastic be to absorb 70% of the light?

Answer. 11.4 cm

9–6.2 The absorptivity of green light by a plastic is 0.05/cm; that of yellow light is 0.12/cm. Assume equal initial intensities of yellow and green light. (a) What is their intensity ratio after 1 mm? (b) 10 cm?

Answer. (a) $I_{gr}/I_{ye} = 1.007$ (b) $I_{gr}/I_{ye} = 2$

[*Note:* Since the ratio changes with thickness, the observed color changes toward the green in thicker sections.]

CHAPTER 10

CERAMICS AND THEIR CHARACTERISTICS

10–1 CERAMIC COMPOUNDS (AX)

In Chapter 1 we described ceramic materials as being *those materials which contain compounds of metallic and nonmetallic elements.* Such compounds may be relatively simple, with only two kinds of atoms present—for example, MgO—or they may be more complex—for example, mica*—with a minimum of five different types of elements.

Both the above examples contain metallic atoms which have given up electrons to produce positively charged cations, e.g., the Mg^{2+} in MgO. Likewise, nonmetallic elements are present which (1) received electrons to develop negatively charged anions (O^{2-} in both examples), or (2) shared electrons with other atoms, e.g., electrons shared between the oxygen and hydrogen in the $(OH)^-$ found in mica.

AX structures (NaCl-type). The simplest ceramic compounds have equal numbers of only two kinds of atoms. We call them AX compounds, A being a metallic atom, and X being a nonmetallic atom. Our previous example, MgO, falls in this category, as do LiF, MnS, TiN, and NaCl, among others.

Each of these compounds has the structure shown in Fig. 10–1.1, where A^+ is the metallic ion and X^- is the nonmetallic ion. The unit cell is a cube with the expected right angles and equal edge dimensions. Note the similarity of this unit cell to the fcc unit cell of Fig. 3–2.2. There is an ion at each corner, and at the center of each face. Both are fcc. Note also a difference between Figs. 10–1.1 and 3–2.2. The interstices of this fcc lattice are occupied by the opposite ion in Fig. 10–1.1, so that each positive ion is surrounded by six immediate neighboring negative ions.† The *coordination number* is six.

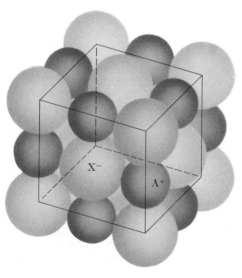

Fig. 10–1.1 AX structure (NaCl). Many ceramic compounds have this structure; for example, LiF, MgO, MnS, TiC, and NaCl. Each positive ion, A^+, is coordinated with six neighboring negative ions. Each negative ion, X^-, is coordinated with six neighboring positive ions. Note that nearest like ions are $\sqrt{2}$ times further apart than the nearest unlike ions. Like ions do *not* touch. In addition, note that the negative ions form an fcc pattern, i.e., fcc lattice. The same is true for the positive ions if the crystal is extended in three dimensions.

---

* The composition of the simplest "white" mica is $[KAl_3Si_3O_{10}(OH)_2]_2$. We do not have to remember this; however, it has one characteristic, that of *cleavage*, which will be important to us.

† To visualize this, imagine that the structure of Fig. 10–1.1 is extended beyond the single unit cell.

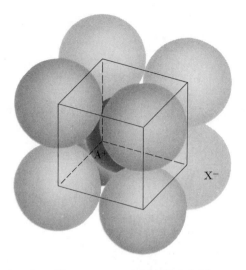

Fig. 10–1.2 AX structure (CsCl). This structure rather than the NaCl structure forms in those compounds with large positive ions because each positive ion may coordinate with more negative ions (eight rather than six). Also each negative ion is coordinated with eight positive ions. Note that this structure is *not* bcc. The center of the cube is not equivalent to the corners because different types of ions are present in the two locations.

Conversely, each negative ion has direct contact with six neighboring positive ions. A close examination of Fig. 10–1.1 reveals that the ions at the face centers and the corners do not touch in this "hard-ball" illustration. This is not surprising, because these ions have like charges and would therefore repel one another. There are, however, strong coulombic attractions between the oppositely charged ions that are in contact.

AX structures (CsCl-type). Figure 10–1.2 shows another AX compound. The prototype for this structure is CsCl, in which each Cs^+ ion is surrounded with eight Cl^- anions (and if we extend the structure, we see that each Cl^- ion is surrounded by eight Cs^+ cations). Observe the similarity between this structure and that of the bcc metal in Fig. 3–2.4. In addition, however, be sure to note one major difference. In Fig. 3–2.4 the position at the center of the unit cell is *identical* to that at the corner. This is not the case in Fig. 10–1.2, because here the center of this cube has one type of ion and the corner has another, of opposite sign. Figure 10–1.2 is *not* bcc; it is *simple cubic* because only the corners of the unit cells are equivalent.

AX structures (ZnS-type). Figure 10–1.3 presents the third AX compound. This structure will return to our attention in Chapter 12 because it is one which is often encountered in semiconducting materials. Each metal atom in Fig. 10–1.3 is coordinated with four nonmetal atoms (and each nonmetallic atom with four metallic atoms). An examination of Fig. 10–1.3 reveals that this compound, like the ones in Fig. 10–1.1 and Fig. 3–2.2, is fcc, with equivalent locations at the corners of the cube and at the center of each face. It is more obvious

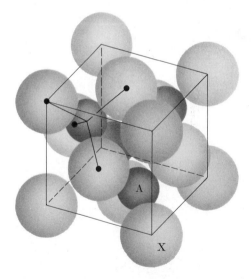

Fig. 10–1.3 AX structure (sphalerite, ZnS). This structure forms when (a) the positive ion is very small and can be coordinated with only four larger negative ions, or (b) the atoms are bonded covalently, and each requires four neighbors. The latter is the situation in semiconducting compounds, which we shall study in Chapter 12. Note that, if the crystal is continued, each type of atom forms an fcc lattice.

Table 10–1.1

RADII OF SELECTED IONS

Ion	Approximate* radius, Å			Ion	Approximate* radius, Å		
	CN = 4	CN = 6	CN = 8		CN = 4	CN = 6	CN = 8
Al^{3+}	0.46	0.51		Mn^{2+}	0.73	0.80	
Ba^{2+}		1.34	1.38	Na^+		0.97	
Be^{2+}	0.32	0.35		Ni^{2+}	0.63	0.69	
Ca^{2+}		0.99	1.02	O^{2-}	1.28	1.40	1.44
Cl^-		1.81	1.87	$(OH)^-$		1.40	
Cr^{3+}		0.63		P^{5+}		0.35	
Cs^+		1.67	1.72	S^{2-}	1.68	1.84	1.90
F^-	1.21	1.33		Si^{4+}	0.38	0.42	
Fe^{2+}	0.67	0.74		Sr^{2+}		1.12	
Fe^{3+}	0.58	0.64		Ti^{4+}		0.68	
K^+		1.33		Zn^{2+}	0.67	0.74	
Li^+		0.68		Zr^{4+}		0.79	0.82
Mg^{2+}	0.60	0.66	0.68				

* The radii are approximate. In LiF, for example, the center-to-center distance between Li^+ and F^- ions can be accurately measured as 2.01 Å; this equals ($r_{Li^+} + R_{F^-}$). Beyond this it is necessary to approximate what fraction of that distance is in each ion. That approximation leaves a small probable error to the radii values. (Patterned after Ahrens.)

here than in Fig. 10–1.1 that the metallic atoms contact only non-metallic atoms, and vice versa. The structure of Fig. 10–1.3 is called the *sphalerite structure*, because the mineral ZnS is the prototype.*

Radius ratios. Since energy is released as ions of unlike charges approach each other, ionic compounds generally have high coordination numbers, i.e., as high as possible without introducing the strong mutual repulsion forces between ions of like charges. This is illustrated in CsCl, which is the prototype for Fig. 10–1.2. Table 10–1.1 presents one widely used set of ionic radii. The Cs^+ has a radius of 1.7 Å. This is large enough to permit eight Cl^- ions to surround each Cs^+ ion without direct "contact" of negative ions with one another. The minimum radius ratio (r/R) possible for eight neighbors without interference is 0.732 (Table 10–1.2). Very few AX compounds have this ratio, however, because positive ions are usually reduced in size after losing an electron (and the anions are larger than those of the corresponding nonmetallic atoms). As a result, the structure of Fig. 10–1.2 is not widely encountered in engineering materials.

* Structural names are often used to identify structures. This enables us to associate identical structures with one prototype, and to differentiate between different structures of a given composition.

In contrast, structures with a coordination number of six are widespread, because many ceramic compounds have r/R of at least 0.414, the minimum required for CN = 6.

We shall observe shortly that the silicon of SiO_2 has CN = 4 because an Si^{4+} ion is too small to have six neighboring oxygens. Another factor favors a CN of only 4 in SiO_2, and the CN of only 4 in ZnS (Fig. 10–1.3). In each of these, there is considerable electron sharing (i.e., a zinc atom and a sulfur atom share a pair of electrons) rather than outright ionization.* As we recall from Table 7–1.1, the maximum number of shared bonds under normal conditions is four. Thus, with sharing, the probability of a CN = 4 is increased over what it would be on the basis of radius ratios alone.

Table 10–1.3 summarizes the relationships of the three most common AX structures.

Table 10–1.3

SELECTED AX STRUCTURES

Prototype compound	Lattice of A (or X)	CN of A (or X) sites	Sites filled	Minimum r_A/R_X	Other compounds
CsCl	Simple cubic	8	All	0.732	CsI
NaCl	fcc	6	All	0.414	MgO, MnS, LiF
ZnS	fcc	4	$\frac{1}{2}$	0.225	β-SiC, CdS, AlP

Example 10–1.1. X-ray data show that the unit cell dimensions of cubic MgO are 4.211 Å. It has a density of 3.6 g/cm^3. How many Mg^{2+} ions and O^{2-} ions are there per unit cell? [The structure of MgO is that of NaCl (Fig. 10–1.1).]

Solution

$$\rho = \frac{mass/u.c.}{volume/u.c.}$$

$$3.6 \ g/cm^3 = \frac{n(24.3 \ g/0.6 \times 10^{24} \ Mg^{2+}) + n(16.0 \ g/0.6 \times 10^{24} \ O^{2-})}{(4.211 \times 10^{-8} \ cm)^3/u.c.}$$

$$n = 4Mg^{2+} \ ions/u.c. = 4O^{2-} \ ions/u.c.$$

Comments. Note: Since MgO is an AX compound, the number of Mg^{2+} ions must equal the number of O^{2-} ions. ◀

---

* Shared electrons provide particularly strong interatomic (covalent) bonds. The carbon atoms of diamonds are bonded by shared electrons.

Table 10–1.2

COORDINATION NUMBERS VERSUS MINIMUM RADII RATIOS

Coordination number	Radii ratios, r/R*	Coordination geometry
3-fold	≥ 0.155	
4	≥ 0.225	
6	≥ 0.414	
8	≥ 0.732	
12	1.0	—

* r — smaller radius
R — larger radius

Example 10–1.2. a) Based on the ionic radii of Table 10–1.1, what is the distance between the *centers* of nearest neighboring Cl^- ions in NaCl? In CsCl?

b) Assume that the "hard-ball" model applies in ionic materials such as NaCl and CsCl. What is the closest approach of the Cl^- ion "surfaces"?

Solution. a) Examine the structure of Fig. 10–1.1, where A is Na^+ and X is Cl^-. The closest approach of *centers* is $(r_{Na^+} + R_{Cl^-})\sqrt{2}$. Therefore

$$D_{Cl^- - Cl^-} = (0.97 \text{ Å} + 1.81 \text{ Å})\sqrt{2} = 3.93 \text{ Å}.$$

Repeat for CsCl in Fig. 10–1.2; use the CN = 8 radii of Table 10–1.1:

$$D_{Cl^- - Cl^-} = (1.72 \text{ Å} + 1.87)(2)/\sqrt{3} = 4.14 \text{ Å}.$$

b) Distance between *surfaces*: In NaCl,

$$d_{Cl^- - Cl^-} = 3.93 \text{ Å} - 2(1.81 \text{ Å}) = 0.31 \text{ Å}.$$

In CsCl,

$$d_{Cl^- - Cl^-} = 4.14 \text{ Å} - 2(1.87 \text{ Å}) = 0.40 \text{ Å}.$$

Comments. The ionic radius ratios are:

$$r_{Na^+}/R_{Cl^-} = 0.97 \text{ Å}/1.81 \text{ Å} = 0.54,$$

$$r_{Cs^+}/R_{Cl^-} = 1.72 \text{ Å}/1.87 \text{ Å} = 0.92.$$

The Cs^+ ions can have eight neighbors because the ionic radius ratio is > 0.732 (Table 10–1.2). ◀

Example 10–1.3. Determine the minimum r/R for a compound with CN = 6.

Solution. Refer to Fig. 10–1.1 and Example 10–1.2. The minimum ratio occurs if the anions, X, have radii such that they just touch one another. In such a case,

$$R + R = (r + R)\sqrt{2},$$

$$2R = r\sqrt{2} + R\sqrt{2},$$

$$r/R = 0.414.$$

Comments. If the radius ratio, r/R, is less than 0.414, a 4-fold coordination is mandatory. If it is greater than 0.414, however, 6-fold coordination— that is, CN = 6—is not mandatory. If some other factor, such as covalent bonds (shared electrons) becomes important, as it did in ZnS (Fig. 10–1.3), there may be 4-fold coordination with $r/R > 0.414$. ◀

10–2 CERAMIC COMPOUNDS (A_mX_p) Ⓜ

If the valences of the two ions are not equal, the above AX structures are not applicable. Of course, many compounds exist with valence ratios other than 1:1. For example, Al_2O_3 has a 2:3 ratio; each aluminum ion is +3 and each oxygen ion is −2. Likewise in ZrO_2, another ceramic compound, the valences are +4 and −2. The list could be extended to cover many other compounds in many other ceramic materials.

AX_2 structures (CaF_2-type). The prototype for the most common of these structures is CaF_2. In addition, ZrO_2 and UO_2 are commercial ceramic compounds of this type. The Ca^{2+} ion has a radius of about 1 Å. The F^- is not very large, since it is in the first row of the periodic table (Fig. 1–2.2) and has only one excess electron. Therefore the radius ratio, r/R, is 0.8, and each positive ion can be surrounded by eight negative ions (Table 10–1.2). This determines the structure (Fig. 10–2.1), and as in CsCl, the cations of CaF_2 have a coordination number CN = 8. However, there are only one-half as many Ca^{2+} ions as F^- ions, since electrical neutrality must be maintained; one-half of the cation sites are therefore vacant. This happens to be an important feature of UO_2, a nuclear fuel with this structure, because these unoccupied sites can accumulate fission fragments from nuclear reactions without introducing excessive strains in the solid.

A_2X_3 structures (Cr_2O_3-type). Aluminum oxide, Al_2O_3, is the most widely used compound in ceramic materials of the nonclay variety. The prime constituent of the spark plug insulator in Fig. 1–2.3, for example, is Al_2O_3. This compound is also used in many other electrical ceramics (Fig. 10–2.2) and abrasion-resistant materials. The prototype for the Al_2O_3 crystal is Cr_2O_3. We may describe Al_2O_3 and Cr_2O_3 crystals as follows. (1) The O^{2-} ions form a hexagonal pattern (cf. Fig. 3–3.2). (2) Cr^{3+} ions (or Al^{3+} ions) are in 6-fold sites among the O^{2-} ions; i.e., these cations have CN = 6. (3) Since the A:X ratio is 2:3, only two-thirds of the 6-fold sites are occupied. (4) Finally, since one-third of those sites are vacant, the structure adjusts slightly.

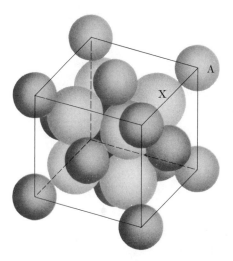

Fig. 10–2.1 AX_2 structure (CaF_2). Since the radius ratio is about 0.8, each positive ion can have eight F^- neighbors. (Observe the atom on the front face, which has four additional neighbors in the next unit cell.) However, there must be only one-half as many positive ions, Ca^{2+}, as negative ions, F^-; therefore half the 8-fold sites are vacant, i.e., the site at the center of the unit cell and the sites at the center of each edge.

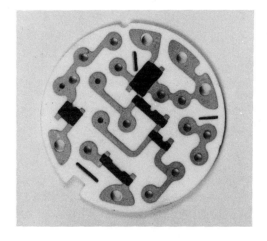

Fig. 10–2.2 Ceramic substrate (Al_2O_3). The *wafer*—onto which conductors, resistors, and capacitors are printed—is one of the newer applications of electronic ceramics. Al_2O_3 makes an excellent material for this purpose because it is one of the strongest electrical insulators. (Courtesy A.C. Spark Plug Division)

(This latter adjustment is noticeable to the crystallographer, but will not affect the properties of interest to us.)

Example 10-2.1. What is the packing factor of the ions in CaF_2 if we assume spherical atoms with radii as shown in Table 10-1.1? Assume hard-ball ions of Table 10-1.1.

Solution. There are $8F^-$ ions (CN = 4) and $4Ca^{2+}$ ions (CN = 8) per unit cell. The distance $(r_{Ca^{2+}} + R_{F^-})$ is one-fourth of the cube diagonal:

$$a = \frac{4(1.02 + 1.21)}{\sqrt{3}} = 5.15 \text{ Å},$$

$$PF = \frac{4(4\pi/3)(1.02 \text{ Å})^3 + 8(4\pi/3)(1.21 \text{ Å})^3}{(5.15 \text{ Å})^3}$$

$$= \frac{(17.7 + 59.4)}{136.5} = 0.56.$$

Comments. The actual lattice dimension, a, is 5.47 Å because the eight F^- ions around the center vacant site have some repulsion for each other, and this causes a decrease in the packing factor. ◀

10-3 CERAMIC COMPOUNDS ($A_mB_nX_p$) Ⓜ ⓜ

Ceramic compounds often contain two distinct kinds of metal atoms. For example, materials of the $BaTiO_3$-type are used for phonograph cartridges and for pressure transducers; these devices require materials which interchange mechanical and electrical forces. Also materials of the $NiFe_2O_4$-type are used for magnetic deflection yokes of TV tubes.

Ionic materials of the $A_mB_nX_p$ type must have positive and negative charges that are balanced; the two examples above have

$$(Ba^{2+} + Ti^{4+} = 3O^{2-}) \quad \text{and} \quad (Ni^{2+} + 2Fe^{3+} = 4O^{2-}),$$

respectively. This places on the structure another restriction in addition to the one of radii ratios.

$BaTiO_3$-type structures. Figure 10-3.1 shows this structure. Above 120°C it is cubic, with the Ba^{2+} ions at the corner of each unit cell (CN = 12). The Ti^{4+} ions are at the center of each unit cell (CN = 6). The oxygen ions are at the center of each face. They are closely coordinated with two Ti^{4+} ions and somewhat less closely associated with four Ba^{2+} ions. We shall see in Section 10-8 that below 120°C there is a slight relative shift in these ions, a shift that has important consequences for dielectric behavior. The high-temperature form of Fig. 10-3.1 will serve our present purposes.

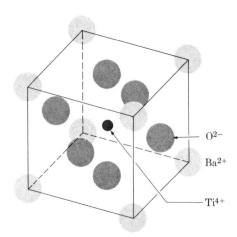

Fig. 10-3.1 Ternary compound ($BaTiO_3$ above 120°C). Each unit cell cube has a Ba^{2+} ion at the corner, a Ti^{4+} ion at its center, and O^{2-} ions on each face. Of course, we could have located the cube to corner on the Ti^{4+} ions (Example 10-3.1).

O^{2-}

Ba^{2+}

Ti^{4+}

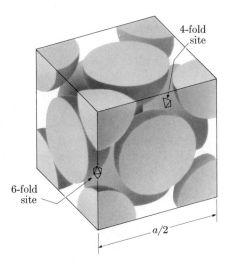

Fig. 10–3.2 Ternary compound ($NiFe_2O_4$). Only the O^{2-} ions are shown. The Ni^{2+} ions are in 6-fold sites, i.e., interstitial positions among six oxygen ions (see Fig. 10–1.1). Half the Fe^{3+} ions are in 6-fold sites; half are located in the positions among four O^{2-} ions, that is, 4-fold sites as in Fig. 10–1.3. Since only three positive ions are present for each four O^{2-}, not all the interstitial sites are occupied. The majority of soft ceramic magnets have this structure. [Only $(a/2)^3$, or one-eighth of the unit cell, is shown.]

$NiFe_2O_4$-type structures. Magnetic ceramics can best be described by Fig. 10–3.2, which shows only the oxygen ions. Among these O^{2-} ions are 6-fold sites (cf. Fig. 10–1.1). The number of these sites equals the number of O^{2-} ions; however, in $NiFe_2O_4$ only half are occupied. The Ni^{2+} ions and half the Fe^{3+} ions sit in these 6-fold sites (CN = 6). Among the O^{2-} ions, there are also 4-fold sites (cf. Fig. 10–1.3). There are twice as many of these sites as there are O^{2-} ions. In $NiFe_2O_4$, one-eighth of these 4-fold sites (CN = 4) are occupied by the remaining half of the Fe^{3+} ions.

Example 10–3.1. Redraw Fig. 10–3.1 so that the Ti^{4+} ion is at the corner of the unit cell.

Solution. See Fig. 10–3.3. We may describe this compound as having Ba^{2+} ions at the cell center, Ti^{4+} ions at the cell corners, and O^{2-} ions at the center of each cell edge.

Comments. Figures 10–3.1 and 10–3.3 are equal alternatives. Each presentation provides the same 1/1/3 ion ratio for Ba/Ti/O.

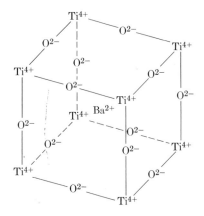

Fig. 10–3.3 $BaTiO_3$ with unit cell corners at the Ti^{4+} ion. See Example 10–3.1 and compare with Fig. 10–3.1.

	Fig. 10–3.1	Fig. 10–3.3
Ba^{2+}	8/8	1
Ti^{4+}	1	8/8
O^{2-}	6/2	12/4 ◄

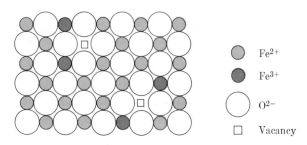

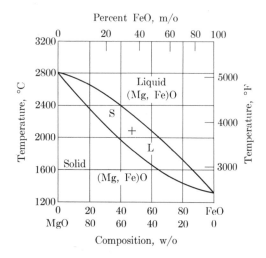

Fig. 10–4.1 Defect structure ($Fe_{1-x}O$). This structure is the same as NaCl (Fig. 10–1.1), except for some iron ion vacancies. Since a fraction of the iron ions are Fe^{3+} rather than Fe^{2+}, the vacancies are necessary to balance the charge. The value of x ranges from 0.05 to 0.16, depending on temperature and the amount of available oxygen.

Fig. 10–4.2 Oxide solutions (FeO-MgO). In both solid and liquid, Fe^{2+} and Mg^{2+} ions can substitute for each other. This is the same as saying that FeO and MgO replace each other in solution. (See Example 10–4.1.)

10–4 SOLID SOLUTIONS Ⓜ ⓜ

Substitutional solutions. Just as in the case of metals, ceramic compounds may have interstitial or substitutional solid solutions (Section 3–4). For ceramic materials, we shall consider only substitutional solid solutions. Both size *and* charge of the replacement atoms must be comparable for substitutional solid solutions. Thus, from Table 10–1.1, we may expect extensive replacement of Mg^{2+} ions ($r = 0.66$ Å) by Ni^{2+}, Fe^{2+}, and Zn^{2+} ions, since they are all divalent and have radii of 0.69 Å, 0.74 Å, and 0.74 Å, respectively. In contrast Li^{+} ($r = 0.68$ Å) cannot replace Mg^{2+} unless some adjustment in charges is made.

One of the better-known ceramic solid solutions is ruby, in which a fraction of one percent of Cr^{3+} ions replace Al^{3+} ions in Al_2O_3. This solid solution is used not only as a gem but also in lasers.

Defect structures. Ⓞ Compounds also have a type of compositional variation not found in metals: *nonstoichiometric*, or *defect structures*. This is best described by iron oxide (Fig. 10–4.1). Iron oxide, $Fe_{1-x}O$, in which the predominant positive ion is Fe^{2+}, has the same basic structure as MgO or NaCl (Fig. 10–1.1). Almost invariably, however, some ferric (Fe^{3+}) ions are present. Since two Fe^{3+} ions have six charges, they replace three Fe^{2+} ions, and in the process leave an ion vacancy, ☐. These vacancies affect diffusion rates (Section 4–8). More important for electrical considerations, Fe^{3+} ions permit electronic conduction. An electron can jump from an Fe^{2+} ion to an Fe^{3+} ion, transporting charge along an electric field. This is the basis of the conductivity through ceramic semiconductors (Chapter 12).

Example 10–4.1. The phase diagram of Fig. 10–4.2 shows a solid solution of MgO-FeO. What is the Mg^{2+}/Fe^{2+} ion ratio in the solid (Mg, Fe)O which is in equilibrium with the liquid oxide at 1600°C?

Solution. Basis: 100 g solid = 35 g MgO and 65 g FeO at 1600°C. From Fig. 1–2.2, the molecular weights are:

MgO: 40.3 amu = 40.3 g/mole,

FeO: 71.8 amu = 71.8 g/mole.

Moles MgO = 35 g/(40.3 g/mole) = 0.87 moles   or 49% of solid is MgO

Moles FeO = 65 g/(71.8 g/mole) = 0.91 moles   or 51% of solid is FeO
                                                      1.78 moles   or 100% solid.

Thus 0.49 MgO/0.51 FeO = 0.49 Mg^{2+}/0.51 Fe^{2+} = 0.96 ion ratio.

Comments. Phase diagrams for ceramics are used in the same way phase diagrams are used for metals (Chapter 5). No new principles are involved. ◀

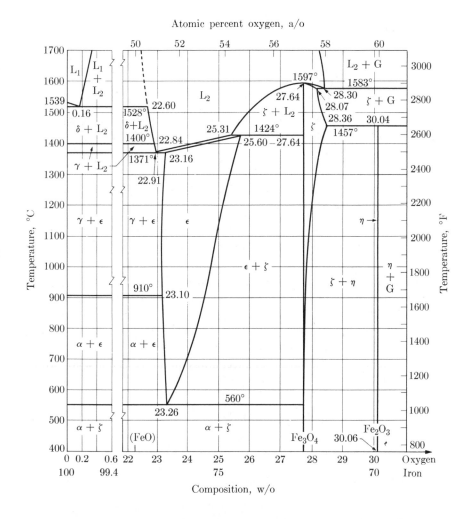

Atomic percent oxygen, a/o

Fig. 10–4.3 The FeO system. Rockhounds will recognize Fe_2O_3 (η) as hematite, and Fe_3O_4 (ζ) as magnetite. The ϵ-phase, which is not normally found as a mineral, has a defect structure, with the composition $Fe_{1-x}O$. (See Fig. 10–4.1.) (ASM *Handbook of Metals,* Metals Park, O.: American Society for Metals)

Example 10–4.2. ⊚ Figure 10–4.3 shows the ceramic part of the Fe-O diagram. The compounds ($Fe_{1-x}O$) or ϵ, Fe_3O_4 or ζ, and Fe_2O_3 or η are present. Note from the top abscissa that the ($Fe_{1-x}O$) composition range has somewhat more than 50 a/o O^{2-}. This occurs because some Fe^{3+} ions are present. What is the Fe^{2+}/Fe^{3+} ion ratio when ε contains 52 a/o O^{2-}?

Solution. Basis: 100 ions $= 52\ O^{2-} + y\ Fe^{2+} + z\ Fe^{3+}$;

$y + z = 48$.

Based on a charge balance ($-$ equal $+$),

$2(52) = 2y + 3z$.

Solving simultaneously, we obtain

$$104 = 2(48 - z) + 3z = 96 + z,$$

$$z = 8 \quad \text{and} \quad y = 40;$$

$$y/z = 40/8 = 5/1 = Fe^{2+}/Fe^{3+}.$$

Comments. Note another feature of the Fe-O phase diagram. The ϵ-phase, that is, $Fe_{1-x}O$, decomposes on cooling by a eutectoid reaction at 560°C:

$$\epsilon(76.74\ Fe) \underset{\text{heating}}{\overset{\text{cooling}}{\rightleftharpoons}} \alpha(100\ Fe) + \zeta(72.36\ Fe). \qquad (10\text{–}4.1)$$

Magnetite, that is, Fe_3O_4 or ζ, and hematite, that is, Fe_2O_3 or η, are stable at normal temperatures. ◀

10–5 SILICATE STRUCTURES Ⓜ ⓜ Ⓞ

The two elements most prevalent in the earth's crust are oxygen and silicon.* It is therefore to be expected that natural raw materials for ceramic products would be predominantly silica (SiO_2) and silicates (MO-SiO_2).

Silicon atoms bond with four oxygens (Fig. 10–5.1). This *tetrahedron*** is the basic structure in all silicates, being a very stable unit. The silica tetrahedron may be a tetravalent ion, SiO_4^{4-}, if SiO_2 is present with sufficient metallic oxides such as Na_2O, K_2O, MgO, CaO, or FeO to supply the necessary oxygens; for example,

$$2MO + SiO_2 \rightarrow 2M^{2+} + SiO_4^{4-}. \qquad (10\text{–}5.1)$$

With an MO/SiO_2 ratio somewhat less than 2/1, a silicate *bimer* forms; for example,

$$3MO + 2SiO_2 \rightarrow 3M^{2+} + Si_2O_7^{6-}. \qquad (10\text{–}5.2)$$

As shown in Fig. 10–5.2, one oxygen is part of two tetrahedra. Such atoms are commonly called *bridging* oxygens.

Chain silicates (and silicones). With increasing silica content (and decreasing amounts of metallic oxides), polymerization produces a chain, as shown in Fig. 10–5.3. In this figure we see the relationship between linear polymers and certain silicates.

Figure 10–5.3(b) is a chain with organic radicals, R, along the side. These are called *silicones.* Figure 10–5.3(c) shows a *silicate* chain

---

*   The surface of the earth is estimated to contain 47 w/o O and 28 w/o Si; iron is third, with 5 w/o.

**   This term is used because a four-apexed shape (Fig. 10–5.1) is a tetrahedron, or literally a *four-sided shape.*

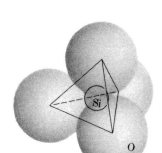

Fig. 10–5.1 SiO_4 tetrahedron. This is the basic structural unit of all silicates. The Si—O bonds are very strong. Depending on the number of positive ions present, this tetrahedron may be a tetravalent ion, SiO_4^{4-}, or it may polymerize with other tetrahedra.

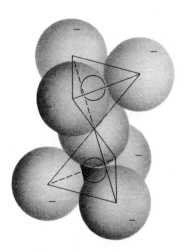

Fig. 10–5.2 Double tetrahedra, $Si_2O_7^{6-}$. This bimer has four oxygens coordinated with each silicon. The center oxygen is coordinated with two silicons, and acts as a *bridging* oxygen.

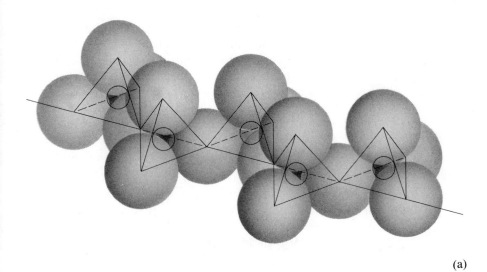

(a)

(b)

(c)

Fig. 10–5.3 Tetrahedral chains. (a) Basic (SiO_3) chain. The coordinations are the same as those described in Fig. 10–5.2. (b) Silicone (siloxane). The nonbridging oxygens—i.e., those that do not join two tetrahedra—have side radicals. These may be —H, —OH, —CH$_3$, —C$_2$H$_2$, — ⬡ , etc. (c) Silicates. Metal atoms produce metal ions and supply an extra electron to each nonbridging oxygen.

The silicones are linear polymeric molecules, $(SiO(OR)_2)_n$. The silicates have linear polymeric negative ions, $(SiO_3)_n^{2-}$, and positive ions.

which is ionically bonded by metallic ions to other tetrahedral chains. Figures 10–5.3(b) and (c) show the same backbone in their tetrahedral chains. The silicone forms molecules with relatively weak bonds to adjacent molecules. Therefore these silicones have the characteristics of plastics. The silicate has stronger ionic bonds between chains than silicones do; therefore the silicate chains cannot slide by one another, do not kink, and are crystallizable. Although ionic bonds are relatively strong, they are cleavable. Therefore these silicates are fibrous. Asbestos, for example, has a silicate chain somewhat like that of Fig. 10–5.3(c).

Sheet silicates. We saw in Fig. 10–5.3 that the silicates can polymerize into chains with the bridging oxygens jointly shared by adjacent tetrahedra.

If three oxygens of each tetrahedron are jointly shared with other tetrahedra (Fig. 10–5.4), a *sheet structure* results. This structure is the basis of materials such as mica, talc, clay, and other sheetlike silicates. The underside of the sheet in Fig. 10–5.4 explains the easy cleavage of mica. No bonds or charges connect onto that side, which makes possible easy splitting.

● Si

◯ O

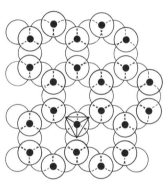

(a)

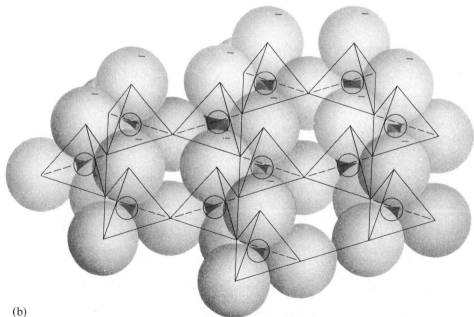

Fig. 10–5.4 Tetrahedral sheets. Three corners of each tetrahedron have a jointly shared—i.e., bridging—oxygen. A sheet structure results which serves as the basis for mica, talc, clay, and other sheetlike silicates. (a) Projected view. (b) Perspective view. Like the chain structures (Fig. 10–5.3), the nonbridging oxygens have a negative charge.

(b)

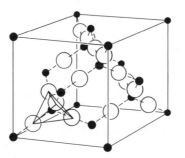

Fig. 10–5.5 Tetrahedral network. Every oxygen of pure silica (SiO_2) is bridging, which gives a rigid network structure. Although the packing factor is less than 0.5 (Example 10–5.1), this structure does not melt until it reaches 1710°C (3110°F). This particular structure is called cristobalite. Quartz, the commonest form of silica, is somewhat more complex. but, as in cristobalite, each silicon in quartz is among four oxygen atoms and each oxygen bridges two silicon atoms.

Network silicates. With pure SiO_2 there are no metal ions, and every oxygen atom is a bridging atom between two silicon atoms (and every silicon atom is among four oxygen atoms, as shown in Fig. 10–5.5). This gives a network-like structure. Silica (SiO_2) can have several different crystal structures, just as carbon can be in the form of graphite or diamond. The structure shown in Fig. 10–5.5 is a high-temperature form. A commoner structure of silica is quartz, the predominant material found in the sands of many beaches.* Just as the form of SiO_2 in Fig. 10–5.5 does, quartz contains SiO_4 tetrahedra, but with a more complex lattice.

Another common natural silicate is feldspar. The pink phase granite is $KAlSi_3O_8$, which may be visualized as a network silicate with one of four silicons replaced by an Al^{3+} ion. The latter, however, has only three charges as compared to four for silicon. Thus K^+ is present to balance out the charges. Fortunately most network structures are quite open, so that there is space for extra ions to be present (cf. Fig. 10–5.5). In fact, the K^+ ion may be viewed as an interstitial ion.

Example 10–5.1. Quartz (SiO_2) has a density of 2.65 g/cm³. (a) How many silicon atoms (and oxygen atoms) are there per cm³? (b) What is the packing factor, given that the radii of silicon and oxygen are 0.38 Å and 1.17 Å, respectively?

Solution

a) $SiO_2/cm^3 = \dfrac{2.65 \text{ g/cm}^3}{(28.1 + 32.0) \text{ g/}(6 \times 10^{23} \text{ } SiO_2)}$

$\qquad = 2.64 \times 10^{22} \text{ } SiO_2/cm^3$

$\qquad = 2.64 \times 10^{22} \text{ } Si/cm^3$

$\qquad = 5.28 \times 10^{22} \text{ } O/cm^3.$

---

*   Typically they are tan in color because of iron oxide stains. The white sands of coral beaches are generally $CaCO_3$.

b) $V_{Si}/cm^3 = (4\pi/3)(2.64 \times 10^{22}/cm^3)(0.38 \times 10^{-8}\ cm)^3 = 0.006$

$V_O/cm^3 = (4\pi/3)(5.28 \times 10^{22}/cm^3)(1.17 \times 10^{-8}\ cm)^3 = \underline{0.354}$

Packing factor $= \overline{0.36}$

Comments. Although there is considerable open space within this structure, most single atoms (except for helium) must diffuse through SiO_2 as ions. Thus their charges prohibit measurable movements. ◀

10–6 GLASSES Ⓜ

The main glasses used commercially are silicates. They have the SiO_4 tetrahedra described in the preceding section, plus some modifying ions. They are *amorphous*, i.e., noncrystalline. Figure 10–6.1 contrasts the structure of a crystal with that of an amorphous solid of the same composition. Atoms in both have the same first neighbors, i.e., each oxygen atom is between two silicon atoms and each silicon is among four oxygens (the fourth is envisioned as being above *or* below the plane of the paper). However, in glass, differences arise in more-distant neighbors because the glass, unlike the crystalline silica, does not have a regular long-range pattern.

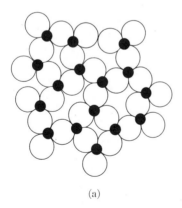

(a)

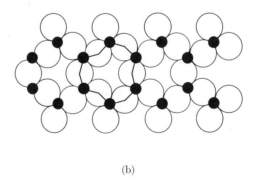

(b)

Fig. 10–6.1 Structure of silica glass. (a) Amorphous (glassy) silica. (b) Crystalline silica. In both glassy and crystalline silica, each silicon atom (black) is in a tetrahedron among four adjacent oxygen atoms (open circles). (The fourth oxygen of each tetrahedron has been removed for clarity.) Each oxygen bridges two silicon atoms. The crystalline silica has long-range order which is absent in the glassy silica.

When only silica (SiO_2) is present, the glass is very rigid. Fused silica, for example, is extremely viscous even when it gets into the temperature range in which it is a true liquid. In polymer terms, its *network* structure is highly cross-linked. Fused silica is very useful in some applications because it has a low thermal expansion. However, its high viscosity makes it extremely difficult to shape.

Network modifiers. Most silicate glasses contain *network modifiers*. These are oxides such as CaO and Na_2O which supply cations (positive ions) to the structure. As shown in Fig. 10–6.2, the addition of Na_2O

to a silica glass introduces (1) two Na^+ ions, and (2) nonbridging oxygens. These are oxygens which are attached to only one silicon. As in the chain structure of Fig. 10–5.3(c), these oxygens carry a negative charge.

Commercial glasses contain these network modifiers because their presence markedly lowers the high-temperature viscosity of the glass by reducing the effective cross-linking, to make the glass more "thermoplastic." Therefore this glass is much easier to shape into the desired product.

Commercial glasses. We read in Section 1–1 that there are as many as 10,000 varieties of glass. Some of these differences arise from minor variations in composition of one of a dozen different components, to produce specific properties such as index of refraction, color dispersion, viscosity, etc. Commercial glasses may be divided into relatively few general categories, however, as indicated in Table 10–6.1.

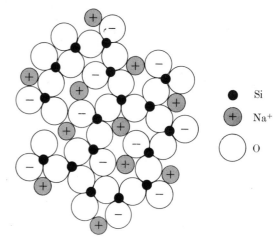

Fig. 10–6.2 Structure of soda-silica glass. The addition of Na_2O to a silica glass decreases the number of bridging oxygen atoms. Each nonbridging oxygen carries a charge. (As in Fig. 10–6.1, the fourth oxygen of each tetrahedron has been removed for clarity.) This glass is less viscous at high temperatures than the silica glass of Fig. 10–6.1 (a).

Table 10–6.1

COMMERCIAL GLASS TYPES*

Type	Major components, %								Comments
	SiO_2	Al_2O_3	CaO	Na_2O	B_2O_3	MgO	PbO	Other	
Fused silica	99								Very low thermal expansion, very high viscosity
96% silica (vycor)	96				4				Very low thermal expansion, high viscosity
Pyrex type	81	2		4	12				Low thermal expansion, low ion exchange
Containers	74	1	5	15		4			Easy workability, high durability
Plate	73	1	13	13					High durability
Window	72	1	10	14		2			High durability
Lamp bulbs	74	1	5	16		4			Easy workability
Lamp stems	55	1		12			32		High resistivity
Fiber (E-glass)	54	14	16		10	4			Low alkali
Thermometer	73	6		10	10				Dimensional stability
Lead glass tableware	67			6			17	K_2O 10	High index of refraction
Optical flint	50			1			19	BaO 13, K_2O 8, ZnO 8	Specific index and dispersion values
Optical crown	70			8	10			BaO 2, K_2O 8	Specific index and dispersion values

* Data adapted from various sources, but primarily from A. K. Lyle, "Glass Compositions," *Handbook of Glass Mfg.* (F. V. Tooley, editor). New York: Ogden Publishing Co.

The choice of a glass depends on striking a balance between the requirements of processing and the properties of the finished product. For example, a fused-silica glass has both a very low thermal expansion and a very high viscosity. Thus, even though there are many applications in which the low thermal expansion would be desirable, costs of processing fused silica may exclude its use because of the difficulty involved in shaping it into the desired product.

Example 10–6.1. A glass contains 80 w/o SiO_2 and 20 w/o Na_2O. What fraction of the oxygens is nonbridging?

Solution. Basis: 100 g = 80 g SiO_2 + 20 g Na_2O.

80 g SiO_2/[(28.1 + 2 × 16.0) g/mole] = 1.33 mole SiO_2

$\qquad\qquad\qquad\qquad\qquad\qquad\qquad\qquad$ or 80.6 m/o SiO_2.

20 g Na_2O/[(2 × 23.0 + 16.0) g/mole] = 0.32 mole Na_2O

$\qquad\qquad\qquad\qquad\qquad\qquad\qquad\qquad$ or 19.4 m/o Na_2O.

80.6 SiO_2 = 80.6 Si + 161.2 O

$$19.4\ Na_2O = \frac{19.4\ O + 38.8\ Na^+}{80.6\ Si + 180.6\ O + 38.8\ Na^+}$$

Note from Fig. 10–6.2 that there is one nonbridging oxygen for each Na^+ added.

Fraction nonbridging oxygens = 38.8/180.6 = 0.215.

Comments. Calcium forms a double charge and provides electrons to two nonbridging oxygens. Therefore each Ca^{2+} plus the accompanying oxygen of CaO breaks open a bridging oxygen and forms two nonbridging, charge-carrying oxygens. ◀

Example 10–6.2. A soda-lime glass contains 13 w/o Na_2O, 13 w/o CaO, and 74 w/o SiO_2 (cf. a plate glass of Table 10–6.1). Soda ash (Na_2CO_3) and limestone ($CaCO_3$) are used as sources of the Na_2O and CaO, respectively. As they are heated, CO_2 is evolved, leaving Na_2O and CaO. How many pounds of each are required for 1000 lb of SiO_2?

Solution. Based on the above composition:

1000 lb SiO_2 = 175 lb Na_2O and 175 lb CaO

$$Na_2CO_3 \rightarrow Na_2O + CO_2 \qquad\qquad\qquad\qquad\qquad (10\text{–}6.1)$$

$$\frac{x}{2(23) + 12 + 3(16)} = \frac{175}{2(23) + 16}$$

$$x = 300\ \text{lb}\ Na_2CO_3$$

$$CaCO_3 \rightarrow CaO + CO_2 \qquad\qquad (10\text{--}6.2)$$

$$\frac{y}{40.1 + 12 + 3(16)} = \frac{175}{40.1 + 16.0}$$

$$y = 312 \text{ lb } CaCO_3$$

Comments. In the process, $125 + 137$, or 262 lb of CO_2 will be evolved per 1000 lb of SiO_2. ◀

10–7 REFRACTORIES AND ABRASIVES ⒠

High-temperature (refractory) and hard (abrasive) materials are grouped together because both properties depend on the extra strong metal–nonmetal bonds found in certain ceramic materials. For example, Al_2O_3 has a hardness of 9 on the mineralogist's Mohs scale (diamond = 10 is nature's hardest). It also has a melting point of more than 2000°C (>3600°F), as shown in Table 10–7.1. As an abrasive material, Al_2O_3 is called *emery*, and is a well-known component of grinding wheels. As a refractory material, it not only withstands high temperatures but is chemically inert to oxidation and many fluxing situations.

Table 10–7.1

HARDNESS OF ABRASIVE* MATERIALS

Material	Composition	Mohs hardness number	Approximate Knoop hardness	Melting temperature, °C
⬆ Diamond	C	10	8000	>3500
Boron carbide	B_4C	—	3500	2450
Silicon carbide	SiC	—	3000	>2700
Titanium carbide	TiC	—	2800	3190
Tungsten carbide	WC	—	2100	2770
Corundum	Al_2O_3	9	2000	2050
Topaz	$SiAl_2F_2O_4$	8	1500	Dissociates
⬇ Quartz	SiO_2	7	1000	Transforms
Orthoclase	$KAlSi_3O_8$	6	600	Dissociates
Apatite	$Ca_5P_3O_{12}F$	5	500	~1400
Fluorite	CaF_2	4	200	1330
Calcite	$CaCO_3$	3	150	Dissociates
Gypsum	$CaSO_2(OH)_4$	2	50	Dissociates
Talc	$Mg_3Si_4O_{10}(OH)_2$	1	20	Dissociates

* Mohs hardness ≥ 7.

Table 10–7.2

SELECTED PROPERTIES OF COMMON REFRACTORY MATERIALS*

Brick	Composition	PCE† or softening point	True specific gravity	Approx. weight 9 in. brick	Porosity	Linear coefficient of expansion, °C^{-1}
High-heat-duty firebrick	Al_2O_3: 35–42% SiO_2: 52–60%	31–33 (3060–3170°F)	2.60–2.70	7.5–8 lb	8–24%	(21–1300°C) 5.4 × 10^{-6}
Superduty	Al_2O_3: 40–44% SiO_2: 49–56%	33–34	2.60–2.70	8–8.3 lb	9–16%	
High alumina	Al_2O_3: 50–60–70 or 80%	35–41 (3200–3390°F)	2.75–3.45	8–8.3 lb	17–31%	
Bonded alumina	Al_2O_3: 90%	40–41	3.55–3.65	9.9 lb	13–25%	7.0–10.0 × 10^{-6}
Kaolin	Al_2O_3: 44–45% SiO_2: 51–53%	33–34 (3060–3200°F)	2.60–2.70	8.1–8.8	7–18%	(21–1610°C) 4.3 × 10^{-6}
Bonded mullite	Al_2O_3: 60–75% SiO_2: 18.0–34.0%	38–39 (3340°F)	3.00–3.20	9.0 lb	15–26%	(200–1600°C) 4.5 × 10^{-6}
Bonded A–Z–S	Al_2O_3: 70% ZrO_2: 20% SiO_2: 10%	34–35	3.65	11.0 lb	15–18%	3.89 × 10^{-6} in./in./°F (80–2700°F)
Zircon	$ZrSiO_4$	Above 2015°C	4.6	12–15 lb	0–30%	(21–1550°C) 4.2–5.5 × 10^{-6}
Silica	SiO_2: 95–96%	31–33 (3060–3170°F)	2.30–2.40	6.6 lb	20–28%	(21–1550°C) 43 × 10^{-6} 20–300°C 3 × 10^{-6} 300°–1200°
Recrystallized silicon carbide	SiC: 98–99%	Dissociates at 2250°C	3.17	9 lb	18–23%	(21–1000°C) 4.5 × 10^{-6}
Fused alumina	Al_2O_3: 88–90%	2050°C 3720°F	3.90	11 lb	21%	(21–800°C) 8.1 × 10^{-6}
Chrome	Cr_2O_3: 30–33% Al_2O_3: 27–29% SiO_2: 5–8% FeO: 5–7%	1950–2200°C 3540–3990°F	3.80–4.10	12 lb	18–25%	(21–1540°C) 10.4 × 10^{-6}
Magnesite (burned)	MgO: 92–98% Fe_2O_3: 0.1–2.0	2200°C 3990°F	3.40–3.60	10.4 lb	16–24%	(21–1700°C) 14.7 × 10^{-6}
Magnesite (unburned)	MgO: 92–96% Fe_2O_3: 0.1–2.0%	2200°C 3990°F	3.40–3.60	10.5 lb	15–24%	(21–1700°C) 14.7 × 10^{-6}

* Courtesy of Charles Taylor Sons Co., a subsidiary of N.L. Industries, Cincinnati, Ohio.
† Pyrometric cone equivalent.

Table 10–7.2 (Continued)

Thermal conductivity, CGS units	Specific heat	Spalling resistance	Deformation under load, 25 lb/in²	Constancy of volume	Slag resistance
(200–1000°C) 0.00339	(25–1000°C) 0.254	Good	3–6% at 2460°F	0.5 to 2% shrinkage at 2250°F	Rapidly attacked by basic slags, particularly those high in Fe_2O_3; moderate resistance to acid slags
Approximately the same as high-heat-duty firebrick		Varies	2–8% at 2640°F	1% max. shrinkage at 2900°F	
Approximately the same as high-heat-duty firebrick		Superior to good	1–6% at 2640°F		More resistant to most alkalies than firebrick; readily attacked by Fe_2O_3
Approximately the same as H.H.S. duty		Excellent	0–2% at 2900°F	0% change at 3000°F	Good except in extreme alkaline or acid slags
(200–1000°C) 0.0045	(250–1000°C) 0.254	Good	1.0% at 2800°F	5% shrinkage at 3000°F	Approximately same as high-heat-duty firebrick; slightly better resistance to alkalies
(600–900°C) 0.0044	(20–1000°C) 0.20	Superior to excellent	0.0% at 2460°F, 1.5% at 2900°F	No shrinkage at 3000°F; 2.5% at 3200°F	Relatively insoluble in most slags and glasses, particularly those high in lime, alkalies, or fluorides; attacked by strongly basic slags or those high in Fe_2O_3
13 Btu/hr/ ft²/°F/in.		Superior	0–0.12%	0.3–0.6% below 3100°F	Alkaline resistance excellent to good
(200–1000°C) 0.0046	(20–51°C) 0.18	Varies	2–6% at 2910°F	No appreciable change at 1550°C	Slightly attacked by acid slags; strongly attacked by basic slags and fluorides
(200–1000°C) 0.0045	(25–1000°C) 0.265	Excellent above 1200°F	0.0% at 1500°C, 50 lb/in²	Reversible expansion to 3170°F; no shrinkage	Resists acid slags and dust; attacked by basic slags, fluorides and Fe_2O_3
(200–1350°C) 0.0330	(20–1000°C) 0.285	Superior	0.0% at 1500°C 50 lb/in²	No shrinkage at 1500°C; up to 10% permanent expansion due to oxidation	Attacked by basic slags, particularly those high in iron oxide
(200–1000°C) 0.0095	(20–1000°C) 0.20	Poor–Fair	2% at 1500°C, 50 lb/in²	No shrinkage at 1500°C	Properties similar to 90% Al_2O_3
(200–1000°C) 0.0040	(20–1000°C) 0.217	Fair	Improved chrome brick show no deformation at 1425°C	No shrinkage at 1450°C; 1.3% at 1550°C	Neutral properties, resistant to both acid and basic slags; poor against strong acid slags
(200–1000°C) 0.0087	(20–1000°C) 0–278	Fair	Not recommended‡ where high hot strength required	Shrinkage at temperatures above 1600°C	Highly resistant to basic slags; readily attacked by acid slags
(200–1000°C) 0.0087	(20–1000°C) 0.278	Fair	3.5% deformation at 1480°C‡	Some shrinkage at 1500°C; 2.6% at 1550°C	Highly resistant to basic slags; readily attacked by acid slags

‡ Fails at 3000°F.

Another commercially available refractory and abrasive material is silicon carbide, SiC. It is somewhat harder than Al_2O_3 and has a higher melting temperature, but suffers from the fact that it oxidizes slowly to CO, a gas, and SiO_2, a glass. Fortunately the latter forms a surface layer which is relatively impervious to oxygen, thereby protecting the SiC from rapid oxidation.

Table 10–7.2 lists refractory materials for making industrial furnaces (for refining metals, generating power, industrial incinerating, heat-treating, glass melting, etc.) together with their properties. The difficulties of specifying refractories for furnace applications are not so much that of knowing the properties of the materials, as of defining the service conditions. Very often they are combinations of high temperatures, mechanical stresses, and corrosive environments.

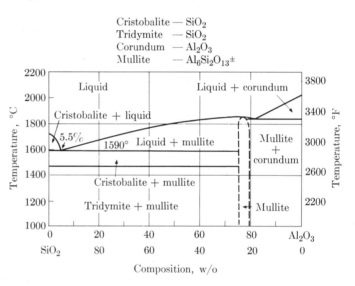

Fig. 10–7.1   The Al_2O_3-SiO_2 system. (Adapted from Muan and Osborn, *Phase Equilibria Among Oxides in Steelmaking*, Reading, Mass.: Addison-Wesley)

Example 10–7.1. A fireclay is a relatively pure clay with a composition of $Al_2Si_2O_5(OH)_4$. When it is used to produce refractory brick it undergoes the dissociation reaction:

$$Al_2Si_2O_5(OH)_4 \rightarrow Al_2O_3 + 2SiO_2 + 2H_2O\uparrow. \qquad (10–7.1)$$

Based on Fig. 10–7.1, and assuming equilibrium, what phases will form at 1500°C (2732°F)? What is the maximum weight percent of each?

Solution. The H_2O produces a gas, leaving Al_2O_3 and SiO_2:

Al_2O_3: $2(27.0) + 3(16.0) = 102.0$ amu $=$ 46 w/o

SiO_2: $2[28.1 + 2(16.0)] = \underline{120.2}$ amu $= \underline{54}$ w/o

                            222.2 amu   100 w/o

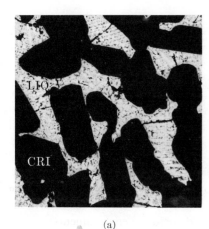

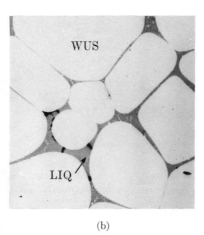

(a)   (b)

Fig. 10–7.2 Photomicrographs of ceramics.
(a) CRIstobalite (SiO$_2$) grains in a matrix of iron oxide-silica LIQuid for 18 hours at 1480°C (2700°F).
(b) WÜStite (~ FeO) grains in a matrix of iron oxide-silica LIQuid for 64 hours at 1200°C (2190°F). The liquid in (a) is saturated with SiO$_2$; the liquid in (b) is saturated with FeO. (c) The FeO-SiO$_2$ system. The dots locate the phase compositions shown in the photomicrographs of (a) and (b). (Phase diagram adapted from Bowen and Schairer, *Amer. Jour. Sci.*)

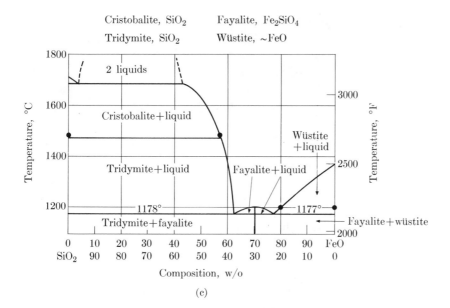

(c)

At 1500°C, we see from the equilibrium diagram that the phases are cristobalite (SiO$_2$) and mullite (~ Al$_6$Si$_2$O$_{13}$).

Fraction cristobalite = (0.76 − 0.46)/(0.76 − 0) = 0.4.

Fraction mullite   = 0.46/0.76   = 0.6.

Comments. The cristobalite crystallizes very slowly because it contains 100% network-forming oxide, SiO$_2$. The equilibrium value of 40 w/o cristobalite requires several days to form at 1500°C. Figure 10–7.2 is the basis for Study Problems. ◀

Fig. 10–8.1 High-voltage insulators. As shown on this test stand, these insulators must be designed with a high surface resistivity, as well as a high volume resistivity. (R. Russell, *Brick and Clay Record*)

10–8 ELECTRICAL CERAMICS Ⓜ ⓜ

Insulators. Ceramics are well known for their use as electrical insulators (Fig. 10–8.1). One of the best ways to tie up the outer electrons of atoms so they can't transfer charge is to transfer them from the metal ions to oxygen atoms, forming oxygen ions, O^{2-}. Let us refer again to the spark plug in Fig. 1–2.3. The Al^{3+} ions of Al_2O_3 have been stripped of the electrons that would carry charge in metallic aluminum. Those electrons are held firmly by the oxygen ions, O^{2-}. In other insulating materials, Mg^{2+} ions lose their electrons to O^{2-} in MgO, and silicon and oxygen rigidly share electrons within the SiO_4 tetrahedra (Fig. 10–5.1). As a result, compositions of MgO-Al_2O_3-SiO_2 form some of the best electrical insulators.

The "triaxial" diagram of Fig. 10–8.2 is a convenient way of presenting *ternary compositions*, i.e., compositions of three components. The corners are 100% of the indicated component—MgO, Al_2O_3, or SiO_2 in this case. Compositions along the sides represent combinations of two of the components, and zero content of the third component that appears at the far corner of the triangle. Compositions within the triangle include all three components. Any particular component increases from zero to 100% according to the lever rule (Section 5–3).

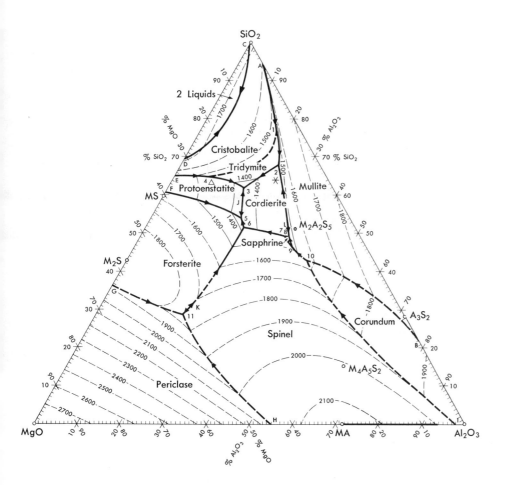

Compound	Melting temp., °C	Phase name	Composition Al$_2$O$_3$	MgO	SiO$_2$
A-Al$_2$O$_3$	~2020	Corundum	100	—	—
M-MgO	~2800	Periclase	—	100	—
S-SiO$_2$	1723	Cristobalite	—	—	100
S-SiO$_2$		Tridymite	—	—	100
S-SiO$_2$		Quartz*	—	—	100
MS	D†	Enstatite*	—	40	60
M$_2$S	~1900	Forsterite	—	57	43
A$_3$S	~1850	Mullite	72	—	28
MA	~2135	Spinel	71.5	28.5	—
M$_2$A$_2$S$_5$	D†	Cordierite	35	14	51
M$_4$A$_5$S$_2$	D†	Sapphrine	65	20	15

*Name of 25°C polymorph
†D = decomposes

Invariant Points

Ternary					Binary			
	Temp., °C	Al$_2$O$_3$	MgO	SiO$_2$		Temp., °C	Al$_2$O$_3$	MgO	SiO$_2$
1	1470	18	5.5	76.5	A	~1590	6	—	94
2	1440	22.5	9.5	68	B	~1840	78	—	22
3	1335	17.5	20.5	62	C	1703	—	1	99
4	1470	8.5	26.5	65	D	1703	—	31	69
5	1365	21	25	54	E	1543	—	35	65
6	1370	23	25.5	51.5	F	1557	—	39	61
7	1453	33.5	17.5	49	G	~1860	—	64	36
8	1460	34.5	16.5	49	H	~1850	55	45	—
9	1482	37	17	46	I	~1925	98	2	—
10	1578	42	15	43	J	~1367	20	24	56
11	~1710	20	51	29	K	~1710	21	49	30
					L	1465	32	15	53

Fig. 10–8.2 The MgO-Al$_2$O$_3$-SiO$_2$ system (partial). The contours indicate the lowest temperatures at which the materials are completely liquid. One of the common types of electrical porcelains lies near the point marked with a triangle. A triaxial diagram such as this is "read" by noting that each corner represents a single oxide; the edge of the triangle is a mixture of two oxides. Ternary mixtures of three components lie within the triangle. (After E. F. Osborn and A. Muan, *Phase Equilibrium Diagrams of Oxide Systems*, Columbus, O.: American Ceramic Society)

Thus a rather common electrical porcelain made out of talc $[Mg_3Si_4O_{10}(OH)_2]$ and clay $[Al_2Si_2O_5(OH)_4]$ is represented by point Δ in Fig. 10–8.2. (The two raw materials decompose on heating to $33MgO\text{-}67SiO_2$ and $46Al_2O_3\text{-}54SiO_2$ respectively.)

Table 10–8.1 shows properties of selected ceramic insulators. The dielectric strength is the voltage gradient, volts/in. (usually represented as volts/mil) which produces electrical breakdown. We discussed the dielectric constant in Sections 2–9 and 9–5. The final item, *loss factor*, is listed, but will not be detailed, since it is not within the scope of this text. In brief, however, the loss factor, tan δ, in combination with the relative dielectric constant, κ, is a measure of the electrical energy lost with ac voltage. Just as you would expect, this will be of interest to electrical engineers in later courses.

Table 10–8.1

PROPERTIES OF CERAMIC DIELECTRICS

Material	Resistivity (volume), ohm · cm	Dielectric strength, volts/mil*	Relative dielectric constant, κ		Loss factor, tan δ	
			60 Hz	10^6 Hz	60 Hz	10^6 Hz
Electrical porcelain	10^{13}–10^{15}	40–200	6	—	0.010	—
Steatite insulators	$>10^{14}$	200–400	6	6	0.005	0.003
Zircon insulators	$\sim 10^{15}$	250–400	9	8	0.035	0.001
Alumina insulators	$>10^{14}$	250	—	9	—	<0.0005
Soda-lime glass	10^{14}	250	7	7	0.1	0.01
E-glass	$>10^{17}$	—	—	4	—	0.0006
Fused silica	$\sim 10^{20}$	250	4	3.8	0.001	0.0001

* Volts per 0.001 in. This value varies somewhat with service conditions.

Piezoelectrics. ◎ The electrical ceramics of Figs. 1–2.3, 10–2.2, and 10–8.1 serve as insulators. Their primary function is to remain inert to the electric circuit and applied voltages. A number of electrical ceramics, however, have an active function within the circuit. The most obvious is the dielectric spacer between capacitor electrodes. The charge on the capacitor is proportional to the relative dielectric constant (Eq. 9–5.3). The dielectric principles described for polymers in Section 9–5 apply equally for dielectric ceramics. In this section we shall direct our attention to *piezoelectric* ceramics, i.e., *pressure-electric* ceramics.

Recall from Section 10–3 that we said that the cubic structure shown for $BaTiO_3$ in Fig. 10–3.1 existed above 120°C, and that there is

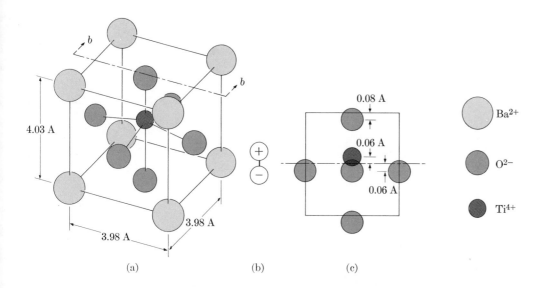

(a)   (b)   (c)

Fig. 10–8.3 Noncubic $BaTiO_3$ (cf. Fig. 10–3.1). As the temperature drops below 120°C, there is a relative shift of the positive and negative ions. The centers of positive and negative charges do not remain at the center of the unit cell. An electric dipole (b) results; its length may be altered by external forces.

a slight shift below that temperature. The shift is revealed in Fig. 10–8.3. Relative to the Ba^{2+} ions, the center Ti^{4+} shifts upward 0.06 Å, the side O^{2-} ions shift downward by the same amount, while the top and bottom O^{2-} shift downward 0.08 Å. This shift provides the engineer with a very useful structure because the center of positive charge is raised relative to the center of negative charge. Specifically, the center of positive charges is 0.04 Å above the center of the unit cell as defined by the Ba^{2+} ions. [The average of $(4^+)(0.06$ Å$)$ and $(2^+)(0.00$ Å$)$ equals $(6^+)(0.04$ Å$)$]. The center of negative charges is 0.067 Å below the center of the unit cell. [The average of $(4/2)(2^-)(0.06$ Å$)$ and $(2/2)(2^-)(0.08$ Å$)$ equals $(6^-)(0.067$ Å$)$].* Thus the centers of positive and negative charges are separated by 0.107 Å. Admittedly this distance is not great; however, when it is multiplied by a large number of unit cells, its effect can be useful.

Figure 10–8.4(a) shows a schematic presentation of several unit cells, each with the centers of positive charges above the centers of negative charges (distances are markedly exaggerated). As in Fig. 9–5.6(b), this displacement collects extra negative charges on the upper electrode and accumulates positive charges on the lower electrode. The amount of electrode charge is proportional to the electric dipole moment within the dielectric material. The dipole moment is equal to the product of the charge at the end of the electric dipoles and the distance between the centers of charge, in this case, 0.107 Å.

---

* 4/2 and 2/2 are used since the oxygen ions are on unit cell faces, and are therefore shared by adjacent unit cells.

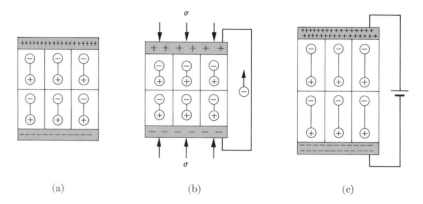

Fig. 10–8.4 Piezoelectricity. (a) The dipoles within the material produce a charge difference between the two ends. (b) Pressure shortens the dipoles, and therefore reduces the charge density on the two electrodes (or introduces a voltage difference if the two ends are not shorted). (c) An external voltage alters the dipole lengths.

(a)   (b)   (c)

Now apply pressure (Fig. 10–8.4b). The distance between the centers of charge is reduced. This reduces the electric dipole moment so that (1) electrons flow from the one electrode to the other if the two are connected, or (2) if not connected, a voltage change occurs between the two electrodes. This voltage change can be detected and amplified in an electric circuit. By proper calibration, the engineer can design a pressure gage. The electrical engineer can use a material of this type in a phonograph cartridge. As the stylus follows the groove, it places a varying pressure on the piezoelectric material in the cartridge. A corresponding variation occurs in the voltage. This signal is amplified and eventually fed into the speaker.

Returning to Fig. 10–8.4, let us operate the mechanical-electrical transducer in reverse. In Fig. 10–8.4(c), a voltage is placed across the electrodes. The extra charges on the plates "stretch" the dipoles within the unit cells, causing the material to strain as if it were stretched in tension. (Or if the applied voltage is reversed, the electric dipole is compressed to a shorter length.) An alternating voltage can produce mechanical vibrations. Ultrasonic generators utilize such materials.

Piezoelectric materials include not only $BaTiO_3$, but also $PbTiO_3$ and $PbZrO_3$, among a number of other such ceramics which have been discovered since the engineer has seen uses for transducers of this type.

Example 10–8.1. ◎ Polarization, $\mathscr{P}$, is the electric dipole moment per unit volume. What is the maximum polarization of $BaTiO_3$ (Fig. 10–8.3)?

Solution. In this section of the text we calculated that the electric dipole moment arm, d, is 0.107 Å. Since (1) the dipole moment, μ, is the product of the moment arm and the charge, Q,

$$\mu = Qd, \tag{10–8.2}$$

(2) each valence charge is 1.6×10^{-19} coul, and (3) there are six charges per unit cell:

$$\mu = (6)(1.6 \times 10^{-19} \text{ coul})(0.107 \times 10^{-8} \text{ cm})$$

$$= 1.03 \times 10^{-27} \text{ coul} \cdot \text{cm}.$$

$$\mathscr{P} = \frac{1.03 \times 10^{-27} \text{ coul} \cdot \text{cm}}{(4.03 \times 3.98 \times 3.98 \times 10^{-24} \text{ cm}^3)} = 1.6 \times 10^{-5} \frac{\text{coul} \cdot \text{cm}}{\text{cm}^3}.$$

Comments. When there is maximum polarization, called *saturation* polarization, we assume that each of the unit cells has its center Ti^{4+} ions displaced in the same direction.

The electrical engineer commonly uses meters rather than centimeters. The above answer would thus be 0.16 coul $\cdot$ m/m^3. ◀

Example 10–8.2. ◎   Refer to Fig. 10–8.5 and the second paragraph of this section. (a) Show graphically that point $\times$ represents 50% A, 40% B, and 10% C. (b) What is the composition of the point marked $\triangle$? (c) The point marked $\square$?

Solution

a) From your geometry course:

$$\overline{a\times}/\overline{aA} + \overline{b\times}/\overline{bB} + \overline{c\times}/\overline{cC} = 1.00. \tag{10–8.1}$$

By measurement in Fig. 10–8.5(b):

$$\overline{a\times}/\overline{aA} = 0.50, \qquad \overline{b\times}/\overline{bB} = 0.40, \qquad \overline{c\times}/\overline{cC} = 0.10.$$

b) For point $\triangle$:

$$\overline{a\triangle}/\overline{aA} + \overline{b\triangle}/\overline{bB} + \overline{c\triangle}/\overline{cC} = 0.40 + 0.20 + 0.40.$$

Composition: $40A + 20B + 40C$.

c) For point $\square$:

$$\overline{a\square}/\overline{aA} + \overline{b\square}/\overline{bB} + \overline{c\square}/\overline{cC} = 0 + 0.60 + 0.40.$$

Composition: $60B + 40C$.

Comments. Note that Eq. (10–8.1) applies whether the triangle is equilateral or not. Every possible composition may be represented by a point, and every point represents a composition. ◀

Example 10–8.3. ◎   Refer to Fig. 10–8.2.

a) Locate the composition of talc after dissociation and loss of water.

$$Mg_3Si_4O_{10}(OH)_2 \rightarrow 3MgO + 4SiO_2 + H_2O\uparrow. \tag{10–8.2}$$

b) Repeat for clay (Eq. 10–7.1).
c) Determine the location on the $MgO\text{-}Al_2O_3\text{-}SiO_2$ diagram of an 8 talc-2 clay mixture to be used for an electrical insulator.

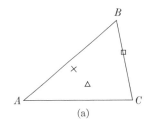
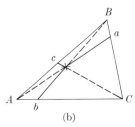

Fig. 10–8.5   Refer to Example 10–8.2. Each point represents a specific composition. Equation (10–8.1) applies.

Solution. Basis: 100 g starting material = 80 g talc + 20 g clay.
a) From Eq. (10–8.2),

$$\frac{80 \text{ g talc}}{(3)(24.3) + 4(28.1) + (10)(16.0) + 2(17.0)} = \frac{y \text{ g MgO}}{3(24.3 + 16.0)}$$

$$= \frac{z \text{ g SiO}_2}{4(28.1 + 32.0)} .$$

$y = 25.4$ g MgO = 33.4 w/o ⎱
$z = \underline{50.7}$ g SiO$_2$ = 66.6 w/o ⎰   See the point marked ☐ in Fig. 10–8.6.

  76.1 g dissociated talc

b) From Eq. (10–7.1),

$$\frac{20 \text{ g clay}}{(2)(27.0) + 2(28.1) + 5(16.0) + 4(17.0)} = \frac{w \text{ g Al}_2\text{O}_3}{2(27.0) + 3(16.0)}$$

$$= \frac{x \text{ g SiO}_2}{2(28.1 + 32.0)} .$$

$w = 7.9$ g Al$_2$O$_3$ = 46 w/o ⎱
$x = \underline{9.3}$ g SiO$_2$ = 54 w/o ⎰   See the point marked ⬡ in Fig. 10–8.6.

  17.2 g dissociated clay

c) Total composition: Al$_2$O$_3$:   7.9 g = 8.5 w/o

                MgO: 25.4 g = 27.2 w/o

                SiO$_2$: (50.7 + 9.3) g = 64.3 w/o

                Total: 93.3 g

(See the point marked △ in Fig. 10–8.6.)

Comments. The location of the composition of the electrical insulator is 64.3% of the distance from the MgO-Al$_2$O$_3$ side to the SiO$_2$ apex; 8.5% of the distance from the MgO-SiO$_2$ side to the Al$_2$O$_3$ apex; and a little over one-fourth of the distance from the Al$_2$O$_3$-SiO$_2$ side to the MgO apex.

The total composition may also be located by the lever rule. The total composition lies on a line between dissociated talc, ☐, and dissociated clay, ⬡, such that 76.1 g of dissociated talc (81.5 w/o), and 17.2 g of dissociated clay (18.5 w/o) balance their lever at △. ◀

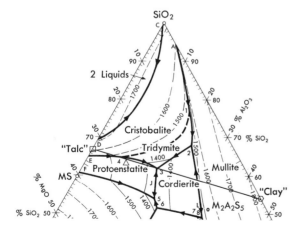

Fig. 10–8.6 Steatite porcelains. Electrical porcelains made out of 80% talc and 20% clay have the composition indicated by the △. (See Example 10–8.3 and Fig. 10–8.2.) The quoted compositions of talc, ☐ and clay, ⬡ , are the compositions after dissociation (Eqs. 10–8.1 and 10–7.1).

10–9 STRUCTURAL CERAMICS ⓔ

The oldest ceramic products are bricks. Sun-dried clay in the form of adobe blocks were used in prehistoric times. These products were improved when a reinforcing material was added. We know from the Bible that straw was used as an additive in early times.*

Present-day bricks are made at the rate of hundreds per minute by extrusion augers, and with close control of temperature during firing. Although the product serves some of the same structural purposes as the early brick, the specifications for their properties are much more stringent.

The ability to stand up under static loading is the most common service requirement for nonmetallic, inorganic (i.e., ceramic) construction materials. Other requirements vary with individual engineering design, but may include demands that the material be able to (a) withstand intermittent dynamic loading (e.g., concrete roadways); (b) that it act as thermal insulation, as in fiber products; (c) that it be inert to various environments; (d) that it resist freezing and thawing; and even (e) that it transmit light (e.g., window glass). Usually cost is also a significant factor; and, of course, installation and maintenance costs must be incorporated with original costs of the material and pro-rated over the period of useful life.

Vitrified products. This category of ceramic products includes a wide variety of clay-base materials such as brick, tile, and whitewares. The processing methods which we shall discuss in Chapter 11 involve the formation of a glassy or vitrified bond within the product. In these cases, the firing temperatures and times are such that melting starts and a silicate glass is formed, which wets the remaining solids. During subsequent cooling, the glass becomes a rigid (but still noncrystalline) solid that gives coherency to the product. In addition, the glass formation reduces porosity and permeability, which gives the product better resistance to environmentally caused deterioration. These consequences of changes in porosity are reflected in Table 10–9.1. Ideally, any vitrified product should contain a significant amount of glass to ensure high strengths and other mechanical properties. However, excessive glass content usually increases the brittleness and thereby reduces the strength. Excessive glass also makes possible viscous flow during firing (Chapter 11), making dimensional specifications more difficult to maintain.

Cements and concretes. In contrast to vitrified ceramic products, which gain their mechanical strengths due to the presence of glass, concretes

---

*   Exodus 5: 7.

Table 10–9.1

PHYSICAL SPECIFICATIONS OF BUILDING BRICK (ASTM C 62–69)

Grade	Minimum compression strength		Maximum H₂O absorption	
	Average of 5 bricks, psi	Of one brick, psi	Average of 5 bricks, %	Of one brick, %
SW*	3000	2500	17	20
MW**	2500	2200	22	25
NW***	1500	1250	No limit	No limit

* SW: Brick intended for use where a high degree of resistance to frost action is desired, and the exposure is such that the brick may be frozen when permeated with water.
** MW: Brick intended for use where exposed to temperatures below freezing, but unlikely to be permeated with water, or where a moderate and somewhat uniform degree of resistance to frost action is needed.
*** NW: Brick intended for use as backup or interior masonry.

gain their strengths because of the formation of hydrated phases, or related compositions such as carbonates. These phases are commonly colloidal in character, i.e., they are made up of extremely fine particles with large surface areas. The chemistry of their formation will be discussed in Chapter 13. Let it suffice here to note that the greatest degree of strength is attained by concrete, which has a maximum packing factor for the aggregated components, and with minimum void space within the cement addition (Section 13–4).

Example 10–9.1. The water absorption referred to in Table 10–9.1 is measured by determining the volume of open pores relative to the total volume. Show how this information can be obtained with three weighings of the material M: (1) dry, M_{dry}, (2) suspended in water, M_{sus}, and (3) saturated with water, M_{sat}.

Answer. Use metric units:

Volume of pores $= [M_{sat} - M_{dry}]/[1 \text{ g/cm}^3\text{H}_2\text{O}] = \text{cm}^3 \text{ H}_2\text{O in pores.}$

Total volume $= [M_{sat} - M_{sus}]/[1 \text{ g/cm}^3 \text{ H}_2\text{O}]$

$=$ water displaced by the volume of (brick + open pores)

$= \text{cm}^3$ total volume;

% porosity $= (M_{sat} - M_{dry})/(M_{sat} - M_{sus}).$

Comments:
Bulk density is mass/(true volume + closed pores + open pores).
Apparent density is mass/(true volume + closed pores).
True density is mass/true volume.
Open porosity is open-pore volume/total volume.
True porosity is (open-pore + closed-pore volume)/total volume. ◀

REVIEW AND STUDY

10–10 SUMMARY

Ceramic materials encompass compounds containing both metallic *and* nonmetallic elements. Due to the attraction of unlike charges, ionic bonds provide the coherence between these elements. This promotes close packing, with the coordination number dependent on the ratio of the radii of the oppositely charged ions. The basic structural types we considered were exemplified by NaCl, CsCl, ZnS, CaF_2, Cr_2O_3, $BaTiO_3$, and $NiFe_2O_4$. Solid solutions exist in ceramics as in metals, but with the added requirement that the positive and negative charges must balance.

Silicates contain partially covalent bonds. As a result, their structures may contain polymer-like units, including chains, sheets, and networks. The majority of commercial glasses are amorphous networks of silicate tetrahedra, modified by alkalies and other metallic oxides.

Some of the more widely used ceramic materials fall in the categories of refractories, abrasives, insulators, piezoelectrics, and structural ceramics.

Terms for Study

Abrasives	Coordination number, CN
Amorphous solids	Covalent bond
AX compounds	Cr_2O_3-type structure
A_mX_p compounds	CsCl-type structure
$A_mB_nX_p$ compounds	Defect structure
$BaTiO_3$-type structure	Density, apparent
CaF_2-type structure	Density, bulk
Ceramic materials	Density, true
Chain silicates	Electric dipole

Electronic conduction	Porosity, open
Glass	Porosity, true
Insulator	Radius ratio, r/R
Ion vacancy, □	Refractories
Ionic compounds	Sheet silicates
NaCl-type structure	Silicones
Network modifiers	Soda-lime glass
Network silicates	Transducer
$NiFe_2O_4$-type structure	Triaxial diagram
Piezoelectrics	ZnS-type structure
Polarization, $\mathscr{P}$	

STUDY PROBLEMS

10–1.1 Estimate the density of MnS, based on the approximate ionic radii of Table 10–1.1. (MnS has the NaCl-type structure.)

Answer. 3.9 g/cm^3

10–1.2 Experimental data indicate that the density of MnS is 3.98 g/cm^3. What is the size of the unit cell edge?

Answer. 5.26 Å [*Note:* This is slightly less than value of $(2r + 2R)$ obtained from Table 10–1.1. As indicated in the footnote to that table, however, the values were obtained indirectly and may be off by 0.01 or 0.02 Å.]

10–1.3 Examine Fig. 10–1.3 closely. What is the radius of the largest "spherical hole" in ZnS if the radii of Table 10–1.1 are assumed?

Answer. 1.03 Å

10.1.4 (a) What is the ratio of the ionic radii in MgO? (b) How far apart are the centers of the nearest Mg^{2+} ions? (c) How far apart are the centers of the *second*-nearest Mg^{2+} ions? O^{2-} ions? (MgO has the NaCl structure of Fig. 10–1.1.)

Answer. (a) 0.47 (b) 2.92 Å (c) 4.2 Å

10–1.5 Calculate the minimum r/R ratio for CN = 8. [*Suggestions:* Refer to Fig. 10–1.2 and Example 10–1.2.]

10–1.6 Using geometry, we can calculate the body diagonal of a cubic unit cell as being equal to $\sqrt{a^2 + a^2 + a^2}$, or $a\sqrt{3}$. Observe in Fig. 10–1.3 that $4(r + R)$ also equals the body diagonal for the ZnS structure $[4(r + R) = a\sqrt{3}]$. The face diagonal of a cube equals $a\sqrt{2}$. Show that the minimum ratio of the radii for CN = 4 in the structure of Fig. 10–1.3 is 0.225.

10–1.7 Examine Fig. 10–1.3 closely. One form of SiC has the same structure. What is the closest approach of two silicon atoms in SiC, given that the radii of silicon and carbon atoms are 1.17 Å and 0.77 Å, respectively?

Answer. $D_{Si-Si} = 3.17$ Å, $d_{Si-Si} = 0.83$ Å

10–1.8 What is the packing factor for the ions in MgO?

Answer. 0.73

10–2.1 What is the coordination number for the F^- ion in CaF_2?

Answer. 4

10–2.2 What is the weight of aluminum in Al_2O_3?

Answer. 53 w/o

10–2.3 The lattice dimension a of fluorite (CaF_2) is 5.47 Å. What is the density of fluorite?

Answer. 3.18 g/cm^3

10–2.4 Can MgF_2 have the same structure as CaF_2?

10–3.1 The density of cubic $BaTiO_3$ is 6.08 g/cm^3. What is the lattice dimension of the unit cell?

Answer. 4.00 Å

10–3.2 If the Ba^{2+} and O^{2-} ions of Fig. 10–3.1 "touch," what is the radius of the center "hole" into which the Ti^{4+} ion must fit?

Answer. 0.60 Å

10–3.3 The chemical formula for magnetite, the mineral lodestone, is Fe_3O_4. Its composition can also be written as $Fe^{2+}Fe_2^{3+}O_4$. This structure is of the $NiFe_2O_4$ type. Its density is 5.2 g/cm^3. What is the center-to-center distance of the oxygen ions? Surface-to-surface?

Answer. $D_{O-O} = 3.0$ Å, $d_{O-O} = 0.2$ Å

10–4.1 Although the radii are the same, an Li^+ ion cannot replace an Mg^{2+} ion directly because of their charge difference. However, an Li^+ can replace an Mg^{2+} ion if other means are available to balance the charge difference. Suggest a possibility which uses an Li^+-Fe^{3+} pair.

10–4.2 What should the weight ratio be for NiO-MgO to have equal numbers of Ni^{2+} and Mg^{2+} ions?

Answer. 65 NiO/35 MgO

10–4.3 An iron oxide contains 25 w/o oxygen (and 75 w/o iron). What is the w/o magnetite (Fe_3O_4) at 1200°C? 600°C? 400°C?

Answer. 0 w/o Fe_3O_4 35 w/o   90 w/o

10–4.4 A ceramic semiconductor must have a 12/1 ion ratio of Fe^{2+}/Fe^{3+}. Is this possible in an equilibrated $Fe_{1-x}O$ phase?

Answer. No; a 12/1 ratio would require 77.1 w/o Fe^{2+} (22.9% O^{2-}). This is above the solubility limit of iron in $Fe_{1-x}O$ (76.9 Fe at 910°C, according to Fig. 10–4.3).

10–5.1 How many negative charges must there be per silicon atom in the structures of the five figures of this section?

Answer. 4 (Fig. 10–5.1)   3 (Fig. 10–5.2)   2 (Fig. 10–5.3c)   1 (Fig. 10–5.4)   0 (Fig. 10–5.5)

10–5.2 The average O—Si—O angle in silica is 109.5°. Estimate the closest oxygen-to-oxygen distance. [With CN = 2, $R_0 \approx 1.17$ Å.]

Answer. 2.53 Å (center-to-center)

10–6.1 An optical glass has an index of refraction of 1.601 and an absorptivity of 0.001/cm. (a) What is the light's velocity within the glass? (b) What percentage of the light is absorbed as it passes through 0.51 cm of the glass? (Refer to the discussions in Section 9–6.)

Answer. (a) 1.873×10^{10} cm/sec (116,000 mile/sec)   (b) 0.56% (absorbed)

10–6.2 A glass contains 75 w/o SiO_2 and 25 w/o CaO. What fraction of the oxygen atoms serve as bridges between pairs of silicon atoms?

Answer. 0.70

10–6.3 A soda-lime glass contains 13 w/o Na_2O, 13 w/o CaO, and 74 w/o SiO_2. What fraction of the oxygens is nonbridging?

Answer. 0.30

10–6.4 Soda ash (Na_2CO_3) is the only carbonate added to the batch for a thermometer glass (Table 10–6.1). How many grams of CO_2 will be evolved per 100 g final glass product?

Answer. 7.1 g

10–6.5 Assume that 44 g of CO_2 (that is, 1 mole) produce 100,000 cm³ of gas at glass-melting temperatures. How many cm³ of CO_2 will be evolved per cm³ of the glass in Study Problem 10–6.2? (The density of the glass is 2.5 g/cm³.)

Answer. 1100 cm³ CO_2/cm³ glass.

10–7.1 Refer to Fig. 10–7.1. What range of ratios of Si/Al atoms can exist in mullite?

Answer. Si/Al = 0.27 at 76 w/o Al_2O_3; 0.21 at 80 w/o Al_2O_3

10–7.2 Refer to Fig. 10–7.2(a). The dark phase is cristobalite ($\rho = 2.3$ g/cm³). The light areas contain a eutectic mixture of cristobalite (SiO_2) and fayalite (Fe_2SiO_4). The eutectic combination has a density of approximately 4.0 g/cm², and was liquid at 1480°C (2700°F). (a) Estimate the volume percentage of cristobalite at 2700°F. (b) The weight percentage. (c) Based on the answer for (b), calculate the composition of Fig. 10–7.2(a).

Answer. (a) 67 v/o   (b) 53 w/o   (c) 72SiO_2-28FeO

10–7.3 Refer to Fig. 10–7.2(b). The light-gray phase is wüstite (FeO). The dark-gray phase was a liquid of FeO and SiO_2 at 1200°C (2190°F). Their densities are 5.7 g/cm³ and 5.0 g/cm³, respectively. (a) Estimate the volume fraction WUS and LIQ that were present at 1200°C. (b) The weight fractions. (c) Based on the answer for (b), calculate the composition of Fig. 10–7.2(b).

Answer. (a) 90 WUS-10 LIQ (b) 91 WUS-9 LIQ (c) $1.8SiO_2$-$98.2FeO$

10–8.1 A piezoelectric crystal has a Young's modulus of 10,000,000 psi. What stress must be applied to reduce its saturation polarization from 1.00×10^{-5} to 0.99×10^{-5} coul·cm/cm³?

Answer. $-100,000$ psi (compression)

10–8.2 Assume that you want to produce an electrical insulator containing $12MgO$-$24Al_2O_3$-$64SiO_2$. (a) Locate this composition on the MgO-Al_2O_3-SiO_2 triaxial diagram. (b) It is to be made by mixing MgO with a silicon-rich clay. What Al_2O_3/SiO_2 ratio should the clay have?

Answer. (a) Point ✳ on Fig. 10–8.2 (b) $27Al_2O_3$-$73SiO_2$

10–9.1 A test cylinder of concrete weighs 29.0 lb dry, 33.2 lb saturated with water, and 20.1 lb suspended in water. (a) What is the open porosity? (b) The apparent density? (c) The bulk density? (The density of water is 62.4 lb/ft³, or 1 g/cm³.)

Answer. (a) 32% (b) 203 lb/ft³ (c) 138 lb/ft³

10–9.2 Five test bricks have the following property data

	Compression strength, psi	Porosity, %
A	2600	22
B	2650	21
C	2550	26
D	2775	14
E	2900	10

What grade of specifications do these building bricks meet?

10–9.3 A brick weighs 6.6 lb dry, 6.9 lb when saturated with water, and 3.8 lb when suspended in water. (a) What is its open porosity? (b) Its bulk density? (c) Its apparent density?

Answer. (a) 9.7 v/o (b) 132 lb/ft³ (c) 147 lb/ft³

10–9.4 An insulating brick weighs 3.9 lb dry, 4.95 lb when saturated with water, and 2.3 lb when suspended in water. (a) What is its open porosity? (b) Its bulk density? (c) Its apparent density?

Answer. (a) 0.40 (b) 92 lb/ft³ (c) 152 lb/ft³

CHAPTER 11

PROCESSING
OF
CERAMICS

11–1 RAW MATERIALS ⓔ ⓜ Ⓜ Ⓞ

Historically clay has been the prime raw material for ceramics. Clays are usually hydroplastic, and therefore can be shaped into desired products. Articles made of dried clay have enough strength to be used in a number of applications. Clay products such as adobe blocks can be used directly for simple walls and comparable structures (provided they have a dry service environment). In prehistoric times it was discovered (probably accidentally, soon after fire began to be used) that clays become even stronger if they are heated to a "red" temperature. Thus the early "materials engineer" came into existence.

Unfortunately nonclay ceramics did not lend themselves as readily to accidental discovery. In fact, most of the modern-day ceramics had to await significant scientific and engineering advances before they could be developed. As a result, until fairly recently, ceramic materials made slow progress. In the past few decades, however, our knowledge of the structure of ceramic compounds and the advances of materials processing has jumped rapidly, until now the science of ceramics is one of the more advanced of the materials sciences.

In this section we shall examine three raw materials which are widely used in ceramic products: clays, carbonates, and Al_2O_3. In general, ceramic raw materials do not melt below 1000°C (or sometimes even 2000°C). Therefore melting is not a widely used manufacturing process (except for glass). Furthermore, none of the widely encountered ceramic phases have the plasticity common to many metals and polymers. Therefore mechanical deformation is not a widely used ceramic process either. These two limitations promote the use of particulate raw materials, i.e., powders, for direct compaction into the final product shapes. This demands much closer selection and control of the raw materials than is necessary for metals and polymers.

Fig. 11–1.1 Clay crystals. A very high magnification (×33,000) shows that clay crystals have a sheet structure. When the clay is wet and water molecules are absorbed between the sheets, clay is soft and plastic. (W. H. East, *J. Amer. Ceramic Soc.*)

Clays. In nontechnical terms, clays are very fine-grained, earthy particles. To the ceramist, a clay is a sheet-structured, layered silicate (Fig. 11–1.1). There are several clay phases. We shall, however, limit our discussion to *kaolinite* [$Al_2Si_2O_5(OH)_4$], which is the simplest and most widespread clay. Even so we shall not try to detail its structure. There are two important features, however, which can be appreciated from Fig. 11–1.1. The structure is a two-dimensional sheet. In fact it could be contended that each flake or sheet of clay in that figure is a large 2-D molecule. These sheets may be stacked into the third dimension. As in the case of polymeric molecules (Section 8–3), small molecules such as those of water may be absorbed between the sheets to plasticize these larger clay structures. This inclusion, shown schematically in Fig. 11–1.2, accounts for the *hydroplasticity* of wet clay. After the plasticized clay has been dried to remove the *inter*layer water, the adjacent sheets come into contact, providing a somewhat stronger bond.

The second important feature of clay arises from the extremely small size of the clay particles (<1 micron). At this size, the surface/volume ratio is high, so that surface adsorption becomes important. Observe, however, there are two types of surfaces in clay. The flat surfaces of the clay sheets are like the surfaces of molecules which

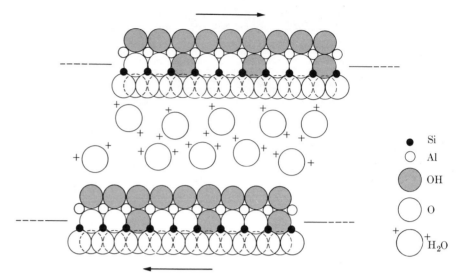

Fig. 11–1.2 Clay plasticized with water (schematic). The "bread" of this sandwich is a cross section of the clay crystals in Fig. 11–1.1. The adsorbed water between these crystals plasticizes the clay, just as micromolecules plasticize polyvinyl chloride (Section 8–3).

have no strong bonds that have been broken. On the other hand, the edges of the sheet expose broken Si-O bonds, and therefore carry a charge and adsorb ions. We shall discuss the importance of these surfaces briefly in the next section, when we talk about suspensions.

Clays dissociate on heating according to:

$$Al_2Si_2O_5(OH)_4 \xrightarrow{\text{heat}} Al_2O_3 + 2\,SiO_2 + 2\,H_2O\uparrow \qquad (11.1–1)$$

This occurs at about 1000°F (500–550°C), and water—usually called the *water of hydration*—escapes. However, water, as such, does not reside within the $Al_2Si_2O_5(OH)_4$ structure. As *dissociation* occurs, some transition structures form, but for the most part an amorphous, noncrystalline, glasslike material develops out of the SiO_2 and Al_2O_3:

$$Al_2O_3 + 2\,SiO_2 \rightarrow \text{glass}. \qquad (11–1.2)$$

This glassy material plays an important role in the firing of clay products (Section 11–3). Eventually mullite, $Al_6Si_2O_{13}$, forms according to the Al_2O_3-SiO_2 phase diagram (Fig. 10–7.1).

Carbonates. Soda ash (Na_2CO_3) and lime ($CaCO_3$) are two widely encountered carbonates that are used for raw materials. Each *calcines*— or dissociates on heating—to release CO_2 and provide quite pure sources of metal oxides. (The BaO for $BaTiO_3$ is obtained from $BaCO_3$.) Thus

$$BaCO_3 \xrightarrow{\text{heat}} BaO + CO_2\uparrow. \qquad (11–1.3)$$

Table 11–1.1

Al_2O_3 PRODUCTION FROM BAUXITE (ALUMINUM ORE)*

Step	Purpose
1. Dry and calcine bauxite at the mine	Reduces weight for shipping Removes any organic material
2. Grind to − 70 mesh	Accelerates the solution in Step 3
3. Dissolve (leach) in an NaOH solution [2–8 hours at 160°C (320°F) and 60 psi].	Forms soluble Na^+ and AlO_2^- ions, $Al(OH)_3 + NaOH \rightleftarrows$ $\qquad AlO_2^- + Na^+ + 2H_2O$
4. Settle any precipitate and insoluble residue, and filter	Removes $Fe(OH)_3$ precipitate and SiO_2
5. Cool to reverse the reaction of Step 3. [Fresh $Al(OH)_3$ particles are added to nucleate further $Al(OH)_3$ precipitation].	Precipitates $Al(OH)_3$
6. Filter and wash. (The NaOH solution is recycled to Step 3.)	Removes $Al(OH)_3$ and salvages NaOH
7. Calcine the $Al(OH)_3$ at 1150°C (2100°F)	Forms Al_2O_3

* Bayer process.

Those raw materials which do not exist in nature in sufficiently pure form for use in ceramic products may be purified by forming carbonates. The desired materials are first dissolved in appropriate solvents, and then precipitated as carbonates in a fine-grained usable form. When this procedure became available, there was a great step forward in the quality of various manufactured products.

Alumina (Al_2O_3). Probably the most widely used nonclay raw material for ceramic products is Al_2O_3. It originates as bauxite, the ore of aluminum metal, with the approximate composition of $Al(OH)_3$. However, it invariably contains undesirable impurities such as SiO_2 and Fe_2O_3. The purification procedure used in one of the major processes (Bayer) includes seven steps, as outlined in Table 11–1.1. Note that the chemistry in simple terms involves dissolving the desired product to remove it from an impure residue, then cooling to precipitate the desired material, in this case $Al(OH)_3$. This product must then be calcined to dissociate it to Al_2O_3 and H_2O:

$$2\,Al(OH)_3 \xrightarrow{\text{heat}} Al_2O_3 + 3\,H_2O\uparrow. \qquad (11\text{–}1.4)$$

The above process requires closer attention than simply dissolving, precipitating, and calcining, because, as we shall see shortly (Section 11–3), very fine particles of Al_2O_3 ($<1\mu$) produce a better product than coarser Al_2O_3. The process engineer must pay close attention to operating temperatures, compositions, pH, etc.

11–2 FORMING PROCESSES (PARTICLES)

Nonglassy ceramics are made by compacting particles into the desired shape and then sintering them into a coherent mass. Three shaping procedures are common: (1) pressure fabrication, (2) hydroplastic forming, and (3) slip casting.

Pressure fabrication. This procedure is widely used in the preparation of electrical ceramics, magnetic ceramics, and other nonclay products. It also serves as the basis for powder metallurgy (Section 6–8); however, in powder ceramics and powder metallurgy, there is a rather significant difference in processing. In many cases, metal particles making up the powders are deformed as they are compacted so as to densify the product. Except for a limited number of materials at high temperatures, ceramic powders do not compact by pressure alone beyond initial contact. To partially overcome this situation, the ceramist employs *sizing*, i.e., he uses coarse and fine particles to eliminate void space. This is shown schematically in Fig. 11–2.1. Maximum density occurs when there is just enough of the fine fraction to fill the spaces among the coarse particles.

Figures 11–2.2 and 11–2.3 sketch the two main procedures for applying pressure. The directional pressing of Fig. 11–2.2 is simpler in operation than hydrostatic pressing (commonly called *isostatic molding*), but is limited to simple shapes that have equal compaction ratios in all longitudinal sections. The hydrostatic molding of Fig. 11–2.3 is more complicated, but is preferred where possible because there is negligible wall friction to produce pressure gradients.

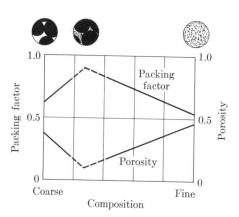

Fig. 11–2.1 Mixed sizing. Maximum density occurs when finer particles fill the spaces between coarser particles.

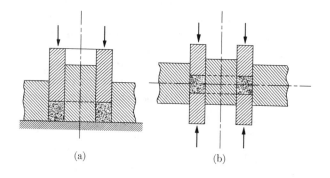

Fig. 11–2.2 Pressure fabrication (magnetic toroid). (a) Single-end pressing. (b) Double-end pressing. More uniform density is attained in double-end pressing because wall friction is less with the neutral plane across the center of the product. (See Fig. 11–2.4.)

(a)   (b)

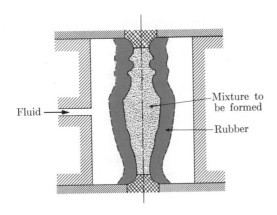

Fig. 11–2.3 Isostatic molding (spark plug manufacture). The hydrostatic fluid provides radial pressure to the product. Maximum compaction is possible with this procedure. (Jeffery, U.S. Patent 1,863,854)

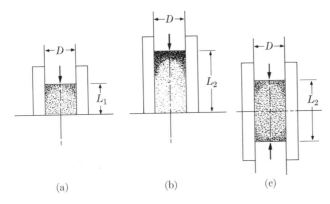

(a)   (b)   (c)

Fig. 11–2.4 Bulk density distributions. (a) $L/D = 0.9$. (b) $L/D = 1.8$. (c) $L/D = 1.8$, double-end pressing. More uniform density is obtained with the neutral plane across the center.

The importance of wall friction is shown in Fig. 11–2.4, in which bulk densities are shaded to indicate various parts of the compact. Because part of the compressive load is transferred to the mold wall, larger L/D ratios provide greater variations in applied pressures within the mold. Therefore the *double-end* compact of Fig. 11–2.4(c) with a smaller effective L/D ratio produces greater homogeneity and average densities than a single-end compact of the same dimensions (Fig. 11–2.4b).

Hydroplastic forming. Ceramic raw materials with layered structures, specifically clays, are amenable to plastic forming. Their moisture content is adjustable, so that they are plastic enough to be extruded without cracking, and also have enough yield strength so that the formed product will maintain its shape during handling and prior to drying. During hydroplastic forming, the layered colloidal particles are oriented by shear stresses. However, a direct comparison of hydroplastic ceramics and the more common metals cannot be made, because most metals have cubic crystals and several equivalent planes for slip. The plastic forming of two-dimensional ceramic particles must be compared more directly with the plastic deformation of zinc, magnesium, or other hexagonal metals with only one major slip plane. Because a preferred orientation is developed, it is expected that anisotropic behavior will occur with respect to drying, shrinkage, and various properties.

Slip-casting. The third forming process for particulate materials utilizes a suspension, or *slip*, of particles in a liquid (commonly water). The water is absorbed from the suspension into a porous mold leaving a dewatered "shell." When the semirigid shell is thick enough, the balance of the suspension is drained (Fig. 11–2.5). After time for additional dewatering, the mold is disassembled and the piece removed.

The primary advantage of slip-casting is that intricate shapes may be formed by this method. Further, it has an economic advantage when production involves only a few items, thus making it a favorite forming process for ceramic artists. It also has industrial applications when thin-walled products are to be made from nonplastic materials, as in the case of high-alumina combustion tubes and thermocouple tubes.

Among the important criteria for slip-casting is that the suspension remain *deflocculated*, i.e., dispersed within a minimum quantity of liquid. It is apparent that if the particles were to settle, the walls of the product would not be uniform. Second, if the water content is high,

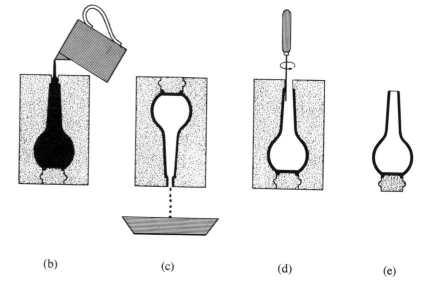

Fig. 11–2.5 Slip-casting. A slip—i.e., a semi-fluid suspension—is poured into a porous plaster mold. The mold absorbs water, producing a solid shell of the product, which is dried and fired. (F. H. Norton, *Elements of Ceramics*, Reading, Mass.: Addison-Wesley)

(a)   (b)   (c)   (d)   (e)

more time is required for processing, and more shrinkage results. The above criteria are controlled through surface charges on the particles. With adsorption and the broken bonds described in the discussion of clay in the preceding section, it is possible to provide all the surfaces with like electrical charges and introduce mutual repulsion between particles. This promotes deflocculation, or dispersion of particles and avoids clumping and settling. The ceramist tries to do more than this, however. By adjusting the pH of the suspension and the adsorbants, he can effect a "card deck" type of structure as the slip is dewatered, rather than a "card house" type of structure (Fig. 11–2.6). The latter retains considerable water, producing a weak shell and permitting excessive shrinkage during drying.

Drying. m̄ The initial product after hydroplastic forming, or after slip-casting, has a high moisture content. When the product is fabricated by means of pressure, it usually contains small amounts of liquid or fugitive lubricants and binders. Although these need not be water, they must be removed.

To be economical, drying, as an engineering process, must be rapid. At the same time, it must not be *so* rapid that the product is damaged by cracks or warping as a result of changes in volume. To design an efficient drying process, one must be familiar with the distribution of the liquid within the product, and also with the rate of movements of the liquid. With this knowledge, one can predict various dimensional and property changes with improved accuracy.

Water is the liquid most commonly used in ceramic production; therefore we shall give it most of our attention. However, the general principles which apply to water also apply to other liquids, such as oil, carbon tetrachloride, etc. Water may be considered a fugitive vehicle because, after it has been used for purposes of suspension or plasticity, it can readily be removed by evaporation, due to its relatively high vapor pressure.

Water distribution can be categorized into several types. (Not all types are present in all ceramic products prior to drying.) These categories include: (1) water of suspension, (2) interparticle or interlayer water, (3) pore water, and (4) adsorbed water. We shall not talk further about the water of suspension, because it must be removed during, or prior to, the formation of the shape of the ceramic product.

Figure 11–2.7(a) shows interparticle (or interlayer) water in products made of claylike materials. After the water of suspension has been removed by filtration, decantation, or by the absorption by the mold, a considerable quantity of water in a relatively free state can remain between the particles. This interlayer film may be as much as

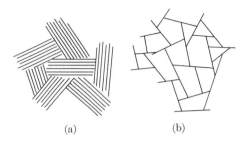

(a)   (b)

Fig. 11–2.6 Flocculation of clay particles. (a) Face-to-face (card-deck) flocculation. (b) Edge-to-face (card-house) flocculation. The former produces a denser product. To obtain this, the ceramist must control particle size and the pH of the slip.

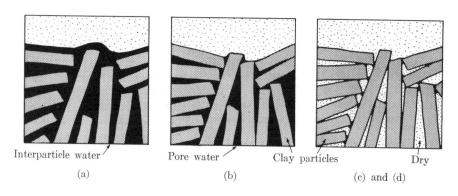

Fig. 11–2.7 Water in wet clay. Shrinkage occurs as the interparticle water is removed (*a → b*). Further drying removes the pore water and introduces porosity (*b → c*). The final adsorbed water on the surfaces of the clay particles is tightly held. (After F. H. Norton, *Elements of Ceramics,* Reading, Mass.: Addison-Wesley)

Interparticle water   Pore water   Clay particles   Dry

(a)   (b)   (c) and (d)

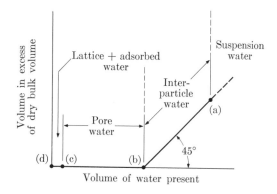

Fig. 11–2.8 Shrinkage versus water content (schematic). The letters refer to Fig. 11–2.7. (After F. H. Norton, *Elements of Ceramics,* Reading, Mass.: Addison-Wesley)

500 Å thick, which is comparable to the colloidal dimensions of some ceramic raw materials. When the water is removed, the layered particles move closer together (probably by capillary action) and the shrinkage of the previously formed product is readily apparent. During this stage of drying, the volume shrinkage directly equals the volume of interparticle water removed [Fig. 11–2.8(a) to (b)].

Figure 11–2.7(b) shows pore water as the water in the interstices among the particles after the particles have come into contact. Removal of this water produces very little if any additional shrinkage [Fig. 11–2.8(b) to (c)]. In fact, in special cases, there is even some evidence that a slight expansion occurs because the capillary forces of the liquid disappear.

Hydroplastically formed products generally contain both inter-particle and pore water because they usually contain clays or other two-dimensional raw materials. This is also true for slip-cast ceramics. Pressure-fabricated ceramics contain very little if any interparticle water, and therefore do not exhibit the large shrinkage depicted in Fig. 11–2.8. This factor constitutes an advantage for pressure molding which partially offsets its other limitations.

Adsorbed water is that water which is held by surface forces. It is usually only a few molecules thick; however, its quantity can be significant in ceramic products manufactured from colloidal-sized particles (see Study Problem 11–2.4). This water usually requires drying temperatures in excess of 100°C ($>$212°F) because the molecules are adsorbed to the surface more tightly than they would be bonded to other water molecules of the liquid during normal evaporation.

The drying process requires a movement of liquid from the interior of the shaped product to the surface, where evaporation can occur. This movement is highly sensitive to temperature; in fact, it increases by a factor of four as the temperature increases from 20°C (68°F) to 100°C. Let us tie this fact in with the data of Fig. 11–2.8. If the surface is heated and dried while the center is still cool, the surface will shrink [(a) to (b) of Fig. 11–2.8] around the center before the latter shrinks. In contrast, if the entire moist product is heated to ~205°F (96°C) in a humid atmosphere before drying is started, moisture can diffuse outward nearly four times as fast while drying occurs. This procedure, called *humidity drying*, produces a much lower moisture (and dimensional) gradient between surface and center. Cracking and warping are significantly reduced by using this procedure.

Example 11–2.1. Silica sand (~ 1.0 mm diameter) and silica flour (~ 0.01 mm) are to be mixed to maximum density. The former has a *bulk* density of 1.6 g/cm^3, the latter 1.5 g/cm^3. (a) How much of each should be used? (b) What is the maximum density possible? [Both silica sand and silica flour (finely crushed and powdered sand) are quartz, with a true density of 2.65 g/cm^3.]

Solution. Basis: 1 cm^3 of sand = 1.6 g sand.

Actual volume sand = $(1.6 \text{ g})/(2.65 \text{ g/cm}^3) = 0.60 \text{ cm}^3$.

    Pore space = 0.40 cm^3.

Fill the pore space of the sand with silica flour.

Weight flour = $(0.40 \text{ cm}^3)(1.5 \text{ g/cm}^3) = 0.6 \text{ g}$.

a) 1.6 g sand to 0.6 g flour.

b) Bulk density = (1.6 g sand + 0.6 g flour)/1 cm^3

      = 2.2 g/cm^3.

Comments. The actual volume of the flour is $(0.6 \text{ g})/(2.65 \text{ g/cm}^3)$ or 0.226 cm^3. Therefore the overall packing factor is $(0.60 + 0.226)/1$ cm^3, or 0.83. ◀

Example 11–2.2. [m] An extruded ceramic insulator which initially weighs 1.90 g/cm^3 shrinks 6.3 l/o (dry basis) during drying. The dry product is 1.75 g/cm^3. (a) What was the weight fraction of interparticle water in the extruded shape? (b) What is the volume fraction of open porosity in the final dried product?

Solution. Basis: 1 cm^3 of dried product = 1.75 g of dried clay.

Volume before drying = $(1.00 \text{ cm} + 0.063 \text{ cm})^3 = 1.20 \text{ cm}^3$.

Weight before drying = $(1.20 \text{ cm}^3)(1.90 \text{ g/cm}^3) = 2.28 \text{ g}$.

Total water = (2.28 g − 1.75 g) = 0.53 g = 0.53 cm^3.

Interparticle water (from Fig. 11–2.8) = $(1.20 \text{ cm}^3 - 1.00 \text{ cm}^3)$

             = 0.20 cm^3 = 0.20 g.

a) Weight fraction interparticle water = 0.20 g/2.28 g = 0.088.

b) Open porosity = (pore volume)/(total volume)

= (0.53 cm³ − 0.20 cm³)/(1.0 cm³) = 0.33.

Comments. The dry weight and dimensions are commonly used as a basis because they are reproducible for sample-to-sample comparison. ◀

11–3 BONDING PROCESSES

Since the majority of ceramic products are made by agglomeration, the mechanism of particle bonding is important. Two prime procedures are used: *sintering* and *cementation*. In this section we shall pay attention only to the former. Sintering is performed at elevated temperatures. Cements will be discussed in Chapter 13, since they are widely used in concrete, which may be viewed as a composite material.

Solid sintering. The use of heat to bond particles is called sintering, or *firing*. In principle the process of sintering is straightforward. Two particles in contact form one boundary rather than two surfaces (Fig. 11–3.1). The driving force for this comes from the surface atoms; having neighbors on only one side, they possess more energy than atoms that have a full complement of neighbors. (Recall that energy is *given off* when atoms become coordinated with other atoms.)

The actual mechanism for sintering is somewhat more complicated. Atoms must move from the points of particle contact and diffuse to the surfaces where contact does not exist. The most common route of diffusion is through the crystalline solid. Another route may be along the boundaries and surfaces, which is the path taken by oxygen ions during the sintering of Al_2O_3. We shall not dwell longer on the mechanism of sintering, except to note that many modern ceramics —such as UO_2 for nuclear fuel elements, and Al_2O_3 envelopes for solium-vapor lights—cannot be produced without an extensive knowledge of the factors which modify sintering.

Firing shrinkage. Extremely high-pressure techniques for powder compacts cannot produce full density, i.e., zero porosity. We depend on sintering to bring about the final pore removal. The centers of the particles move closer together as atoms or ions move from the points of contact to the surfaces where contact does not occur. This produces shrinkage,* which is a "necessary evil" in sintering processes. Allowances must always be made in production for the correct amount of shrinkage. (See Example 11–3.2.) When firing shrinkage is added to

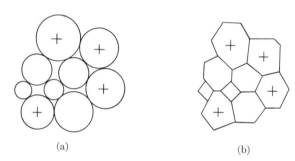

Fig. 11–3.1 Solid sintering (schematic). (a) Particles before sintering have two adjacent surfaces, each with high energy. (b) Grains after sintering have one, lower-energy boundary (cf. Fig. 6–8.2).

---

*   In a few materials with high vapor pressures—for example, NaCl— sintering can occur by volatilization from the free surfaces and condensation adjacent to the points of contact to enlarge the contact area. This does not produce shrinkage, and furthermore it does not reduce the total porosity.

the drying shrinkage described in the preceding section, the ceramic engineer finds it mandatory to maintain extremely close control over his processing variables. These include moisture content, particle sizes, drying rates, pressures, and firing temperatures and times.

Vitreous sintering. Ⓜ ⓜ Ⓞ Siliceous materials normally produce glass during the firing operation. Such materials include the multitude of clay-based ceramics, and, as indicated by Eq. (11–1.1), the dissociation products of clay form a glass. We shall see in the next section that glass responds to viscous deformation. Thus, at firing temperatures, the capillary tension arising from the surface energies gradually shrinks the product, reducing porosity. This, plus the hydroplastic forming discussed in Section 11–2, enabled the early skilled craftsmen to produce intricate, impermeable ceramic products even though they had negligible scientific knowledge of the accompanying structural changes. However, a knowledge of the structural changes can be both useful and interesting to the engineer.

Figure 11–3.2 is a plot of the corner of the K_2O-Al_2O_3-SiO_2 triaxial diagram. The older electrical porcelains, called *triaxial porcelains*, are made with three main components: clay [$Al_2Si_2O_5(OH)_4$ calcined to $46\ Al_2O_3$-$54\ SiO_2$], feldspar [$KAlSi_3O_8$], and quartz [SiO_2]. Their compositions are indicated by C, F, and Q, respectively. The total composition must lie inside a triangle with those compositions at the apexes. The firing reactions can be deciphered from the

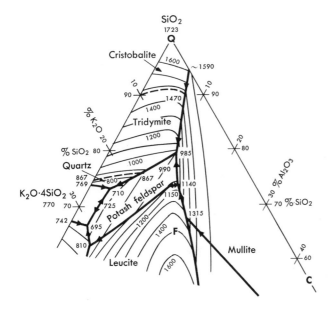

Fig. 11–3.2 Reactions during firing (electrical porcelain). The conventional electrical porcelain contains clay (**C**), powdered quartz (**Q**), and feldspar (**F**) as raw materials. These are located within the K_2O-Al_2O_3-SiO_2 system. The materials react slowly, even at firing temperatures, giving mullite and glass. (See Fig. 11–3.3.) (Adapted from *Phase Equilibria Among Oxides of Steelmaking*, A. Muan and E. F. Osborn, Reading, Mass.: Addison-Wesley (1965), page 90.)

Crystalline phases	
Notation	Oxide formula
Cristobalite ⎫ Tridymite ⎬ **Q**—Quartz ⎭	SiO_2
Corundum	Al_2O_3
Mullite	$3Al_2O_3 \cdot 2SiO_2$
F—Potash feldspar	$K_2O \cdot Al_2O_3 \cdot 6SiO_2$
Leucite	$K_2O \cdot Al_2O_3 \cdot 4SiO_2$
C—Clay (dissociated)	$Al_2O_3 + 2SiO_2$

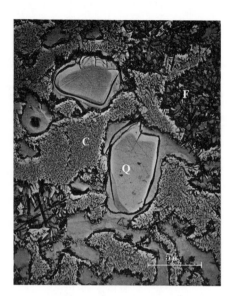

Fig. 11–3.3 Electrical porcelain microstructure (×3000). Although not attained, reactions are moving toward the equilibrium. The former clay (**C**) has formed mullite and glass because it contains about $60SiO_2$-$40Al_2O_3$ (Fig. 10–7.1). The former feldspar (**F**) contains more glass and larger mullite crystals because it contains considerable K_2O. The original quartz grains (**Q**) are only slightly altered around the edge by the glassy liquid. Since the original particle sizes were very small (<1 micron) and the glass was semifluid at high temperatures, the pore space was eliminated during firing. (Courtesy S. T. Lundin, Technological University of Lund, Sweden)

microstructures shown in Fig. 11–3.3.* (1) The clay has changed completely to glass and small growing mullite ($Al_6Si_2O_{13}$) crystals, the phase we would expect on the basis of the Al_2O_3-SiO_2 diagram (Fig. 10–7.1). (2) The feldspar is a flux which produces considerable glassy liquid. Furthermore the K^+ ions, like the Na^+ ions in Fig. 10–6.2, make the glass more fluid than the glass from clay alone. Crystals grow readily in this liquid, to give large mullite needles in original feldspar areas. For this to occur a small fraction of the K^+ ions must diffuse away from these areas. (3) The quartz reacts only as fast as it is fluxed by the diffusing K^+ ions. Consequently it remains almost intact, providing stability in volume during the firing process. The silica-rich liquid around the quartz grains crystallizes extremely slowly, remaining as a crystal-free glass.

It would take a long time for the diffusion and crystallization required for complete homogenization to take place. As a result the phase diagram of Fig. 11–3.2, which assumes equilibrium, cannot be used quantitatively, but can be used to predict the direction of reaction.

Example 11–3.1. ⓔ ⓜ Ⓜ Ⓞ Surface energy is commonly expressed in ergs/cm². The surface energy of an oxide powder is 1000 ergs/cm², that is, 10^{-4} joules/cm². The energy of the grain boundary is 550 ergs/cm². How much energy, Γ, is released per cm³ when a powder (assume cubes) of a 2-micron size is sintered? (Assume that there is no grain growth.)

Solution. Basis: one cubic particle.

Initial surface energy $= 6(2 \times 10^{-4}$ cm)²(1000 ergs/cm²)/(2 $\times$ 10^{-4} cm)³

$$= 3.0 \times 10^7 \text{ ergs/cm}^3 \text{ (or 3 joules/cm}^3).$$

Final boundary energy

$$= (6/2)(2 \times 10^{-4} \text{ cm})^2(550 \text{ ergs/cm}^2)/(2 \times 10^{-4} \text{ cm})^3$$
$$= 0.82 \times 10^7 \text{ ergs/cm}^3.$$
$$\Gamma = (3.0 - 0.8)(10^7) = 2.2 \times 10^7 \text{ ergs/cm}^3$$
$$\text{(or 2.2 joules/cm}^3).$$

Since 1 cal $= 4.185 \times 10^7$ ergs, $\Gamma \cong 0.5$ cal/cm³.

Comments. The (6/2) factor is required because each of the internal boundaries is shared by two grains.

Of course, the assumed cubic shape is not accurate. A corrected surface area value for actual grain shapes within a solid would be about 10% lower. ◀

---

* Since the structural dimensions are in the submicron region, this knowledge had to await modern technology and the development of the electron microscope.

Example 11-3.2. The final dimension for a ceramic component must be 0.75 in. Laboratory tests indicate that the drying and firing shrinkages are 7.3 l/o and 6.1 l/o, respectively, when based on the dried dimensions. (a) What is the required initial dimension? (b) What was the dried porosity if the final porosity is 2.2%?

Solution

a) 0.75 in. $= 0.939\ L_d$,

$$L_d = 0.80 \text{ in.}$$

$$L_i = 1.073\ (0.80 \text{ in.}) = 0.858 \text{ in.}$$

b) Basis: 1 cm³ fired product $= 0.978$ cm³ actual material.

$$1 \text{ cm}^3 \text{ fired} = (0.939)^3\ V_d$$

$$V_d = 1.21 \text{ cm}^3 \text{ dried.}$$

Porosity $= 100\% - \%$ space occupied by material

$$= 100 - 100(0.978 \text{ cm}^3 \text{ actual})/(1.21 \text{ cm}^3 \text{ dried})$$

$$= 19 \text{ v/o.}$$

Comments. This calculation assumes no density changes of the phases. In some cases appropriate corrections would have to be made. ◀

Example 11-3.3. ⓜ Ⓜ Ⓞ An electrical porcelain is made into an insulator from 50 w/o clay [$Al_2Si_2O_5(OH)_4$], 40 w/o potassium feldspar [$K_2Al_2Si_6O_{16}$], and 10 w/o silica flour (SiO_2). What is its fired composition? (a) Solve by computation. (b) Solve graphically.

Solution. Basis: 100 g $= 50$ g of clay (after dissociation we have 43 g of 54% SiO_2 and 46% Al_2O_3) $+ 40$ g $KAlSi_3O_8 + 10$ g SiO_2.

a) Computation (in grams):

	K_2O, g	Al_2O_3, g	SiO_2, g
Dissociated clay:	0	$0.46 \times 43 = 19.7$	$0.54 \times 43 = 23.3$
Feldspar:	$\left(\dfrac{94.2}{556.6}\right)(40) = 6.8$	$\left(\dfrac{102}{556.6}\right)(40) = 7.3$	$\left(\dfrac{360.6}{556.6}\right)(40) = 25.9$
Silica flour:	0	0	10.0
Total	6.8	27.0	59.2
Weight percent:	7.3	29.0	63.7

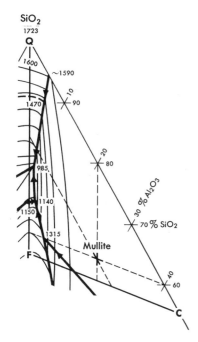

Fig. 11–3.4 Portion of K_2O-Al_2O_3-SiO_2 system. (See Example 11–3.3 and Fig. 11–3.2.)

b) Graphically:

Dissociated clay:	43 g = 46.2 w/o
Feldspar:	40 g = 43.0 w/o
Silica flour:	10 g = 10.8 w/o
	93 g

On Fig. 11–3.4 locate the point within the **F-Q-C** triangular region for the above ratio. This point corresponds to the 7-29-64 point in the overall K_2O-Al_2O_3-SiO_2 triangle, thus checking the computation.

Comments. The clay dissociation follows Eq. (11–1.1), in which 258 amu of $Al_2Si_2O_5(OH)_4$ loses 36 amu of H_2O to give 102 amu of Al_2O_3 and 120.2 amu of SiO_2.

The triaxial diagram is not limited to the corner oxides alone, but can be used for any combination of components that lie within the triangle. ◄

11–4 FORMING PROCESSES (VITREOUS)

Glasses and metals have only limited points in common. Both are melted and deformed during the course of their manufacturing processes. Beyond this point, however, one can't successfully draw comparisons. (1a) Metals are atomic liquids. (1b) Glasses retain much of their polymerized structure well above the melting temperature. (2a) Metals crystallize during cooling, and therefore have specific solidification temperatures, or temperature ranges. (2b) Below their theoretical melting temperatures, glasses are supercooled liquids, which continuously increase in viscosity until they become rigid at their glass transition temperature (Fig. 11–4.1). (3a) Metals deform by dislocation movements and resulting plastic slip along specific crystal planes. (3b) Glasses deform by viscous flow; consequently the flow is not direction-sensitive, but is highly time-temperature-sensitive.

Fig. 11–4.1 Glass transition temperature, T_g. Both organic glasses (Fig. 7–4.1) and the normal silicate glasses have a shrinkage slope discontinuity at T_g. Above this temperature the atoms can rearrange themselves by thermal agitation and respond to applied forces. Below this temperature, the materials are rigid. The transition temperature is somewhat lower when there is slow cooling because there is more time for rearrangements of the atoms.

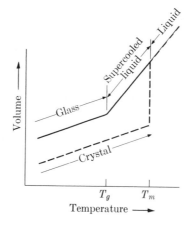

Table 11–4.1

VISCOSITIES OF NONCRYSTALLINE MATERIALS

Materials	Viscosity, poises*	Remarks
Air	0.00018	20°C
Pentane, C_5H_{12}	0.0025	20°C
Water	0.01	20°C
Phenol, C_6H_5OH	0.1	20°C
Syrup (60% sugar)	0.56	20°C
Oil, machine	1 to 6	20°C
Polymers at T_g	$\sim 10^{13}$	Glass temperature
Polymers at $T_g + 10°C$	$\sim 10^{10}$	
Polymers at $T_g + 20°C$	$\sim 10^{8}$	
Window glass at 515°C	10^{13}	
Borosilicate glass at 500°C	10^{13}	Annealing point
Fused silica glass at 1000°C	10^{13}	
Window glass at 700°C	10^{7}	
Borosilicate glass at 900°C	10^{7}	Working range (viscous side)
Fused silica at 1550°C	10^{7}	

* Poise = g/cm · sec

Glass viscosity. Glass is a vitreous material. Its viscosity, like the viscosity of polymers or of true liquids, is a measure of the amount of shear stress required for a specified shear velocity gradient. More briefly stated for our purposes, a liquid which flows readily has a low viscosity, η (Table 11–4.1). Note in this table that the viscosity drops off rapidly as the temperature T is raised. The relationship is

$$\log_{10} \eta = A + B/T, \qquad (11\text{–}4.1)$$

where A and B are constants for any given glass. This is shown graphically for several glasses on the semilog plot of Fig. 11–4.2. Note that the lower axis of the graph is the reciprocal temperature in accordance with Eq. (11–4.1); furthermore, the temperature which is used must be the absolute temperature in °K. [°K = °C + 273 = (°F + 460)/(1.8°F/°K).] Three viscosity values are important to us in vitreous deformation of glass: the *working range*, the *annealing point*, and the *strain point.**

Working range. This range is defined as temperatures between viscosity values of $10^{3.5}$ and 10^7 poises (Fig. 11–4.2). It is a wide range because

* Glass manufacturers also pay attention to the *melting range* (log η = 1.5 to 2.5) and to the *softening point* (log η = 7.5).

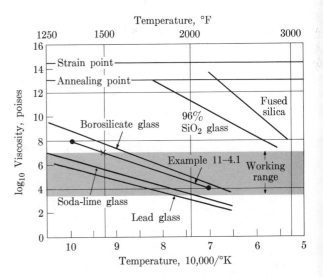

Fig. 11–4.2 Viscosity versus temperature. Since viscous flow is sensitive to thermal activation—i.e., sensitive to thermal agitation—the viscosity follows Eq. (11–4.1).

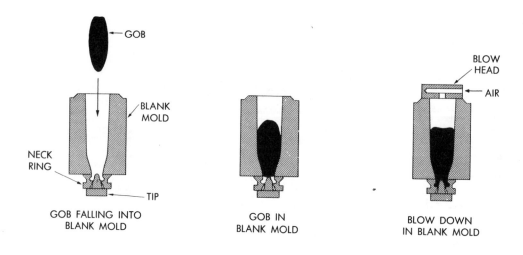

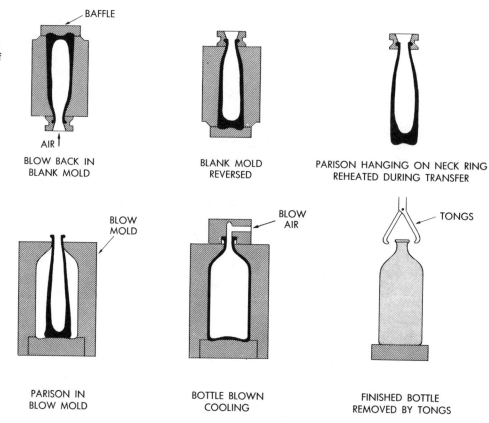

Fig. 11-4.3 Steps in the production of a glass bottle. Air is blown into the center of a "gob," or large drop of viscous glass. The blowing and mold manipulations are all automatic mechanical steps. (F. H. Norton, *Elements of Ceramics,* Reading, Mass.: Addison-Wesley)

working procedures vary from (1) *hot-pressing*, used to make products like the two halves of a glass building block, which requires a very viscous glass ($\eta \cong 10^7$ poises), to (2) the *blowing of thin glass envelopes for light bulbs* ($\eta \cong 10^4$ poises) at several thousand per hour. Figure 11–4.3 shows the procedure for producing glass containers; this process is the forerunner of the blow-molding of plastics (Fig. 8–4.3d). Intermediate glass viscosities are required here.

Annealing point. At 10^{13} poises, the movements of atoms within a glass are very sluggish; however, the atoms do move, and stresses can be relaxed within approximately 15 minutes (cf. Section 9–2). Therefore the temperature at which a glass has 10^{13} poises is called the *annealing point*. Almost all glass products are annealed after they are shaped, to remove those residual tensile stresses which might induce premature cracking.

Strain point. At $10^{14.5}$ poises, that is, 3.1×10^{14} poises, the movements of atoms in a glass are so sluggish that rapid cooling does not introduce new stresses. This is pertinent to the annealing process. If a glass is annealed at the annealing point and then quickly cooled, new residual stresses could form. However, if glass is cooled from the annealing point to the strain point slowly enough so that stresses are not introduced, subsequent cooling can be rapid.

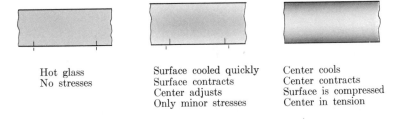

Hot glass
No stresses

Surface cooled quickly
Surface contracts
Center adjusts
Only minor stresses

Center cools
Center contracts
Surface is compressed
Center in tension

Fig. 11–4.4 Dimensional changes and induced stresses in tempered glass.

Tempered glass. (e) Residual stresses are commonly avoided, as we said above. However, if they are compressive and correctly distributed, the engineer can use them to advantage. Consider the rear window of an automobile. As the final step of production, it is heated above the annealing point, but not high enough to start deformation ($\sim 10^8$ poises). Then the glass is cooled rapidly. As the glass surface contracts, the interior is still hot enough for the atoms to adjust (Fig. 11–4.4). The surface, however, soon drops below the strain point; therefore the surface cannot adjust to the contraction at the center. As a result, the surface is placed under compression (and the center under tension). The product of this heat treatment is called *tempered glass*. Figure

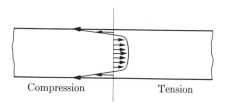

Fig. 11–4.5 Stress profile across the thickness of tempered glass. Since the surface has residual compressive stresses, the surface is protected from the initiation of cracks.

11–4.5 shows the stress pattern across the thickness of the glass. With the surface in *compression*, the glass is very strong because a tensile load must overcome this compression before the surface is sufficiently strained in tension to initiate a crack. The purpose of the annealing which was cited earlier was to *avoid* having residual *tensile* stresses at the surface, where cracking starts.

Example 11–4.1. The viscosity of a glass is 10^8 poises at 727°C and 10^4 poises at 1156°C. At what temperature would it be 10^7 poises?

Solution. Based on Eq. (11–4.1):

$$\log 10^8 = A + \frac{B}{1000°K} = 8;$$

$$\log 10^4 = A + \frac{B}{1429°K} = 4.$$

Solving simultaneously, we obtain:

$$A = 8 - \frac{B}{1000} = 4 - \frac{B}{1429},$$

$$B = 13{,}300.$$

$$A = 8 - \frac{13{,}300}{1000} = -5.3.$$

$$\log 10^7 = -5.3 + \frac{13{,}300}{T} = 7$$

$$T = 1080°K - 273 = 807°C \ (1485°F).$$

Comments. A graphical solution is possible if one uses an appropriate straight line on Fig. 11–4.2. (See dots.) ◀

11–5 CRYSTALLIZATION PROCESSES ⓜ Ⓜ Ⓞ

Interesting new materials have come forth in recent years which utilize *devitrification*, i.e., the crystallization of glass. In addition, a number of newly developed electrical and optical devices require *single crystals* of ceramic compounds. Each of these require attention to crystal nucleation and growth.

Devitrification. Glass is easily formed; it has zero porosity. Historically, it has been used in its amorphous, noncrystalline condition because of the above characteristics, and because it is transparent. When it is used in the above situation, crystallization is to be avoided. Isolated crystals mar the transparency. More important, any volume changes which accompany crystallization introduce possible stresses which can cause delayed cracking within the brittle glass product.

Fig. 11–5.1 Devitrified glass product (centrifugal pump). Nonporous glass is shaped by conventional glass-forming techniques. It is then heated (1) to an appropriate intermediate temperature, to nucleate crystallization, i.e., devitrification. (2) This is followed by a higher-temperature treatment to complete crystallization. The final product is stronger than glass, and is less porous (0%) than most sintered ceramic products. (Courtesy Owens-Illinois Glass Co.)

Glasses have been designed which can be heat-treated at temperatures in the vicinity of the glass transition temperature, and which crystallize uniformly and in a controlled manner from innumerable nuclei. Furthermore, if the composition is selected correctly, there is no change in volume between the glass at its transition temperature and the glass when it is a crystalline solid.

Glass-ceramics, as these devitrified glasses are sometimes called, have two chief merits. (1) The versatile processes of glass shaping can be used in commercial production (Fig. 11–5.1). (2) Glass-ceramics develop nonporous, fully dense ceramic products which do not soften just above the glass temperature, but remain rigid to within a few degrees of the melting temperature. A third advantage is also important. The compositions used in glass-ceramics of necessity have low thermal-expansion coefficients, so that heat-treating does not introduce stresses, a characteristic which is also important in the final product. Therefore we find glass-ceramics such as Pyroceram and Cervit used extensively in temperature-sensitive applications.

Single crystals. Products such as the synthetic ruby laser require single crystals. Ruby is Al_2O_3 containing approximately one percent Cr_2O_3 in solid solution. In order to grow a single crystal, the technician melts previously prepared powders of the correct composition in an oxygen-hydrogen flame. The small molten droplets fall on a preselected seed crystal which gradually grows into a large, single-crystal *boule* (Fig. 11–5.2).

There are several other procedures for growing single crystals, including growing them from water solutions at elevated temperatures

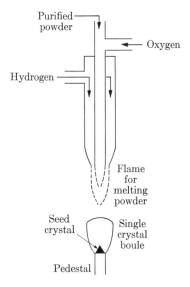

Fig. 11–5.2 Single crystal boules (sapphire, that is, Al_2O_3). Al_2O_3 powder is melted in an oxygen-hydrogen flame. The seed grows to a large single crystal.

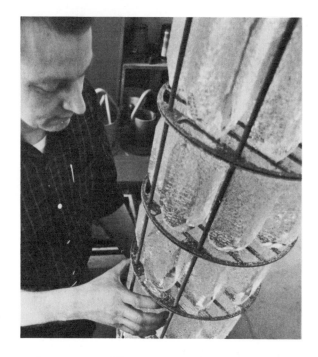

Fig. 11–5.3 Quartz crystals. These single crystals of quartz (SiO_2) were grown in an autoclave by precipitation from alkaline solutions. They will be used for electrical oscillators. (Courtesy Western Electric)

and pressures. Figure 11–5.3 shows quartz crystals grown in autoclaves. Slices of these are used in radio transmission to control the frequency.

REVIEW AND STUDY

11–6 SUMMARY

Ceramic raw materials include not only the clays, which have been used for a long time, but also oxides and other compounds that are the result of modern developments. Most ceramics are agglomerated from powders, so that sintering is the prime method of processing. The engineer must control the process to reduce the final porosity to near zero, and at the same time maintain appropriate control over the drying and firing shrinkage.

Glass is melted and shaped by viscous deformation. Since the viscosity of glass is highly sensitive to temperature, for production control the glass processor focuses his attention on temperature control. Glass production includes subsequent heat-treatment steps of (1) annealing, to remove uncontrolled and unwanted residual stresses, or

(2) tempering, to introduce desirable compressive residual stresses on the surface for strengthening purposes.

Two expanding fields of materials production are (a) devitrification of previously shaped glass products, and (b) the growth of single crystals.

Terms for Study

Absolute temperature, °K
Annealing point
Boule
Calcine
Clay
Deflocculation
Devitrification
Dissociation (by dehydration)
Drying
Double-end pressing
Firing
Flocculation
Glass-ceramics
Glass transition temperature, T_g
Humidity drying

Hydroplastic forming
Isostatic molding
Kaolinite
Porcelain
Pressure fabrication
Sintering, solid
Sintering, vitreous
Slip-casting
Slip (ceramic)
Strain point
Tempered glass
Triaxial composition
Viscosity, η
Vitreous
Working range (glass)

STUDY PROBLEMS

11–1.1 A bauxite ore contains 90% $Al(OH)_3$ and 10% impurities. If 85% of the $Al(OH)_3$ is recoverable, how many pounds of Al_2O_3 are available per 100 lb of ore?

Answer. 50 lb

11–1.2 The clay flakes in Fig. 11–1.1 average 0.5 microns (5×10^{-5} cm) in "diameter" and are 100 Å thick. What is the surface area per gram of the clay, given that the true density is 2.6 g/cm³?

Answer. 800,000 cm²/g

11–1.3 What percent weight loss occurs in the calcining (high-temperature dissociation) of (a) clay? (b) limestone ($CaCO_3$)? (c) $Al(OH)_3$?

Answer. (a) 14 w/o (b) 44 w/o (c) 35 w/o

11–1.4 Magnesite ($MgCO_3$) is a raw material used for making the MgO which is incorporated into high-temperature brick (> 3000°F). How much MgO will 1 ton (2000 lb) of magnesite make?

Answer. 960 lb

11–2.1 Refer to Example 11–2.1. Combinations of (a) 75 g sand–25 g silica flour, (b) 50–50, and (c) 25–75 are thoroughly mixed and compacted. What is the greatest bulk density possible for each of the combinations?

Answer. (a) 2.13 g/cm³   (b) 1.92 g/cm³   (c) 1.68 g/cm³

11–2.2 A dried, slip-cast product was 6% longer before drying. What was the v/o shrinkage (based on dried dimension)?

Answer. 19 v/o

11–2.3 A cylindrical extrusion should have a dried diameter of 1.05 in. A die which is 1.12 in. in diameter is required under normal operating conditions of 20 v/o interparticle water (dried basis). What dried diameter will result if 28 v/o water is used rather than the 20 v/o?

Answer. 1.03 in. diameter

11–2.4 The clay of study problem 11–1.2 is dried at 100°C to remove all the interparticle and pore water. Only adsorbed water remains. It is then heated to 200°C (392°F). An additional 2.4 g of water (adsorbed) were evaporated per 100 g of this "superdried" clay. How many square angstroms were there per molecule of adsorbed water?

Answer. 10 Å²

11–3.1 At elevated temperatures a glass fiber tends to shrink into a sphere as a result of its surface energy. What is the energy decrease per gram if 0.0009-in. glass fibers with a surface energy of 300 ergs/cm² (that is, 3×10^{-5} joules/cm²) spheroidize into 0.01-in. balls? The density is 2.5 g/cm³.

Answer. 1.8×10^5 ergs/g (or 0.018 joules/g)

11–3.2 A magnetic ferrite for an oscilloscope component is to have a final dimension of 0.621 in. Its volume shrinkage during sintering is 32.9% (unfired basis). What initial dimension should the powdered compact have?

Answer. 0.709 in.

11–3.3 A ceramic electrical insulator of Al_2O_3 weighs 1.23 g, occupies 0.47 cm³, and is 0.87 cm long before sintering. After sintering, it is 0.77 cm long. If the true specific gravity is 3.85, that is, 3.85 times as dense as water, what are the porosities (a) before, and (b) after sintering?

Answer. (a) 32 v/o   (b) 2 v/o

11–3.4 The composition of mullite is approximately $Al_6Si_2O_{13}$. This is frequently considered to be $3Al_2O_3 \cdot 2SiO_2$. What is the percentage of (a) Al_2O_3? (b) SiO_2? (c) Al? (d) Si? (e) O?

Answer. (a) 71.8%   (b) 28.2%   (c) 38%   (d) 13%   (e) 49%   (All w/o)

11–3.5 A ceramic wall tile contains 25 w/o mullite ($\sim Al_6Si_2O_{13}$) and 75 w/o glass of an average composition of $8K_2O-18Al_2O_3-74SiO_2$. How much clay, feldspar, and silica flour were required? (Solve graphically.)

Answer. Clay 60% (before dissociation), 30% feldspar, 10% silica flour

11–3.6 A ceramic product was made with 10 w/o magnesite ($MgCO_3$), 30 w/o talc ($Mg_3Si_4O_{10}[OH]_2$) and 60 w/o clay ($Al_2Si_2O_5[OH]_4$). Determine the final $MgO-Al_2O_3-SiO_2$ analysis by computation and graphically.

Answer. $17MgO-28Al_2O_3-55SiO_2$

11–4.1 From Fig. 11–4.2, calculate the values of A and B (of Eq. 11–4.1) for soda-lime glass.

Answer. $A = -4.5, B = 11,000$

11–4.2 The constants of viscosity for a glass are $A = -3.9$ and $B = 18,800$. (a) What is the temperature of the annealing point? (b) The strain point?

Answer. (a) 839°C (1542°F) (b) 748°C (1378°F)

11–4.3 The viscosity of a glass is 3100 poises at 1300°C (2372°F) and 10^8 poises at 800°C (1472°F). What is the viscosity at 1050°C (1922°F)?

Answer. 200,000 poises

CHAPTER 12

SEMICONDUCTORS

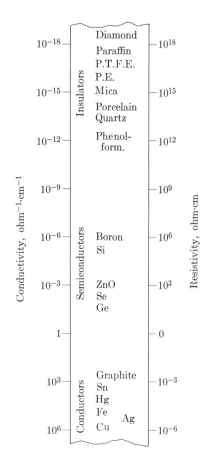

Fig. 12–1.1 Conductivity–resistivity spectrum. Materials in the intermediate range between 10^{-3} and 10^9 ohm · cm have become very important in modern technology as semiconductors.

12–1 CHARGE CARRIERS ⓜ

Metals have high conductivities, i.e., low resistivities, with $\rho < 10^{-3}$ ohm · cm. Insulators have high resistivities, $\rho > 10^9$ ohm · cm (Fig. 12–1.1). Many materials fall in the intervening gap, which covers 12 orders of magnitude. It was not too long ago that the materials, called *semiconductors*, which fell in this gap had little electrical use. People considered them to be neither "good" conductors to carry charges nor "good" insulators to isolate charges. Today we are all well aware that materials such as silicon which fall in this gap are now used extensively as electronic materials. A transistor is but one example of electronic devices which make use of semiconductors. Semiconductors are important enough to deserve a separate chapter in this text.

Semiconducting materials do not fit into our convenient categories of metals, polymers, and ceramics which were outlined in Chapter 1. They overlap these three categories for several reasons: (1) Semiconductors do not release their valence electrons as readily as normal metals do. (2) When covalently shared, the electrons are less strongly held than they are in most polymers. (3) Finally, unlike the case of idealized ionic compounds, the transfer of electrons in semiconducting compounds from cation to anion is not really complete.

It is these intermediate characteristics which are responsible for semiconduction. To illustrate, let us consider iron oxide, which contains both Fe^{2+} and Fe^{3+} ions. The former (ferrous iron) can be changed to ferric iron by releasing an electron:

$$Fe^{2+} \rightleftarrows Fe^{3+} + e^-. \tag{12–1.1}$$

The action is reversible. Thus in the oxide of Fig. 12–1.2, an electron can be pulled away from any of the Fe^{2+} ions.* Each electron has a

---

* In effect, however, an electron is not pulled away unless there is a nearby Fe^{3+} ion to receive it. The Fe^{3+} ion serves as a trap (it contains an *electron hole*) into which the electron can fall.

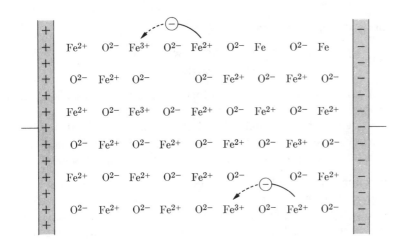

Fig. 12–1.2   Semiconduction in defect compounds $(Fe_{1-x}O)$ (cf. Fig. 10–4.1). Electrons can jump from an Fe^{2+} ion to an Fe^{3+} ion. This produces the same effect as a positive charge moving the opposite direction.

charge of 1.6×10^{-19} coul (or 1.6×10^{-19} amp · sec); therefore if there is a net movement in one direction, charge is carried and conduction occurs.

Electrons.   The *charge carriers* of interest to us in this chapter are the electrons.* The electrons are accelerated as they move toward the positive electrode, and lose velocity as they move away from it. Their *drift velocity*, $\bar{v}$, in an electric field $\mathscr{E}$ is an appropriate average of all movements.

It will be useful to think of *mobility* as it is defined in Eq. (2–8.3),

$$\mu = \bar{v}/\mathscr{E}.$$

Mobility μ is the drift velocity in cm/sec for each volt/cm of voltage gradient,

$$cm^2/volt \cdot sec = (cm/sec)/(volt/cm).$$

Mobility can be determined by experiment, as can the number of charge carriers there are per cm^3, n; therefore we can calculate the *conductivity* σ according to Eq. (2–8.1):

$$\sigma = nq\mu.$$

The term q is the charge per carrier, and since we are considering electrons as carriers, the value of q is 1.6×10^{-19} amp · sec.   The

---

*   In this text we shall not consider the charge carried by diffusing ions. For example, an oxygen ion, O^{2-}, has eight protons (+) and 10 electrons (−). Therefore, as it moves, it carries a charge of $2 \times 1.6 \times 10^{-19}$ coul.

units are repeated from Eq. (2–8.2):

$$\text{ohm}^{-1} \cdot \text{cm}^{-1} = \left(\frac{\text{carriers}}{\text{cm}^3}\right)\left(\frac{\text{amp} \cdot \text{sec}}{\text{carriers}}\right)\left(\frac{\text{cm}^2}{\text{volt} \cdot \text{sec}}\right).$$

Let us illustrate by a very straightforward calculation: A semiconducting material which has 10^{16} electrons available for conduction in each cm^3, and which permits an electron mobility of 2000 (cm/sec)/(volt/cm), has a conductivity of $(10^{16}/\text{cm}^3)(1.6 \times 10^{-19}$ amp $\cdot$ sec) $\times$ (2000 cm^2/volt $\cdot$ sec), or 3.2 amp/volt $\cdot$ cm, that is, 3.2 $\text{ohm}^{-1} \cdot \text{cm}^{-1}$.

Electron holes. Each Fe^{3+} ion of Fig. 12–1.2 has one less negative charge than the majority of the iron ions (an excess positive charge). These ions have a *hole* into which an electron can drop. As an electron moves from an adjacent ferrous ion to this hole, an effective positive charge moves in the opposite direction (Fig. 12–1.2). In every sense, therefore, it is appropriate to refer to electron holes as having positive charges (of 1.6×10^{-19} coul each) which have mobility. Thus the total conductivity is equal to

$$\sigma = n_n q \mu_n + n_p q \mu_p, \tag{12–1.2}$$

where n_n is the number of negative carriers (electrons), n_p is the number of positive carriers (holes), and μ_n and μ_p are their respective mobilities. As a rule the electron mobility, μ_n, is greater than the hole mobility, μ_p, because the electrons can be accelerated to greater velocities in an electric field than can holes which jump from atom to atom. Table 12–1.1 lists typical values for semiconducting elements.

Example 12–1.1. A semiconductor is known to be of the *n*-type; i.e., its charge carriers are predominantly negative, thus they are electrons. It has a resistivity of 20 ohm $\cdot$ cm. Separate experiments indicate that the mobility of the electrons is 100 cm^2/volt $\cdot$ sec. How many electron carriers are there per cm^3?

Solution. Since $\sigma = 1/\rho$, $n = 1/\rho q \mu$.

$n = 1/(20$ ohm $\cdot$ cm)$(1.6 \times 10^{-19}$ amp $\cdot$ sec)$(100$ cm^2/volt $\cdot$ sec)

   $= 3.12 \times 10^{15}/\text{cm}^3$.

Comments. The term *p*-type refers to semiconductors which have more positive carriers (holes) than negative carriers (electrons). The electrical engineer utilizes this notation extensively; hence a *p-n* junction is the interface between a *p*-type semiconductor and an *n*-type semiconductor. ◀

Example 12–1.2. Indium phosphide (InP) is a semiconductor which has equal numbers of negative and positive charge carriers, i.e., electrons and holes. It has the structure of zinc sulphide (Fig. 10–1.3) with a cube edge of 5.87 Å. Its conductivity is 4.0 $\text{ohm}^{-1} \cdot \text{cm}^{-1}$. How many unit cells are there

Table 12–1.1

CHARGE MOBILITIES IN SEMICONDUCTING ELEMENTS (20°C)

Element	Electron mobility μ_n, cm^2/volt $\cdot$ sec	Hole mobility μ_p, cm^2/volt $\cdot$ sec
C (diamond)	1700	1200
Si	1400	500
Ge	3900	1900
Sn (gray)	2000	1000

per carrier, given that the values of μ_n and μ_p are 4000 and 100 cm^2/volt · sec, respectively (Table 12–2.2)?

Solution. From Eq. (12–1.2),

$$n_n = n_p = \sigma/q(\mu_n + \mu_p)$$
$$= (4 \text{ ohm}^{-1} \cdot \text{cm}^{-1})/(1.6 \times 10^{-19} \text{ amp} \cdot \text{sec})(4100 \text{ cm}^2/\text{volt} \cdot \text{sec})$$
$$= 6.1 \times 10^{15}/\text{cm}^3;$$

$$n = n_n + n_p = 1.22 \times 10^{16} \text{ carriers/cm}^3.$$

$$\text{Carriers/unit cell} = (1.22 \times 10^{16} \text{ carriers/cm}^3)(5.87 \times 10^{-8} \text{ cm})^3/\text{uc}$$
$$= 2.5 \times 10^{-6} \text{ carriers/uc}.$$

Unit cells/carrier = 400,000.

Since the carriers are equally divided between conduction electrons and electron holes, there are:

800,000 uc/conduction electron
(and 800,000 uc/hole).

Comments. The last calculation illustrates that a relatively small fraction of the total electrons within the structure account for its semiconductivity, since there are approximately 4×10^{22} atoms/cm^3.

Since this material has equal numbers of negative and positive carriers, it cannot be considered either *n*-type or *p*-type. The term, *intrinsic*, is commonly applied to InP, meaning that it is a semiconductor in its pure state. ◀

12–2 INTRINSIC SEMICONDUCTORS ⓜ

Intrinsic semiconductors are materials which do not depend on impurities for their conductivities.

Energy gap. We envision an ionic compound as transferring valence electrons from positive to negative ions. It then takes considerable energy to remove those electrons from the negative ions. Likewise, the pair of electrons which are shared between two carbon atoms require an extremely high energy to be freed so that they can conduct charge toward the positive electrode. Since the electrons in the two examples just cited are not easily freed to conduct charge, the materials are *insulators*. However, all electrons possess energy as they move about their orbits.

The materials scientist describes insulators on the basis of the sketch in Fig. 12–2.1. Let us take carbon (diamond) as an example. Its crystal structure, which is shown in Fig. 12–2.2(a), is directly comparable to that of ZnS (Fig. 10–1.3) except that all the atoms are carbon. Furthermore, each carbon has a coordination number of 4, and each neighboring pair of atoms shares a pair of electrons. The

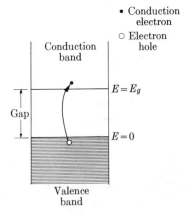

Fig. 12–2.1 Energy gap (insulator). In an insulator, there is an energy gap between the filled energy band for valence electrons and the empty energy band for conduction electrons. The empty one could accommodate electrons. The wavelike movements of the electrons do not permit energies to exist between these two bands; thus the energy gap. A valence electron, to become a conduction electron, must jump the gap. It leaves an electron "hole" in the valence band.

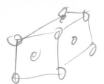

Table 12–2.1
ENERGY GAPS IN SEMICONDUCTING ELEMENTS

Element	Energy gap E_g, eV	At 20°C (68°F)	
		Fraction of valence electrons with energy $> E_g$	Conductivity σ, ohm$^{-1} \cdot$ cm^{-1}
C (diamond)	~6	~$1/30 \times 10^{21}$	$< 10^{-18}$
Si	1.1	~$1/10^{13}$	5×10^{-6}
Ge	0.72	~$1/10^{10}$	0.02
Sn (gray)	0.08	~$1/5000$	10^4

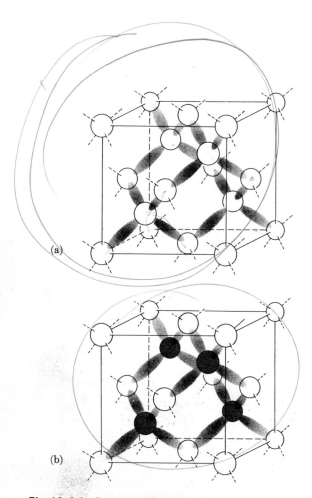

(a)

(b)

Fig. 12–2.2 Crystal structure of familiar semiconductors. (a) Diamond, silicon, germanium, gray tin. (b) ZnS (Fig. 10–1.3), GaP, GaAs, InP, etc. The two are similar, except that two types of atoms are in alternate positions in the semiconducting compounds. All atoms have CN = 4; each material has an average of four valence electrons per atom, and two electrons per bond.

shared electrons have energy as they move between the atoms, but not enough of it to become independent of the atoms. To do this they would need to have their energy raised from the *valence band* to the *conduction band*. In diamond, this would require about 6 eV of energy for each electron. Physicists call the interval between the valence band and the conduction band the *energy gap*. To jump this gap would take much more energy than the typical electron possesses because the thermal energy of an average electron at normal temperatures is only 0.025 eV. Although at any particular instant of time, approximately one-half the electrons have more than the average energy, the probability of an electron having 240 times the average is indeed remote.* As a result, very few electrons can break loose from the bonds to serve as charge carriers. Diamond's conductivity is $<10^{-18}$ ohm$^{-1} \cdot$ cm^{-1}, compared to 10^3 ohm$^{-1} \cdot$ cm^{-1} for graphite.

Silicon, immediately below carbon in the periodic table (Table 1–2.2) also has the structure shown in Fig. 12–2.2(a) except, of course, that every atom is Si. The energy gap** that a valence electron of silicon must jump in order to break loose and become a conductor is 1.1 eV. This is only 45 times the average electron energy of 0.025 eV; therefore the number of electrons which can jump the gap to serve as conductors is increased from one out of more than 10^{22} (for diamond) to 1 out of 10^{13} for silicon. Even so, the conductivity remains low, at about 5×10^{-6} ohm$^{-1} \cdot$ cm^{-1}.

---

* But not absolutely impossible among the many valence electrons per cm^3. A few out of 30 sextillion (3×10^{22}) do have the necessary energy.

** Arbitrarily, some physicists consider an energy gap of 4 eV as a convenient division point between insulators and semiconductors.

Below carbon and silicon in the periodic table are germanium, Ge, and gray tin, Sn. Their energy gaps, E_g, are only 0.72 eV and 0.08 eV, respectively (Table 12–2.1). As a result, their conductivities are markedly higher than that of silicon, because at room temperature a significant fraction of their electrons have energies in excess of the amount needed to jump the energy gap.

Semiconducting compounds. We observed in Fig. 12–2.2 that diamond, C, and zinc sulfide, ZnS, have essentially the same structure. The prime difference between them is the fact that *alternate* atoms of ZnS are zinc (and alternate atoms are sulfur). Every atom has four neighbors, and the average number of valence electrons is four/atom. (Zn contributes two and S contributes six.) There are a variety of binary compounds which produce such AX structures (Section 10–1). The prime requirement is that the atoms share electrons; thus we find this type of compound more common when at least one of the atoms is in, or near, the overlap region between metallic and nonmetallic elements of the periodic table (Fig. 1–2.2.) These compounds are semiconducting. Table 12–2.2 lists several examples.

Table 12–2.2
PROPERTIES OF COMMON SEMICONDUCTORS (20°C)

Material	Energy gap E_g, eV	Mobilities, cm²/volt · sec		Intrinsic conductivity, ohm^{-1} · cm^{-1}	Lattice constant, a
		Electron, μ_n	Hole, μ_p		
Elements					
C (diamond)	~6	1,700	1,200	$< 10^{-18}$	3.57 Å
Silicon	1.1	1,400	500	5×10^{-6}	5.43
Germanium	0.72	3,900	1,900	0.02	5.66
Tin (gray)	0.08	2,000	1,000	10^4	6.49
Compounds					
AlSb	1.6	1,000	400	—	6.13
GaP	2.2	500	20	—	5.45
GaAs	1.4	4,000	300	10^{-8}	5.65
GaSb	0.7	5,000	1,000	—	6.10
InP	1.3	4,000	100	5	5.87
InAs	0.33	30,000	500	100	6.04
InSb	0.17	80,000	500	—	6.48
ZnS	3.6	100	10	—	—
SiC (hex)	3	100	20	—	—

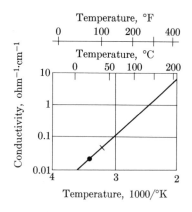

Fig. 12–2.3 Semiconductivity versus temperature (intrinsic germanium: $E_g = 0.72$ eV). This plot is a straight line because it follows Eq. (12–2.1).

Semiconductivity (intrinsic) versus temperature. At room temperature, 20°C (68°F), the average energy of the valence electrons is 0.025 eV. This average increases in proportion to the absolute temperature, in °K. More important, the fraction with abnormally high energies increases markedly.

Let us examine the conductivity of germanium as a function of temperature (Fig. 12–2.3). Materials scientists tell us that theory and experiment check each other to give

$$\log_{10} \sigma = A - \frac{E_g}{0.0004T}. \tag{12–2.1}$$

Thus, if we plot $\log \sigma$ versus $1/T$, we get a straight line, and

$$\log_{10} \sigma_2/\sigma_1 = \frac{E_g}{0.0004} \left[\frac{1}{T_1} - \frac{1}{T_2} \right]$$

$$= \frac{E_g}{0.4} \left[\frac{1000}{T_1} - \frac{1000}{T_2} \right]. \tag{12–2.2}$$

Again T must be the absolute temperature in °K. The simplest means of measuring the energy gap E_g of a semiconductor is to measure its conductivity σ at two different temperatures, and then solve Eq. (12–2.2). (This can also be done graphically.)

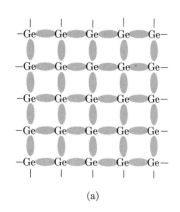
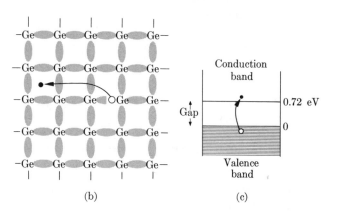

Fig. 12–2.4 Intrinsic semiconductor (germanium). (a) Schematic presentation showing electrons in their covalent bonds (and their valence bands). (b) Electron-hole pair. (Positive electrode at the left.) (c) Energy gap, across which an electron must be raised to provide conduction. For each conduction electron, there is a hole produced among the valence electrons.

Intrinsic semiconduction. Figure 12–2.4 reveals that each conduction electron which breaks loose in a semiconductor, such as pure germanium, also produces an electron hole. For this reason an intrinsic semiconductor has an equal number of *n*-type (electron) and *p*-type (hole) charge carriers.

Photoconduction. Ⓜ Thermal energy raises a relatively small fraction of the electrons out of the valence band of silicon into the conduction band (1 out of 10^{13}, according to Table 12–2.1). In contrast, if an electron is hit by a photon of light, it may readily be energized to the conduction level (Fig. 12–2.5). As an example, a photon of red light (wavelength $\lambda = 6.6 \times 10^{-4}$ cm) has 1.9 eV of energy, more than enough to cause an electron to jump the 1.1 eV energy gap in silicon. Thus the conductivity of silicon increases markedly when it is exposed to light.

Recombination. Ⓜ Ⓞ The reaction to produce an *electron-hole pair*, as shown in Fig. 12–2.5, may be written as

$$E \rightarrow n + p \qquad\qquad (12\text{–}2.3\text{a})$$

where E is energy, n is the conduction electron, and p is the hole in the valence band. In this case the energy came from light.

Since all materials are more stable when they reduce their energies, electron-hole pairs recombine sooner or later:

$$n + p \rightarrow E. \qquad\qquad (12\text{–}2.3\text{b})$$

In effect, the electron drops from the conduction band back to the valence band, just the reverse of Fig. 12–2.4(c). Were it not for the fact that light or some other energy source continually produces additional electron-hole pairs, the conduction band would soon become depleted.

The time required for recombination varies from material to material. However, it follows a regular pattern because within a specific material every conduction electron has the same probability of recombining within the next second (or minute). This leads to the relationship,

$$N = N_0 e^{-t/\lambda}, \qquad\qquad (12\text{–}2.4\text{a})$$

which we usually rearrange to

$$2.3 \log_{10} (N_0/N) = t/\lambda. \qquad\qquad (12\text{–}2.4\text{b})*$$

In these equations, N_0 is the number of electrons in the conduction band at a particular moment of time (say, when the source of light is turned off). After an additional time t, the number of remaining conduction electrons is N. The term λ is called the *relaxation time*, and is characteristic of the material. (Compare this discussion with that for stress relaxation in Section 9–2.)

Luminescence. Ⓜ Ⓞ The energy released in Eq. (12–2.3b) may appear as heat. It may also appear as light. When it does, we speak of

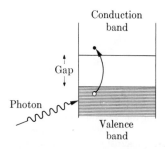

Fig. 12–2.5 Photoconduction. A photon (i.e., light energy) raises the electron across the energy gap, producing a "conduction electron + valence hole" pair, forming charge carriers. Recombination (Eq. 12–2.3b) occurs when the electron drops back to the valence band.

---

*   Equation (12–2.4) can be derived through calculus (by those who wish to do so) from the information stated above:

$$dN/dt = -N/\lambda. \qquad\qquad (12\text{–}2.5)$$

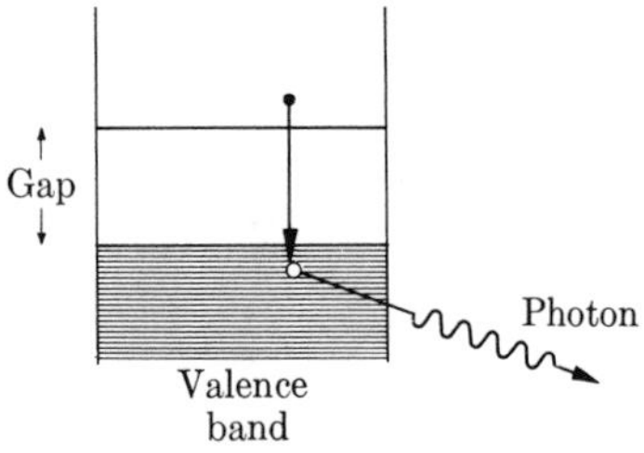

Fig. 12–2.6 Luminescence. Each millisecond, a fraction of the electrons energized to the conduction band return to the valence band. As the electron drops across the gap, the energy may be released as a photon of light.

luminescence (Fig. 12–2.6). Sometimes we subdivide luminescence into several categories. *Photoluminescence* is the light emitted after electrons have been activated to the conduction band by light photons. *Chemoluminescence* is the word used when the initial activation is due to chemical reactions. Probably *electroluminescence* is best known, because this is what occurs in a TV tube, in which a stream of electrons scans the screen, activating the electrons in the phosphor to their conduction band. Almost immediately, however, the electrons recombine, emitting energy as visible light.

Since the recombination rate is proportional to the number of activated electrons, the intensity I of luminescence also follows Eq. (12–2.4):

$$2.3 \log_{10} (I_0/I) = t/\lambda. \tag{12–2.6}$$

For a TV tube, the engineer chooses a phosphor with a relaxation time such that light continues to be emitted as the next scan comes across. Thus our eyes do not see a light-dark flickering. However, the light intensity from the previous trace should be weak enough so that it does not compete with the new scan that follows one-thirtieth of a second later. (See Example 12–2.4.)

Example 12–2.1. Each gray tin atom has four valence atoms and the unit cell size (Fig. 12–2.2a) is 6.49 Å. Separate calculations (Study problem 12–2.1) indicate that there are 2×10^{19} conduction electrons per cm³. What fraction of the electrons have been activated to the conduction band?

Solution. Basing our calculations on Fig. 12–2.2, we find that there are 8 tin atoms per unit cell.

$$\text{Valence electrons/cm}^3 = \frac{(8 \text{ atoms/uc})(4/\text{atoms})}{(6.49 \times 10^{-8} \text{ cm})^3} = 1.17 \times 10^{23}/\text{cm}^3.$$

$$\text{Fraction activated} = \frac{2 \times 10^{19}}{1.17 \times 10^{23}} \cong 0.0002.$$

Comments. In this problem, and elsewhere in this chapter, we refer to gray tin. It has the structure of diamond and is stable below 13°C (55°F). White tin, the more commonly encountered variety, is metallic and stable above 13°C, although it may exist in the unstable form at lower temperatures. ◀

Example 12–2.2. The resistivity of germanium at 20°C (68°F) is 50 ohm · cm. What is its resistivity at 40°C (104°F)?

Solution. Based on Eq. (12–2.2) and an energy gap of 0.72 eV (Table 12–2.2):

$$\log_{10} \sigma_{40°}/\sigma_{20°} = \log_{10} \rho_{20°}/\rho_{40°} = \frac{0.72}{0.4} \left[\frac{1000}{293} - \frac{1000}{313} \right]$$

$$= 1.8(3.42 - 3.20) = 0.396.$$

Therefore

$$\rho_{40°} = \rho_{20°}/10^{0.396} = 50 \text{ ohm} \cdot \text{cm}/2.5$$

$$= 20 \text{ ohm} \cdot \text{cm}.$$

This problem may also be solved graphically (Fig. 12–2.3).

Comments. It is possible to measure resistance changes (and therefore resistivity changes) of <0.1%. Therefore one can measure temperature changes of a small fraction of a degree. (See Study Problem 12–2.3.)

The units of Eq. (12–2.2) are

$$\log \frac{\text{ohm}^{-1} \cdot \text{cm}^{-1}}{\text{ohm}^{-1} \cdot \text{cm}^{-1}} = \frac{\text{eV}}{\text{eV}/°\text{K}} \left[\frac{1}{°\text{K}} - \frac{1}{°\text{K}} \right]. \blacktriangleleft$$

Example 12–2.3. Ⓜ Ⓞ The energy of a light photon was described in Eq. (9–4.1) as

$$E = h\nu,$$

or alternatively,

$$E = hc/\lambda,$$

where c is the velocity of light, 3×10^{10} cm/sec; λ is the wavelength of light; and h is a constant whose value is 4.13×10^{-15} eV $\cdot$ sec when energy is expressed in electron volts. (a) What is the minimum frequency of light necessary to cause photoconduction in gallium phosphide, GaP, if all the energy comes from the photons? (b) What color light is this?

Solution. The energy gap is 2.2 eV (from Table 12–2.2).

a) $\nu = 2.2$ eV$/(4.13 \times 10^{-15}$ eV $\cdot$ sec$)$

   $= 5.34 \times 10^{14}$/sec.

b) $\lambda = c/\nu = (3 \times 10^{10}$ cm/sec$)/(5.34 \times 10^{14}$/sec$)$

   $= 5620$ Å.

This is in the yellow part of the visible spectrum.

Comments. Longer wavelengths (shorter frequencies) can also activate those electrons which possess considerable thermal energy at the moment they are hit. However, the number of electrons which have thermal energy of more than a few tenths of an eV is extremely limited. Therefore, for all intents and purposes, red light does not produce photoconduction in GaP because the red light has longer wavelengths (less energy) than yellow light.

This problem uses λ for wavelength; this is a standard symbol. We shall observe in the next example that λ is used for relaxation time, also a standard symbol. Fortunately none of our problems require wavelength and relaxation time in the same calculation. ◀

Example 12–2.4. Ⓜ Ⓞ The scanning beam of a television tube covers the screen with 30 frames per second. What must the relaxation time for the activated electrons of the phosphor be if only 20% of the intensity is to remain when the following frame is scanned?

Solution. Refer to Eq. (12–2.6):

$$2.3 \log (1.00/0.20) = (0.033 \ \text{sec})/\lambda,$$

$$\lambda = 0.02 \ \text{sec}.$$

Comments. We use the term *fluorescence* when the relaxation time is short compared to the time of our visual perception. If the luminescence has a noticeable afterglow, we use the term *phosphorescence*. ◄

12–3 EXTRINSIC SEMICONDUCTORS ⓜ Ⓜ

***N*-type semiconductors.** Impurities alter the semiconducting characteristics of materials by producing excess electrons or excess electron holes. Consider, for example, some silicon containing an atom of phosphorus. Phosphorus has five valence electrons rather than the four which are found in pure silicon. In Fig. 12–3.1(a), the extra electron is present independently of the electron pairs which serve as bonds between neighboring atoms. This electron can carry a charge toward the positive electrode (Fig. 12–3.1b). Alternatively, in Fig. 12–3.1(c), the extra electron—which cannot reside in the valence band because it is already full—is located near the top of the energy gap. From this position—called a *donor* level E_d—the extra electron can easily be activated into the conduction band. Regardless of which model is used, Fig. 12–3.1(a) or 12–3.1(c), we can see that atoms from Group V of the periodic table (Fig. 1–2.2) can supply negative, or *n*-type, charge carriers to semiconductors.

***P*-type semiconductors.** Group III elements have only three valence electrons. Therefore, when such elements are added to silicon as impurities, electron holes come into being. As shown in Fig. 12–3.2(a)

Fig. 12–3.1 Extrinsic semiconductors (*n*-type). A Group V atom has an extra valence electron beyond the average of four sketched in Fig. 12–2.2. This fifth electron can be pulled away from its parent atom with very little added energy, and "donated" to the conduction band, to become a charge carrier. We observe the donor energy level, E_d, as being just below the top of the energy gap. (a) An *n*-type impurity. (b) Ionized phosphorus atom. (Positive electrode at left.) (c) Band model.

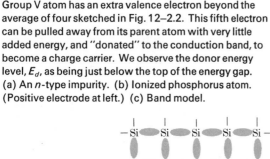

(a)

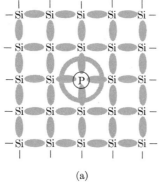

(b)

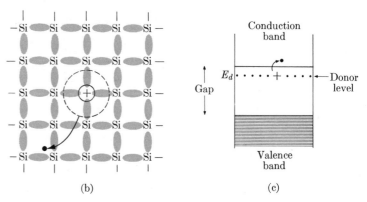

(c)

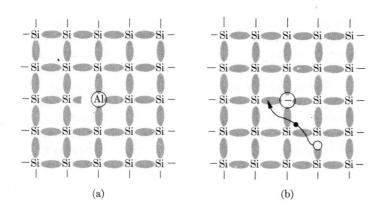

(a)   (b)

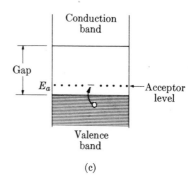

(c)

and (b), each aluminum atom can accept one electron. In the process a positive charge moves toward the negative electrode. Using the band model (Fig. 12–3.2c), we note that the energy difference for electrons from the valence band to the *acceptor* level, E_a, is much less than the full energy gap. The electron holes remaining in the valence band are available as positive carriers for *p*-type semiconduction.

Donor exhaustion (and acceptor saturation). If the energy difference for electrons from the donor level to the conduction band is small in comparison with the size of the energy gap, all the electrons in the donor level can be raised to the conduction band, even at room temperatures. This is illustrated in Fig. 12–3.3. At absolute zero, beyond the left edge of the figure, all the electrons are either in the valence band or in the donor levels, and Fig. 12–3.1(a) applies. As the temperature is raised, more and more of the donor electrons jump to the conduction band and carry charge. When the thermal energy of the electrons is approximately equal to this small energy difference $(E_g - E_d)$, essentially all the donor electrons have jumped to the conduction band and this reservoir of charge carriers arising from

Fig. 12–3.2 Extrinsic semiconductors (*p*-type). A Group III atom has one less valence electron than the average of four sketched in Fig. 12–2.2. This atom can accept an electron from the valence band, thus leaving an electron hole as a charge carrier. The acceptor energy level, E_a, is just above the bottom of the energy gap. (a) A *p*-type impurity. (b) Ionized aluminum atom. (Negative electrode at right.) (c) Band model.

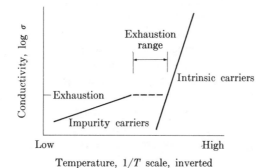

Temperature, $1/T$ scale, inverted

Fig. 12–3.3 Donor exhaustion. Intrinsic (right-hand curve) and extrinsic (left-hand curve) conductivity require energies of E_g and $(E_g - E_d)$, respectively, to raise electrons into the conduction band. At lower temperatures, donor electrons provide most of the conductivity. Exhaustion occurs when the donor electrons have entered the conduction band, and before the temperature is raised high enough for valence electrons to jump the energy gap. The conductivity is nearly constant in this temperature range.

impurities is exhausted. As a result there is a conductivity plateau; then, when the temperature is increased appreciably, the thermal energy becomes sufficient to cause electrons to jump across the full energy gap from the valence band to the conduction band.

Donor exhaustion of *n*-type semiconductors has its parallel in acceptor saturation of *p*-type semiconductors. The reader is asked to make the various comparisons. Donor exhaustion and acceptor saturation are important to materials and electrical engineers, since they provide a region of essentially constant conductivity. This means that it is less necessary to compensate temperature changes in electrical circuits than it would be if the log σ versus $1/T$ characteristics followed an ever-ascending line.

Example 12–3.1. Silicon, according to Table 12–2.2, has a conductivity of only 5×10^{-6} ohm$^{-1} \cdot$ cm^{-1} when pure. An engineer wants it to have a conductivity of 2 ohm$^{-1} \cdot$ cm^{-1} when it contains aluminum as an impurity. How many aluminum atoms are required per cm^3?

Solution. Since the intrinsic conductivity is negligible compared with 2 ohm$^{-1} \cdot$ cm^{-1}, assume that all the conductivity comes from the holes:

$$n_p = (2 \text{ ohm}^{-1} \cdot \text{cm}^{-1})/(1.6 \times 10^{-19})(500 \text{ cm}^2/\text{volt} \cdot \text{sec})$$
$$= 2.5 \times 10^{16}/\text{cm}^3.$$

Comments. Each aluminum atom contributes one acceptor site, i.e., electron hole. Therefore 2.5×10^{16} aluminum atoms are required per cm^3. This is, of course, a large number; however, it is still small compared with the number of silicon atoms per cm^3. (See Study Problem 12–3.1a.) ◀

12–4 SEMICONDUCTING DEVICES Ⓜ ⓜ

There are many electronic devices that use semiconductors. We shall consider but a few.

We have already seen (Example 12–2.2 and Study Problem 12–2.3) that the resistivity and therefore the resistance of semiconductors varies markedly with temperature changes (Figs. 12–2.3 and 12–3.3). Some ceramic semiconductors are available which can detect temperature changes of approximately 10^{-6} °C! Such a resistance device acts as a thermometer, and is called a *thermistor*.

Because many semiconducting materials have low packing factors, they have a high compressibility. Experiments show that as the volume contracts, the size of the energy gap is measurably reduced; this of course increases the number of electrons which can jump the energy gap. Thus pressure can be calibrated against resistance for *pressure gages*.

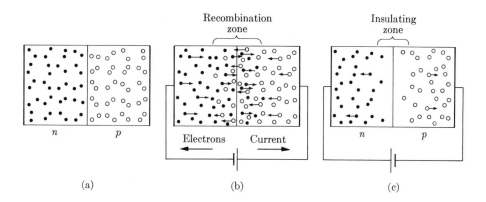

(a)   (b)   (c)

Rectifying junctions. Suppose that a *p*-type semiconductor and an *n*-type semiconductor are joined. Their junction can serve as a *rectifier*; i.e., it is an electrical valve that lets current pass one way and not the other. Figure 12–4.1 presents a brief description, somewhat over-simplified. In part (a) of the figure, we see positive carriers in the *p*-half and negative carriers in the *n*-half. With the voltage in one direction (Fig. 12–4.1b), the current feeds through the device from right to left (electrons from left to right). However, with a reverse voltage (Fig. 12–4.1c), the conduction electrons are attracted to the positive electrode and the holes are displaced in the opposite direction, leaving a carrier-free zone at the junction. The conductivity of this zone is low. If a greater voltage is applied, it simply widens the "insulating" zone. Current passes in only one direction, as shown in Fig. 12–4.2.

Fig. 12–4.1 An *n-p* junction (rectifier). (a) No voltage. (b) Forward bias. Charge is transported across the junction. Electrons and holes recombine beyond the junction. (c) Reverse bias. The carriers do not carry charge across the junction. The holes (in the *p*-type) and the conduction electrons (in the *n*-type) are pulled away from the junction by the electric field.

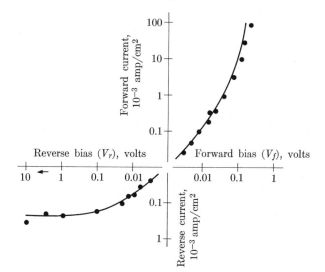

Fig. 12–4.2 Characteristics of a rectifier junction. Forward bias gives a high current density, amp/cm². A reverse bias forms a high resistance. As the reverse voltage is increased from 0.1 to 10 volts, there is only a slight increase in reverse current.

As the holes move across the junction with a forward bias (Fig. 12–4.1b), they recombine with the electrons in the *n*-type material according to Eq. (12–2.3b). Likewise, the electrons combine with the holes as the electrons move into *p*-type material. These reactions do not occur immediately, however. In fact, an excess number of positive and negative carriers may move measurable distances beyond the junction. The numbers of excess, unrecombined carriers are an exponential function of the voltage *V*. These facts do not affect the rectifier significantly, but they are important to transistors.

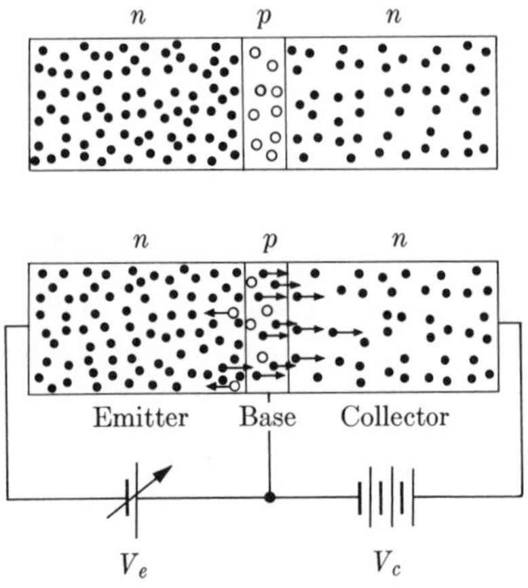

Transistors. ◎ Transistors have revolutionized engineering design in the field of communications. Let us explore their operation by means of a simplified model. A transistor has two junctions in series. They may be *p-n-p* or *n-p-n*. The former is somewhat more common; however, we shall consider the *n-p-n* transistor, since it's a little easier for us to envision the movements of electrons than the movements of holes. The principles behind each type are the same, though.

A transistor consists of an *emitter*, a *base*, and a *collector* (Fig. 12–4.3). For the moment, consider only the *emitter junction*, which is biased so that electrons move into the base (and toward the collector). As discussed a moment ago, the number of electrons that cross this junction and move into the *p*-type material is an exponential function of the emitter voltage, V_e. Of course, these electrons at once start to combine with the holes in the base; however, if the base is narrow, or if the relaxation time is long (λ of Eq. 12–2.4), the electrons keep on moving through the thickness of the base. Once they are at the second or *collector junction*, the electrons have free sailing, because the collector is an *n*-type semiconductor. The total current that moves through the collector is controlled by the emitter voltage, V_e. As the emitter voltage fluctuates, the collector current, I_c, changes exponentially; written logarithmically:

$$2.3 \log_{10} I_c \cong A + V_e/B, \tag{12–4.1}$$

where A and B are constants. Thus, if the voltage in the emitter is increased even slightly, the amount of current is increased markedly. It is because of these relationships that a transistor serves as an amplifier.

Example 12–4.1. ◎ A transistor has a collector current of 4.7 milliamperes when the emitter voltage is 17 millivolts. At 28 millivolts, the current is 27.5 milliamperes. Given that the emitter voltage is 39 millivolts, estimate the current.

Solution. Based on Eq. (12–4.1),

$$2.3 \log_{10} 4.7 \cong A + 17/B = 1.55,$$

$$2.3 \log_{10} 27.5 \cong A + 28/B = 3.31.$$

Fig. 12–4.3 Transistor (*n-p-n*). The number of electrons crossing from the emitter-base junction is highly sensitive to the emitter voltage. If the base is narrow, these carriers move to the base-collector junction, and beyond, before recombination. The total current flux, emitter to base, is highly magnified, or *amplified*, by fluctuations in the voltage of the emitter.

Emitter   Base   Collector

V_e V_c

Solving simultaneously, using *milli*units, we have

$$A \cong -1.175 \quad \text{and} \quad B \cong 6.24.$$

At 39 millivolts,

$$2.3 \log_{10} I_c \cong -1.175 + 39/6.24 \cong 5.075,$$

$$I_c \cong 160 \text{ milliamp.}$$

Comment. The electrical engineer modifies Eq. (12–4.1) to take care of added current effects. These, however, do not change the basic relationship: The variation of the collector current is much greater than the variation of the signal voltage. ◀

12–5 MAKING OF SEMICONDUCTORS ⓜ ◎

As we have seen, impurities markedly alter the properties of semi-conductors. In general, however, impurities are to be avoided except for those specific elements which are used as *dopants*. The normal practice is to refine the semiconductor to the highest degree of purity, then add controlled amounts of the desired impurity.

Let us use silicon as our example. Its purest form in nature is as an oxide of 99+% purity. The oxygen must be removed to give silicon "metal"; however, this is not good enough, because the impurities that are present must be less than 1 part per billion. One of the more common impurities is aluminum. Of course, aluminum is one of the critical ones because it is a Group III element and produces electron holes.

Zone refining. ⓔ Our knowledge of phase diagrams shows us the basic principle involved in removing aluminum from silicon. The same procedure removes many of the other impurities. Figure 12–5.1 shows the Al-Si phase diagram. The solid solubility of Al in Si is approx-imately 1 w/o at 600°C (1110°F); however, this drops to zero at the melting point. Assume that the solubility line is straight between these two points. Thus the solid solubility C_S would be about 0.04 w/o Al at 1400°C. At the same temperature, the liquid solubility C_L is 5 w/o aluminum, about 125 times as great. The ratio C_S/C_L, called the *distribution coefficient k*, continues right up to the melting point,

$$k = C_S/C_L, \tag{12–5.1}$$

where C_S is the solute composition in the solid, and C_L is the solute composition in the liquid. Thus, for silicon plus aluminum, $k = 0.008$.

Now consider a 99Si-1Al melt. As the mixture cools and starts to solidify, C_L is 1 w/o Al when the very first solid forms just under 1430°C. The solid composition is (1 w/o)/125, or only 0.008 Al! Admittedly, if we let equilibrium occur during cooling, the solid would

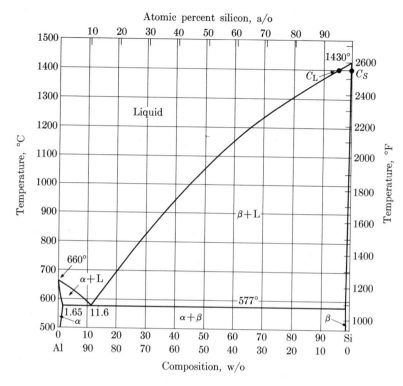

Fig. 12–5.1 The Al-Si system. Near the melting point of silicon, the ratio of aluminum in the solid, C_S, to the aluminum in the liquid, C_L, is approximately 0.008; that is, $k = C_S/C_L = 0.008$. (See text.) (ASM *Handbook of Metals,* Metals Park, O.: American Society for Metals)

gradually become enriched until the full 1 w/o Al would be dissolved as a solid solution in silicon at 600°C. However, if we could isolate some of this early solid, we would have silicon of "four nines" purity, that is, >99.99 w/o Si. The next step is obvious. If we melted the silicon with only 0.008 w/o Al, and let it solidify, the solid which formed this time would contain only (0.008 w/o)/125, or 0.000064 w/o Al! Then we could try additional cycles, until we achieved "*nine* nines" purity (1 ppb impurity).

The materials engineer had to work out a feasible way to utilize the above process. The result is *zone refining* (Fig. 12–5.2). A rod of semipure silicon is placed inside a moving induction coil. The

Fig. 12–5.2 Zone refining. Each pass by the *R-F* induction coil reduces the impurities in the metal by a factor of k, the ratio of solid solute, C_S, to liquid solute, C_L, composition. (See Fig. 12–5.1.)

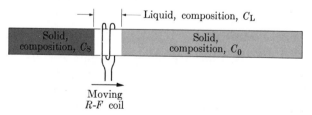

high-frequency (*R-F*) current melts a short zone of the rod.   The initial liquid at the molten (left-hand) end has the overall composition, but as the coil moves to the right, the silicon solidifies at the stern end of the molten zone in purer form.   This process continues until the molten zone comes out at the right-hand end.   Admittedly the very last molten material contains accumulated aluminum that was swept to the end of the rod.   This part is quite high in aluminum content; therefore the last material to solidify is highly impure silicon and is discarded.   Two, three, or more passes of the induction coil down the rod are made; each time more aluminum is swept to the discard end.

Junctions. There are several different methods by which *n*- and *p*-type regions are produced in a single semiconductor crystal.   These are summarized in Fig. 12–5.3.

In an *alloy junction* (Fig. 12–5.3a), a few micrograms of dopant, such as aluminum, are melted on the surface of a single crystal of *n*-type silicon.   Melting causes some silicon to dissolve, and when it cools it resolidifies with aluminum in solid solution.   A similar procedure on the opposite side of a small wafer gives a *p-n-p* junction.   An *n-p-n* junction would require an initial *p*-type silicon with a Group V alloying element added later.

For a *grown junction* (Fig. 12–5.3b), a crystal is solidified from an *n*-type melt.   At the desired stage of growth, a Group III dopant is added to the melt.   The amount must be sufficient not only to cancel the Group V elements of the *n*-type, but there must be enough extra to give the desired impurity level in the *p*-type material.   After the crystal has grown another fraction of a millimeter, the necessary amount of a Group V element is added to the melt, to reverse the type from *p* to *n*.   As before, a *p-n-p* junction could be made by reversing the dopants.   Obviously, processing procedures require close control if one is to achieve the compositions and dimensions desired.

The third and last process we shall describe involves an *n*-type silicon wafer which is oxidized to give a protective SiO_2 coating.   Part of the silicon wafer is masked off; then the protective coating is removed

Fig. 12–5.3 Types of transistors. (a) Alloy; the junction is made by melting the alloy. (b) Grown; the junctions are made during the original crystal growth. (c) Planar (or diffusion); the junction is made by vapor deposition and diffusion. (See text.)

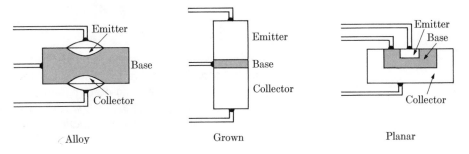

Alloy   Grown   Planar

from the rest of it before the wafer is heated in a boron-rich vapor. The boron (Group III) diffuses into the silicon to form a *p*-type region. The oxidation, masking, and heating process is repeated again, but with a phosphorus-rich vapor to form the second *n*-type region, which usually serves as the emitter (Fig. 12–5.3c). Finally, contacts can be added by repeating the cycle once more, this time with gold or aluminum. A transistor of this type is called a *planar* transistor, simply because all preparation and contacts are on one flat surface or plane. This type of junction lends itself to mass production because the above steps can be automated for the manufacture of transistors and integrated circuits of modern solid-state technology products.

Example 12–5.1. (e) A long rod of sterling silver (92.5Ag-7.5Cu) is melted in a zone refiner. What is the composition of the first solid to form in the first pass? The second pass? The fourth pass?

Solution. Basing our answer on Fig. 5–2.3, we find that a 92.5-7.5 liquid starts to form α at 900°C. The solid contains 97Ag-3Cu; $k = 3Cu/7.5Cu = 0.4$.

Pass	Starting liquid, w/o Cu	Solidifying α, w/o Cu
1	7.5	3
2	3	1.2
3	1.2	0.5
4	0.5	0.2

Comments. Zone refining is most efficient if the molten zone is very short compared with the length of the rod. The materials engineer using this procedure must consider this and other factors.

For another factor which determines the effectiveness of zone refining as a production process, compare Study problem 12–5.1 with this example. ◀

REVIEW AND STUDY

12–6 SUMMARY

Semiconductors contain limited numbers of charge carriers; the numbers may be varied and controlled by chemical composition, electric fields, light, and other parameters. The charge carriers may be electrons (negative carriers) which have been activated out of the valence band into the conduction band. They may also be holes created by "missing" electrons in the valence band which carry a positive charge.

In intrinsic semiconductors there are equal numbers of negative and positive carriers because each activated electron leaves a hole behind. Extrinsic semiconductors are either *n*-type or *p*-type, depending on whether the dopants have raised the average number of valence electrons to more than four, or reduced their average number per atom to less than four.

Semiconductors are used in many devices. Some of these, such as thermistors and photoconductors, depend on the variations in resistivity that can be introduced by variations in temperature, light, and pressure. The more widely used devices have one or more *n-p* junctions which serve as "electron valves." Rectifying diodes and transistors are made by the millions for present-day technologies.

The production of semiconductors has required the development of many new processing techniques, such as zone refining.

Terms for Study

Acceptor	Intrinsic semiconductor
Acceptor saturation	Luminescence
Base	Mobility, μ
Charge carriers	*n*-type
Collector	*p*-type
Conduction band	*p-n* junction
Conduction electron, n	Photoconduction
Distribution coefficient, k	Recombination
Donor	Rectifier
Donor exhaustion	Relaxation time, λ
Dopant	Semiconducting compounds
Drift velocity, $\bar{v}$	Semiconductors
Electron hole, p	Thermistor
Electron-hole pairs	Transistor
Emitter	Valence band
Energy gap, E_g	Valence electron
Extrinsic semiconductor	Zone refining
Hole (*see* electron hole)	

STUDY PROBLEMS

12–1.1 A p-type germanium semiconductor contains 10^{18} electron holes/cm^3. What is its conductivity?

Answer. 300 ohm$^{-1} \cdot$ cm^{-1}

12–1.2 Silicon has eight atoms per unit cell. Each unit cell is cubic, with a dimension of 5.43 Å. An impurity has been added to it to make it n-type. Only one impurity atom is added per every billion (10^9) silicon atoms; however, each impurity atom provides one negative carrier. (a) What is the concentration of impurities? (b) What is the conductivity?

Answer. (a) 5×10^{13}/cm^3 (b) 0.011 ohm$^{-1} \cdot$ cm^{-1}

12–1.3 How many electron carriers (and electron holes) does intrinsic silicon, that is $n_n = n_p$, require to give the same conductivity, $\sigma = 0.011$ ohm$^{-1} \cdot$ cm^{-1}, determined in Study problem 12–1.2?

Answer. 3.7×10^{13}/cm^3 of each

12–1.4 A semiconductor contains 10^{16} holes/cm^3 and has a resistivity of 7 ohm $\cdot$ cm. It is placed in an electric field of 15 volts/mm. What is the drift velocity of the electron holes toward the negative electrode?

Answer. 13,500 cm/sec ($\sim$ 300 mi/hr)

12–2.1 Gray tin has a conductivity of 10^4 ohm$^{-1} \cdot$ cm^{-1} derived from equal numbers of negative carriers (electrons) and positive carriers (holes). Their mobilities are indicated in Table 12–2.2. How many electrons have been raised to the conduction band?

Answer. 2×10^{19}/cm^3

12–2.2 According to the data in Tables 12–1.1 and 12–2.1, how many electron holes are there per cm^3 in pure germanium? (Pure germanium is intrinsic; therefore it has equal numbers of holes and conduction electrons.)

Answer. 2.2×10^{13}/cm^3

12–2.3 The resistance of a certain silicon wafer is 1031 ohms at 25.1°C. With no change in measurement procedure, the resistance decreases to 1029 ohms. What is the temperature change? The energy gap of silicon is 1.1 eV. [*Hint:* $(1/T_1 - 1/T_2) = (T_2 - T_1)/T_2 T_1 \cong (T_2 - T_1)/T_1^2$ when $T_2 \approx T_1$.]

Answer. +0.03°C

12–2.4 A semiconductor has a conductivity of 1.11 ohm$^{-1} \cdot$ cm^{-1} at 10°C and 1.72 ohm$^{-1} \cdot$ cm^{-1} at 17°C. What is its energy gap?

Answer. 0.88 eV

12–2.5 To what temperature must you raise InP in order to make its resistivity half what it is at 0°C? Its energy gap is 1.3 eV. (See hint for Study problem 12–2.3.)

Answer. 7°C

12–2.6 When an electron and hole pair recombine in silicon, luminescence appears, due to the 1.1 eV energy released. Is this luminescence visible?

Answer. No; $\lambda = 11,250$ Å. Visible light extends to only about 7000 Å. This luminescence would be in the infrared zone of the spectrum.

12–2.7 What energy is required for an electron-hole recombination to give blue light in a TV tube? (Wavelength of blue light = 4800 Å.)

Answer. 2.6 eV

12–2.8 Suppose that the relaxation time of the electrons in a given phosphor is 500 milliseconds. How long will it take 50% of the electrons to drop across the energy gap?

Answer. 350 milliseconds (0.35 sec)

12–2.9 The relaxation time for a phosphor is 750 milliseconds. What is the ratio of light intensities at the 0.5 second and 1.0 second decay points?

Answer. $I_{0.5}/I_{1.0} = 1.95$

12–3.1 Silicon has a density of 2.33 g/cm³. (a) What is the concentration of silicon atoms per cm³? (b) Phosphorus is added to silicon to make it an *n*-type semiconductor with a conductivity of 1 mho/cm and an electron mobility of 1400 cm²/volt · sec. What is the concentration of donor electrons per cm³?

Answer. (a) 5×10^{22} atoms/cm³ (b) 4.5×10^{15} donor electrons/cm³

12–3.2 How many silicon atoms are there for each aluminum atom in Example 12–3.1?

Answer. 2×10^6 Si/Al

12–3.3 Extrinsic germanium used for transistors has a resistivity of 2 ohm · cm and an electron hole concentration from impurities of 1.6×10^{15} holes/cm³. (a) What is the mobility of the holes in the germanium? (b) What impurity elements could be added to the germanium to create more holes?

Answer. (a) 1950 cm²/volt · sec (b) Al, In, Ga

12–3.4 Extrinsic germanium is formed by melting 3.22×10^{-6} g of antimony (Sb) with 100 g of germanium. (a) Will the semiconductor be *n*-type or *p*-type? (b) Calculate the concentration of antimony (in atoms/cm³) in germanium. (The density of germanium is 5.32 g/cm³.)

Answer. (a) *n*-type (b) 8.5×10^{14} Sb/cm³

12–3.5 $Fe_{1-x}O$, which has an Fe^{2+}/Fe^{3+} ratio of 10, may be shown to have 4.35×10^{21} Fe^{3+} ions per cm³ (and 4.35×10^{22} Fe^{2+} ions). Its conductivity is 1 ohm⁻¹ · cm⁻¹. (a) Is this oxide *n*-type or *p*-type? (b) What is the mobility of the carriers?

Answer. (a) *p*-type (b) 0.0014 cm²/volt · sec

12–4.1 Refer to Example 12–4.1. If the emitter voltage is doubled from 17 to 34 millivolts, by what factor is the collector current increased?

Answer. 15.4 (= 72.4 milliamps/4.7 milliamps)

12–4.2 The weakest usable signal in the transistor of Example 12–4.1 is 8 millivolts. Approximately what collector current is developed?

Answer. ~1.1 milliamps

12–5.1 Assuming fully efficient operation, how many passes by an *R-F* induction coil are required to refine a rod of 90Si-10Al material to eight-nines purity, that is, <0.000001 w/o Al, at the starting end of the zone-refined rod?

Answer. 4 passes

COMPOSITE MATERIALS

13–1 MATERIALS SYSTEMS ◎

Engineering designs call for various components—gears, armatures, bearings, etc.—as revealed in the cutaway view of the electric motor in Fig. 1–1.1. Often each component consists of only one basic material. A motor housing, for example, may be a die-cast aluminum alloy. It is more and more the case, however, that a component is designed as a system of several materials. A tire contains two or three kinds of rubber and one or more kinds of cord, plus a steel bead (Fig. 13–1.1). The combination of these materials serves as an integrated materials system. Other examples include carburized steel, in which the surface is altered to another composition, and glass-reinforced plastics, GRP, a composite of two distinct materials. One material, for example, the glass in GRP, cannot serve its function without the other material, the plastic.

In this final chapter we shall pay specific attention to how combinations of materials can provide useful characteristics not obtainable by using one material alone. We shall also give attention to certain complex materials which are emerging as important factors in today's technological advances.

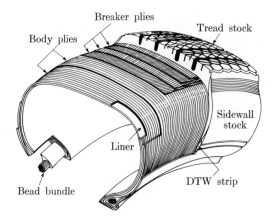

Fig. 13–1.1 A materials system (a tire). Several kinds of rubber, cord, and wire are combined into a system in which each material has a functional role. The behavior of each material is affected by the behavior of the other materials. (Courtesy of H. Howe, Uniroyal Tire)

13–2 REINFORCED MATERIALS

We are all familiar with reinforced concrete. The steel component is strong in tension. The concrete component is strong in compression, and also can be cast on the site into large or complex shapes—a great advantage. As suggested above, neither the concrete nor the steel could be used alone with the same desired results, or at the same cost, as can be obtained when they are used together. Their joint use, however, requires appropriate technical considerations. Although most of these considerations are best discussed in a course in reinforced concrete aimed at civil engineers, let us cite the types of factors which must receive attention. (1) The steel must be used in that part of the

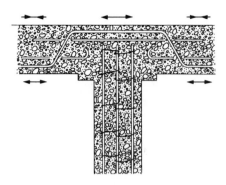

Fig. 13–2.1 Reinforced materials (concrete). The materials are selected to match the design requirements. The steel is used for tension; the concrete for compression and for rigidity.

structure subject to tension. Therefore stress analysis and calculations are required (Fig. 13–2.1). (2) The steel and concrete must be sufficiently well bonded so that they act in concert. If the steel were to slip within the concrete, the mutual benefits would be lost.

Glass-reinforced plastics are another familiar type of reinforced material. Like steel in concrete, the glass serves the plastic as a reinforcing material. Since it is more rigid than the plastic, it carries the load. Consider, for example, a "glass" fishing rod, which contains a bundle of aligned glass fibers bonded together within a plastic matrix. Assume a straight tensile load. The glass and plastic must stretch the same amount, since they are bonded together. However, it takes a greater stress to strain the glass 0.1% than to strain the plastic 0.1%. The glass therefore carries a larger proportion of the load than its volume fraction. *For a reinforcing material to be effective, it must have a greater modulus of elasticity than the matrix.*

Example 13–2.1. ⓔ A steel wire (0.05 in. in diameter) is coated with aluminum (combined diameter = 0.10 in.). The two materials have elastic moduli listed in Appendix B and yield strengths of 40,000 psi and 10,000 psi, respectively. (a) If this composite is loaded in tension, which will yield first? (b) How much load, F, can the composite carry in tension without plastic deformation? (c) What is the modulus of elasticity, $\bar{Y}$, of this composite material?

Solution

$$\text{Area}_{st} = \pi(0.05 \text{ in.})^2/4 = 0.002 \text{ in}^2,$$

$$\text{Area}_{Al} = \pi(0.10 \text{ in.})^2/4 - 0.002 \text{ in}^2 = 0.006 \text{ in}^2.$$

Using the notations of Section 2–2,

$$\sigma_{st}/Y_{st} = \varepsilon_{st} = \varepsilon_{Al} = \sigma_{Al}/Y_{Al}$$

$$\sigma_{st} = \sigma_{Al}(30,000,000 \text{ psi})/(10,000,000 \text{ psi}) = 3\sigma_{Al}.$$

a) With only a 3/1 stress ratio, the aluminum is stressed 10,000 psi when the steel is stressed only 30,000 psi.

b) $F_{total} = F_{Al} + F_{st}$

$$= (10,000 \text{ psi})(0.006 \text{ in}^2) + (30,000 \text{ psi})(0.002 \text{ in}^2),$$

Load = 120 lb.

c) $\bar{\sigma}/\bar{Y} = \bar{\varepsilon} = \varepsilon_{Al} = \sigma_{Al}/Y_{Al} = [F/(A_{Al} + A_{st})]/\bar{Y}.$

$$\bar{Y} = \frac{120 \text{ lb}}{(0.006 \text{ in}^2 + 0.002 \text{ in}^2)}\left(\frac{10,000,000 \text{ psi}}{10,000 \text{ psi}}\right)$$

$$= 15,000,000 \text{ psi}.$$

Comments. One may show from solution (c) that for $\bar{Y}$, the combined modulus of elasticity,

$$\bar{Y} = f_{st}Y_{st} + f_{Al}Y_{Al}.$$

Or more generally,

$$\bar{Y} = f_1Y_1 + f_2Y_2, \tag{13–2.1}$$

where f is the fraction of component 1 and 2 as indicated. ◄

Example 13–2.2. $\boxed{e}$ What is the thermal expansion coefficient of the Al-coated steel wire of Example 13–2.1?

Solution. In the composite, $(\Delta l/l)_{st} = (\Delta l/l)_{Al}$, and in the absence of an external load, $F_{Al} = -F_{st}$. Basis for calculation, $\Delta T = 1°F$, and the data of Appendix B:

$$(\Delta l/l)_{st} = (\Delta l/l)_{Al},$$

$$\alpha_{st}\,\Delta T + \left(\frac{F/A}{Y}\right)_{st} = \alpha_{Al}\,\Delta T + \left(\frac{F/A}{Y}\right)_{Al},$$

$$(\sim 6 \times 10^{-6})(1) + \frac{F_{st}/0.002\text{ in}^2}{30 \times 10^6\text{ psi}} = (\sim 12 \times 10^{-6})(1) + \frac{-F_{st}/0.006\text{ in}^2}{10 \times 10^6\text{ psi}},$$

$$F_{st} \cong +0.18\text{ lb}_f \qquad \text{(tension on heating)};$$

$$F_{Al} \cong -0.18\text{ lb}_f \qquad \text{(compression on heating)}.$$

$$(\Delta l/l) = (12 \times 10^{-6}/°F)(1°F) + \frac{-0.18\text{ lb}/0.006\text{ in}^2}{10 \times 10^6\text{ psi}},$$

$$\tilde{\alpha} = \sim 9 \times 10^{-6}/°F.$$

Comments. The thermal expansion varies slightly with the exact types of steel and aluminum alloys. ◄

13–3 SURFACE-MODIFIED MATERIALS $\boxed{e}$

Coatings. Coatings may be used for purposes of decoration, protection, or wear resistance. They may be metallic, polymeric, or ceramic, as summarized in Table 13–3.1. They are used to modify the surface composition of a material so that the surface has characteristics different from those of the bulk of the material. Only as the two materials of the system are combined is the engineering design complete.

Of course, the coating must have properties capable of meeting the situation in which it is to be used (Table 13–3.1). In addition, the two materials must be compatible. Compatibility involves adherence, which may be of the chemical type, in which bonds connect directly from atoms in the substrate to atoms in the coating. Or may be mechanical: The coating locks onto rough surfaces or around edges.

Table 13–3.1
SELECTED COATINGS

Type*	Composition	Substrate	Prime purpose
Paint	Polymers	Metal	Protection
		Wood	Protection, appearance
Enamel (vitreous)	Glass	Metal	Protection, appearance
Galvanize	Zinc	Metal	Protection
Electroplate	Noble metal	Metal	Appearance
Vapor plate	Metal	Plastic	Appearance, conductivity
Glaze	Glass	Ceramic	Appearance, sealing
Anodize	Al_2O_3	Aluminum	Protection
Vapor plate	Ceramic	Glass	Optical applications
Sizing	Polymer	Fiber	Lubrication, protection
Abrasive	Hard metal	Metal	Wear resistance
	Hard ceramic		

* Coating or process.

In some cases, attention must be given to the thermal expansion or to Young's modulus. Consider, for example, glass coatings on steel. The engineer carefully chooses a vitreous enamel coating which has a thermal expansion (or contraction) less than that of the underlying metal. Thus, after processing, in which the glass is melted onto the steel, the glass is in compression (and the metal is in tension) due to thermal differential contraction. This combination reduces the chances that the protective coating will crack and expose the metal, because any applied tension forces would first have to overcome the residual compression. Of course, glass is strong in the presence of compression, so applied compressive loads do not produce cracks.

Mechanically modified surfaces. Surfaces that have been subjected to compression also offer advantages when one is using ductile materials which must encounter cyclic stresses (see the discussion of *fatigue* in Section 4–5). Two processing procedures are available for metals. One is *shot-peening*, in which hardened steel shot are impinged onto the surface of a softer metal. As the surface (~ 0.005 in. thick) is plastically deformed, residual compressive stresses are introduced. This markedly increases the metal surface's resistance to fatigue. In certain products, the same effect can be achieved in metal fillets by special rolling techniques.

Thermally modified surfaces. Surface heat treatments of steel are common because one can heat the surface of the steel into the austenitic

Fig. 13–3.1 Modification of a surface by flame hardening. Heat is applied only where hardness is desired. Cooling is rapid because the unheated metal underneath serves as a heat sink. (Farrel Company Division, USM Corporation)

Fig. 13–3.2 Modification of a surface by induction hardening. This is a cam shaft. The high-frequency (R-F) currents heated only the surface region. The rapid cooling which followed produced a martensitic surface; the interior remained as a tougher ($\alpha + C$) steel. (Courtesy H. B. Osborn, Jr., Tocco Division, Park Ohio Industries)

range without doing the same for the center of the material. Two techniques are available. (1) The older method is *flame hardening* (Fig. 13–3.1); it is still widely used when one is making large products. (2) The second is *induction hardening*. Since high-frequency induction concentrates the current in the surface, it is particularly adapted to surface heating (Fig. 13–3.2). Either of these heating procedures provides a means of forming a hard martensitic surface (Section 6–6). In fact, quenching rapidly enough to get martensite is not a problem when only the surface is heated, because the center of the metal conveniently serves as a heat sink rather than a source of added heat that must be removed during cooling.

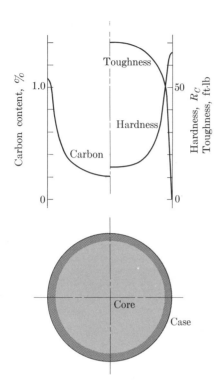

Fig. 13–3.3 Modification of a surface: case hardening by carburization. Carbon is diffused into the case to give a surface zone which is harder than the core. The latter remains tough and ductile.

Chemically modified surfaces. The compositions of surfaces can be altered by diffusing desired elements into the material. *Carburization* is the procedure most widely used to achieve this. Carbon is diffused into the surface (0.03–0.08 in.) of steel, for example. As shown in Fig. 6–6.1, the higher the carbon content, the harder the surface. The advantage of this *case hardening* (and the previously discussed induction hardening), is that not only does the surface become hard enough to resist wear, but the bulk of the steel behind the surface remains tough and ductile (Fig. 13–3.3). These surface-altered materials are as near to being composites as if two distinct materials were forged together.

A second way of altering the surface of steel chemically is by the process of *nitriding*. For this, the original steel must contain about 1 w/o aluminum. When nitrogen diffuses into the steel, it combines with the aluminum to form aluminum nitride,

$$\underline{Al} + N \rightarrow AlN. \tag{13–3.1}$$

The latter is very hard, giving the steel a surface-zone hardness of R_C 75. Note from Fig. 6–6.1 that this is harder than any of the martensite which is formed by quenching.

Glass also can be chemically altered at the surface to modify its properties. Recall from Section 11–4 that tempered glass is strong because the surface has been placed under compression. This was done by introducing residual stresses. A soda-lime glass (Section 10–6) contains 12–15 w/o Na_2O, or $\sim 6 \times 10^{21}$ Na^+ per cm^3. According to Table 10–1.1, the Na^+ ionic radius is 0.98 Å. If such a glass is heated in a bath of molten K_2SO_4, there can be *ion exchange* between the Na^+ of the glass and the K^+ of the melt:

$$Na^+_{glass} + K^+_{melt} \rightarrow Na^+_{melt} + K^+_{glass}. \tag{13–3.2}$$

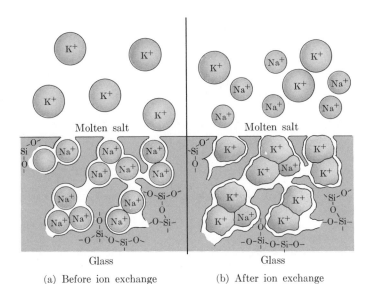

Molten salt | Molten salt

Glass | Glass

(a) Before ion exchange | (b) After ion exchange

Fig. 13–3.4 Modification of a surface: ion stuffing in glass. If a soda-lime glass is heated in molten K_2SO_4, part of the Na^+ ions of the surface zone of the glass are exchanged with larger K^+ ions. These larger ions place the surface under compression, thus strengthening the glass against tensile failure. (Courtesy George McLellan, Corning Glass Co.)

The K^+, which moves into the glass, has an ionic radius of 1.33 Å. It is therefore impossible for all the Na^+ ions to be exchanged because space is simply not available. However, enough K^+ ions can penetrate to place the rigid silicate network in surface compression (Fig. 13–3.4), building up to as much as 50,000 psi. Here, as with temper glass, bending or other service loading must remove the induced compressive stresses resulting from this *ion stuffing* before the surface is placed in tension.

Example 13–3.1. Figure 6–6.3(a) provides a hardness traverse for a 2.0-inch diameter bar of AISI 1040 steel. Based on the hardenability data of Fig. 13–3.5, sketch a similar traverse for an AISI 1020 steel which has been carburized as follows.

Surface	0.90 w/o C
$\frac{1}{16}$ in.	0.80 w/o C
$\frac{1}{8}$ in.	0.40 w/o C
$\frac{1}{4}$ in.	0.30 w/o C
$\frac{1}{2}$ in.	0.20 w/o C

This steel has the same dimensions and is quenched in the same manner as the 1040 steel in Fig. 6–6.3(a).

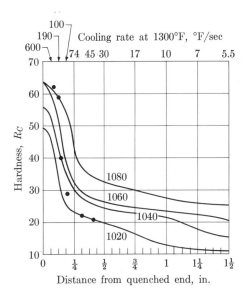

Fig. 13–3.5 Hardenability curves for AISI 10xx steels. (See Example 13–3.1.)

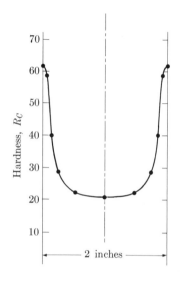

Fig. 13–3.6 Hardness profile (carburized 1020 steel). (See Example 13–3.1.)

Solution

	1040 hardness, R_C	Cooling rate, °F/sec (Fig. 6–6.6)	Equivalent end-quench position, in.	AISI	Hardness, R_C
Surface	53	400	1.5/16	1090	~62
$\frac{1}{16}$ in.	48	~200	2/16	1080	~59
$\frac{1}{8}$ in.	40	~150	2.5/16	1040	40
$\frac{1}{4}$ in.	35	~100	3/16	1030	29
Midradius	28	60	5/16	1020	22
Center	26	37	6.5/16	1020	21

See Fig. 13–3.6.

Comments. Compare with Example 6–6.2.

Surfaces can be selectively carburized by copper plating those areas one wants to be masked from the carbon. ◀

13–4 AGGLOMERATED MATERIALS Ⓜ ⓔ

Cements of various types have been used for centuries. These range from (a) silicate cement, used to bond aggregate in concrete, to (b) organic adhesives, used for bonding sheets into laminates. In this text we shall consider only portland cement and the concrete product.

Portland cement. The most common commercial cement contains compounds such as Ca_2SiO_4, Ca_3SiO_5, and $Ca_3Al_2O_6$. First let us consider the latter: tricalcium aluminate. It hydrates at normal temperatures in the presence of water,

$$Ca_3Al_2O_6 + 6H_2O \rightarrow Ca_3Al_2(OH)_{12}. \tag{13–4.1}$$

This *hydration* reaction is reversible at high temperatures and will dissociate, returning from $Ca_3Al_2(OH)_{12}$ to $Ca_3Al_2O_6$, much as clay dissociates on heating (Eq. 10–7.1). However, that need not concern us at the moment. The hydration of Eq. (13–4.1) is a surface reaction for the cement particles; therefore the reaction occurs more rapidly when the $Ca_3Al_2O_6$ is finely ground. The reaction product adheres tightly to the surface of various types of rock. This enables it to serve as a cement between two adjacent surfaces of sand, gravel, or rock.

The more common portland cement phase is dicalcium silicate, Ca_2SiO_4. In a greatly simplified form, its hydration reaction may be

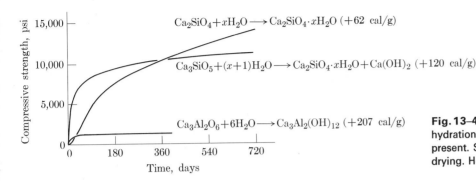

Fig. 13–4.1 Strength of cement components (through hydration). For hydration to proceed, moisture must be present. Strengthening of concrete does *not* occur by drying. Heat is given off during hydration.

shown as

$$Ca_2SiO_4 + xH_2O \rightarrow Ca_2SiO_4 \cdot xH_2O. \qquad (13\text{–}4.2)*$$

The product is almost completely amorphous and therefore provides a stronger bond with the surface of the aggregate than the tricalcium aluminate discussed above. The hydration of dicalcium silicate is slower than the hydration of the tricalcium aluminate of Eq. (13–4.1). See Fig. 13–4.1 for the time required to approach its maximum strength.

Finally, when tricalcium silicate hydrates it releases free lime, $Ca(OH)_2$:

$$Ca_3SiO_5 + (x + 1)H_2O \rightarrow Ca_2SiO_4 \cdot xH_2O + Ca(OH)_2. \qquad (13\text{–}4.4)$$

This occurs relatively rapidly, giving a rapid initial *set*, but the eventual maximum strength of the concrete is less than the product of Eq. (13–4.3).

Each of these reactions emphasizes that cement does not harden by drying, but by the chemical reaction of hydration. In fact it is necessary to keep concrete moist to ensure proper setting. The hydration reactions described above release heat, as shown in Fig. 13–4.1. Engineers take advantage of this in cold climates in which it may be

---

*   This reaction is more correctly stated as:

$$2Ca_2SiO_4 + (5 - y + x)H_2O$$
$$\rightarrow Ca_2[SiO_2(OH)_2]_2 \cdot (CaO)_{y-1} \cdot xH_2O + (3 - y)Ca(OH)_2,$$
$$(13\text{–}4.3)$$

where x varies with the availability of water and y is approximately 2.3. Equations (13–4.2) and (13–4.4) will suffice for our purposes in this book. However, a civil engineer should be aware of Eq. (13–4.3), because it is the basis of the changes in volume of concrete, which vary with humidity.

necessary to pour concrete at temperatures that are slightly below the freezing temperature of water. Conversely, the *heat of hydration* presents a problem when concrete is poured into massive structures such as dams. In these extreme cases, special cooling is required, and a refrigerant is passed through embedded pipes which are left in place and later act as reinforcement.

The variables we have discussed suggest that the engineer may utilize various kinds of cement to obtain different characteristics. For example, a cement with a larger fraction of tricalcium silicate (Ca_3SiO_5) sets rapidly and thus gains strength early.* In contrast, in cases in which the heat of reaction might be a complication, and more setting time is available, a Ca_2SiO_4-rich cement is used. The American Society for Testing and Materials (ASTM) lists the following types of cement: (1) *Type* I, used in general concrete construction in which special properties are not required. (2) *Type* III, used when one wants very strong, early-strength concrete. (3) *Type* IV, used when a low heat of hydration is required. There are also two other types (II and V) for special applications in which sulfate attack is a design consideration.

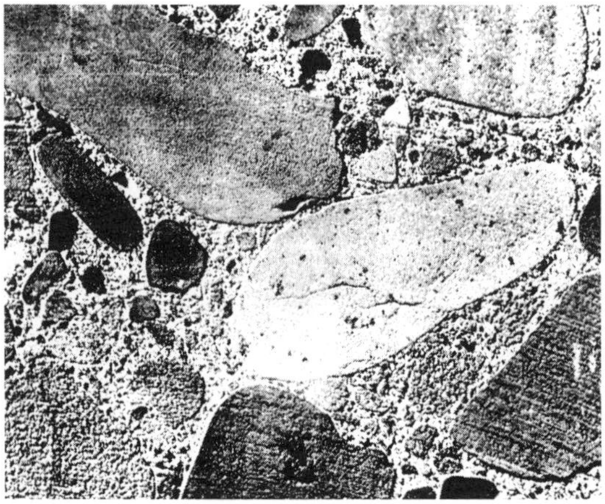

Concrete. In simplest terms, concrete is a composite of gravel with an admixture of sand to fill the pores. The space still remaining in the sand is filled with a "paste" of cement and water (Fig. 13–4.2). The cement hydrates to form a bond within the concrete. Experience has indicated that, in addition to the aggregate and the cement paste, a few percent (by volume) of small air bubbles should be entrained into the concrete. This entrained air improves the workability of the concrete during placement, but more important, these small air bubbles increase the resistance of the concrete to deterioration resulting from freezing and thawing of absorbed water.

Assuming that the aggregate is stronger than the cement, it is the property of the cement paste which governs the properties of concrete. Consequently, concrete of a *low water-to-cement ratio* is stronger than cements with high water-to-cement ratios. With a low ratio, there is more hydrated cement and less excess water in the spaces among the sand and gravel. Many engineers do not fully appreciate the indirect relationship of Fig. 13–4.3.

Example 13–4.1. A *unit mix* of concrete comprises one sack of cement, $2\frac{1}{4}$ ft³ of sand, and $2\frac{3}{4}$ ft³ of gravel. Using the data given below, calculate the number of cubic feet of concrete you would get from this mix if you used 7 gal of water.

Fig. 13–4.2 Concrete (×1.5). This is a composite with sand among the crushed rock or gravel. The pores among the sand are filled with a "paste" of cement and water. It is the water that reacts to produce bonding. (Portland Cement Association)

---

* A *high, early-strength cement* can also be achieved by grinding the cement more finely, as discussed in connection with Eq. (13–4.1).

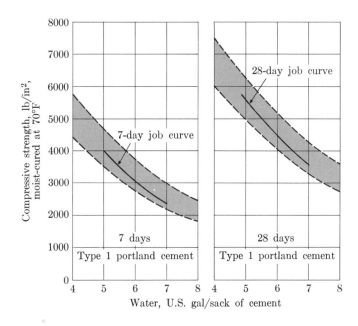

Fig. 13–4.3 Strength of concrete versus water content. There must be enough water to provide workability to the mix and to ensure that the forms are filled without voids. Additional water simply requires extra space and means less bonding between the sand and gravel. (J. N. Cernica, *Fundamentals of Reinforced Concrete,* Reading, Mass.: Addison-Wesley, 1964)

	Bulk density	Specific gravity
Sand	110 lb/ft^3	2.65
Gravel	108 lb/ft^3	2.62
Cement	94 lb/sack	3.15
H$_2$O	62.4 lb/ft^3	1.00
	(8.33 lb/gal)	

Solution

Sand: (110 lb/ft^3)(2.25 ft^3)/(2.65 × 62.4 lb/ft^3)　= 1.50 ft^3

Gravel: (108 lb/ft^3)(2.75 ft^3)/(2.62 × 62.4 lb/ft^3) = 1.82 ft^3

Cement: (94 lb)/(3.15 × 62.4 lb/ft^3)　　　　　　 = 0.48 ft^3

Water: (7 gal)(8.33 lb/gal)/(62.4 lb/ft^3)　　　　= 0.93 ft^3
$$\overline{4.73\ \text{ft}^3}$$

Comments. The standard sack (94 lb) of cement has a bulk volume of 1.00 ft^3.

Compare with Example 11–2.1. There is more sand in the concrete than there is pore space among the gravel. This is necessary because it is essentially impossible to distribute the sand exactly right for maximum packing. Likewise some excess water is necessary to ensure workability. ◀

Example 13–4.2. ⓞ   A concrete mix contains 550 lb of sand and gravel, 94 lb of cement, and 50 lb of water. Assume that the cement is (a) all tricalcium aluminate, $Ca_3Al_2O_6$, (b) all tricalcium silicate, Ca_3SiO_5, (c) all dicalcium silicate, Ca_2SiO_4. What is the temperature rise during hydration, given that the average specific heat of the concrete mix is 0.2 cal/g · °C? Assume that half the heat is lost. (Calories are the metric units more widely used in chemical reactions.)

Solution

Basis: 94 lb cement = (550 + 94 + 50) lb mix,

$$1 \text{ g cement} = 7.4 \text{ g mix.}$$

From data given in Fig. 13–4.1,

a) (7.4 g/g cement)(0.2 cal/g · °C) ΔT = (207 cal/g cement)(0.5),

$$\Delta T = 70°C \text{ with } Ca_3Al_2O_6.$$

b) ΔT = (120 cal)/(1.48 cal/°C)(2) = 40°C with Ca_3SiO_5.

c) ΔT = (62 cal)/(1.48 cal/°C)(2) = 21°C with Ca_2SiO_4.

Comments. Not only does Ca_3SiO_5 release more heat than Ca_2SiO_4; it does it in a relatively short period of time. Therefore much less of the heat escapes from a large structure during the period of setting. ◀

13–5 BIOMATERIALS Ⓜ ⓞ

The engineer uses a number of composite materials of biological origin. Wood is the most familiar, and will be the last material discussed in this book. Leather is another composite. Only as man learned to simulate its subskin structure (corium), which is a composite of fibers, pores, collagen, etc., was he able to make artificial leathers which simulated natural leather well enough to make such products as shoes.

Proteins are complex biological materials. Research work to decipher their structures has found that they contain specific combinations of a limited number of building blocks, or *mers*, which possess unique characteristics depending on their sequence within the molecular chain.

Dental materials illustrate another aspect of biomaterials. The tooth, with its enamel, dentin, cementum, and pulp (Fig. 13–5.1), is certainly a composite. However, for the person interested in materials, it is more; because amalgam fillings, inlays, bridges, false teeth, and other dental materials must be selected so that their properties and characteristics match nature's own tooth without alteration during use or the appearance of mechanical side effects. The science of dental materials is rapidly expanding. Similarly, medical implants which

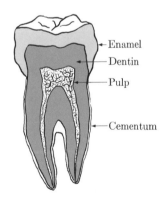

— Enamel
— Dentin
— Pulp
— Cementum

Fig. 13–5.1 Biomaterials (a tooth). This materials system has a hard, wear-resistant surface composed mainly of calcium phosphate. Dentin lends the tooth elastic strength to prevent cracking.

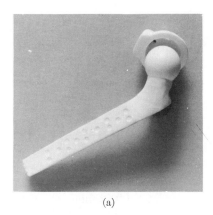
(a)

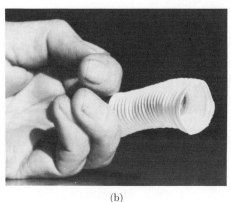
(b)

Fig. 13–5.2 Prosthetic materials. (a) Ceramic hip joint (W. D. Campbell). (b) Artificial aorta of PTFE (J. W. Freeman).

Table 13–5.1

TYPICAL MECHANICAL PROPERTIES OF HUMAN HARD TISSUES*

	Femur bone	Tooth enamel	Tooth dentin
Modulus of elasticity, 10^6 psi	2.82–2.98	6.7–6.9	1.7–2.0
Yield strength, 10^3 psi		28.2–32.5	18–21.5
Ultimate tensile strength, 10^3 psi	13–17.7		
Ultimate compressive strength, 10^3 psi	18–24	37.8–41.8	10.1–44.2
Shear strength,† 10^3 psi	16.8		
Shear strength,‡ 10^3 psi	5.7–13.3		

* From D. S. Crimmins, "The Selection and Use of Materials for Surgical Implants," *J. of Metals*, **21**, 1969, page 41.
† Parallel to long axis.
‡ Transverse to long axis.

replace bone, arteries, organs, or flesh are receiving ever more attention (Fig. 13–5.2 and Table 13–5.1). Literally, as well as figuratively, they require a lifetime guarantee.

Wood. This most familiar material is last in this text, but very important. As an engineering material, it has a high strength-to-weight ratio. It is easily processed, even on the job. And finally, it is a replaceable resource in this day of diminishing raw materials. We are, of course, also aware of its directional properties, which must be taken into consideration wherever it is used. To better understand wood, let's examine its structure.

Wood is a natural polymeric composite. The principal polymeric molecules are those of cellulose (Fig. 8–1.1),

$$
\left[
\begin{array}{c}
\text{CH}_2\text{OH} \\
\end{array}
\right]_n
\tag{13–5.1}
$$

in which the molecular weights range up to 2,000,000 g/mole (as many as 12,000 mers). Since cellulose has many —OH radicals, it develops a fair degree of crystallinity. In addition to its more than 50% cellulose composition, wood contains 10% to 35% *lignin*, a more complex carbohydrate polymer.

Larger than polymeric molecules, wood's next-most-prevalent structural units are biological cells, of which the most extensive are the *tracheids*. These are hollow, spindle-shaped cells which are elongated in the longitudinal direction of the wood. The most visible structural unit is the "grain" of the wood, which is made up of *spring* and *summer* layers. The biological cells of the spring wood are larger and have thinner walls than those of the summer wood (Fig. 13–5.3). In this respect, biological cells are much more variable in structure than the unit cells of crystals.

The above description of wood is, of course, oversimplified.* However, it does indicate the source of the anisotropies in properties that are so characteristic of wood.

In materials which have simple structures, *density* is a structure-insensitive property. This is not the case in a complex material such as wood. First, the amount of spring wood and summer wood varies from species to species. Second, the ratio of cellulose to the more dense lignin varies. As a result of these two factors, the density can range from about 0.15 g/cm³ for *balsa* to 1.3 g/cm³** for the dense *lignum vitae*, so named because of its abnormally high lignin content (and low tracheid content). In addition to the above factors, wood is hydroscopic; hence it absorbs moisture as a function of humidity of

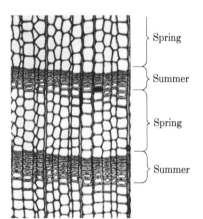

Fig. 13–5.3 Biomaterials (wood). The cells which form with spring growth are larger, but have thinner walls than those which form with summer growth. Cellulose (Eq. 13–5.1) is a major molecular constituent.

Spring

Summer

Spring

Summer

---

*   It does not take into account the wide variety of minor wood chemicals, nor the *vascular rays*, which are rows of single, almost equidimensional cells that radiate from the center of the tree. These become important during deformation.

**   Recall that water has a density of 1.0 g/cm³.

the surrounding atmosphere. Because this is so, most woods, when they are wet, exhibit a net increase in density.

It should come as no surprise that *dimensional changes* which accompany variations in temperature, moisture, and mechanical loading in wood are anisotropic. Wood technologists point out that *thermal expansion* is about 40% greater in the tangential, *t*, direction (Fig. 13–5.4) than in the radial, *r*, direction, and 6 to 8 times greater in the radial direction than in the longitudinal, *l*, direction. Thermal expansion in the longitudinal direction is relatively independent of density, while in the other two directions it depends on density, ρ (in g/cm^3):

$$\alpha_t \cong 70\rho \times 10^{-6}/°C,$$

$$\alpha_r \cong 50\rho \times 10^{-6}/°C,$$

$$\alpha_l \cong 3 \times 10^{-6}/°C.$$

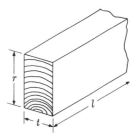

Fig. 13–5.4 Anisotropy of wood. The longitudinal thermal expansion coefficient, α_l, differs from the radial and tangential coefficients, α_r and α_t. The elastic moduli— Y_l, Y_r, and Y_t—are also anisotropic.

As an order of magnitude, longitudinal values for *Young's modulus* are between 1,000,000 and 2,000,000 psi when measured in tension. Tangential values are normally between 60,000 and 100,000 psi, while the radial values are commonly in the 80,000- to 150,000-psi range. We would expect the longitudinal value to be the highest. However, we could also have predicted that the tangential modulus would be higher than the radial. But in Fig. 13–5.3 we did not take into account the vascular rays cited in the accompanying footnote. This feature provides an additional rigidity in the radial direction.

Shrinkage is also anisotropic; longitudinal changes are negligible, but tangential shrinkage is very high (~ 0.25 *l*/o per 1 w/o moisture for Douglas fir). Radial shrinkage is intermediate (~ 0.15 *l*/o per 1 w/o moisture for the same wood) because of the restraining effects of the vascular rays. Effects of shrinkage are summarized in Fig. 13–5.5. The consequences of dimensional distortion on *warpage* are readily apparent.

The *longitudinal tensile strength*, S_l, is upwards of 20 times the radial tensile strength, S_r, because any fracture that takes place must occur across the elongated tracheid cells. Increased densities (for a given moisture content) reflect an increase in the thickness of cell walls, and therefore a proportional increase in the longitudinal strength. The transverse strength increases with the square of the density (again for a given moisture content), because the denser the wood, the less the opportunity for failure parallel to the hollow tracheid cells.

Except in cases in which designs capitalize on the longitudinal strength, the anisotropies of wood are undesirable. Therefore much has been done to modify its structure. Examples include (1) *plywood*, in

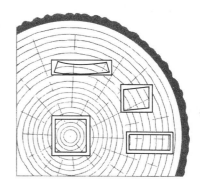

Fig. 13–5.5 Shrinkage of wood. Since the structure is anisotropic (that is, it varies with direction), shrinkage is nonuniform. This may cause warpage if the wood is cut before it has dried completely. (Shrinkage is exaggerated in this sketch.)

which the longitudinal strength is developed in *two* coordinates, and (2) *impregnated wood*, in which the pores are filled with a polymer such as phenol-formaldehyde to provide better bonding across the grain structure. In effect, impregnated wood is a plastic which is reinforced with fibers (i.e., elongated tracheid cells). (3) Finally, the *cellulose* may be extracted from the wood to serve as a raw material for many polymeric products (Section 8–1).

REVIEW AND STUDY

13–6 SUMMARY

Composites are integrated materials in which the composition, microstructure, or stress state is selected to give a better overall product. We are all familiar with reinforced concrete and various surface coatings such as paint and electroplate. In addition, the opportunities are almost unlimited to *design* new combinations of materials for engineering applications. Considerable developmental work is going on in this area.

Surface-modified materials include not only coatings, but mechanical and chemical treatments which induce residual stresses within a material, plus thermal treatments and diffusional coatings to alter the surface properties of various materials.

Portland cement sets, not by drying, but by a hydration reaction which bonds the particles of the aggregate. The strength (and economy) of a concrete depend on the mixing of fine aggregate into the pores of the coarse aggregate, the cement/water ratio, and the components of the cement. The engineer can choose cements with variations in setting times, heats of hydration, etc.

Wood is the most common biomaterial for engineering use. Its anisotropy is a direct consequence of its structure. The bioengineer must focus ever more attention on the properties and characteristics of materials of the future as he uses them for implants, prosthetics, and other biological functions.

Terms for Study

Carburization	Composite
Case hardening	Concrete
Cellulose	Flame hardening
Coatings	Glass-reinforced plastics, GRP

Heat of hydration	Portland cement
Hydration	Setting (cement)
Induction hardening	Shot peening
Ion exchange	Spring wood
Ion stuffing	Types (ASTM) of cement
Lignin	Unit mix (concrete)
Materials system	Warpage (wood)
Nitriding	

STUDY PROBLEMS

13-2.1 A glass-reinforced plastic rod (fishing pole) is made of 67 v/o boro-silicate glass fibers in a polystyrene matrix. What is the modulus of elasticity of the composite?

Answer. 6.8×10^6 psi

13-2.2 Estimate within 10% the electrical resistance of the aluminum-coated steel wire in Example 13-2.1. [*Hint:* Recall from your high school science courses that, when two resistances are in parallel, $1/R = 1/R_1 + 1/R_2$.]

Answer. 0.002 ohm/ft (or 0.00007 ohm/cm)

13-2.3 What is the thermal expansion of the glass-reinforced rod of Study problem 13-2.1?

Answer. 2.2×10^{-6} in./in./°F (or $\sim 4 \times 10^{-6}$/°C)

13-3.1 On the basis of Figs. 6-6.6 and 13-3.5, show a plot of the effect of carbon on surface hardnesses of (a) 1-in. and (b) 3-in. round bars when quenched in oil. (Recall from Section 6-6 that hardness can be determined on the basis of cooling rate.)

Answer. (a) R_C 30 at 0.2C, R_C 57 at 0.8C

13-3.2 A 1020 steel bar 1.6 in. in diameter has been case-hardened, or carburized, so that the case has 0.7% carbon at the surface and 0.4% carbon 0.2 in. below the surface. On the basis of the hardenability data in Fig. 13-3.5, draw a hardness traverse for this steel after oil quenching.

Answer. S, R_C 34; (0.2 in.), R_C 27; M-R, $R_C \sim 21$; C, R_C 20

13-3.3 A glass rod (0.073 in. in diameter) supports a tensile load of 75 lb just prior to fracture. Another rod (0.068 in. in diameter) is made of the same glass, but with ion exchange as described in Eq. (13-3.2). It supports a load of 247 lb when loaded in the same manner. Assume that fracture starts in each case with a similar tension crack from the surface. What residual compressive stress was present in the treated glass?

Answer. 50,000 psi

13-4.1 The bulk volume of a sack of cement is 1.00 ft^3. Using the data of Example 13–4.1, calculate the porosity of dry cement.

Answer. 52 v/o

13-4.2 A box 6 in. $\times$ 4 in. $\times$ 9 in. contains 12 lb of SiO_2 when it is level full. (a) What is the packing factor? (Specific gravity = 2.65). (b) What is the maximum weight of water which could be added to just fill the pores of the sand in this box?

Answer. (a) 0.58   (b) 3.28 lb

13-4.3 A concrete contains a $1:2:3$ unit mix (bulk volume) of cement, sand, and gravel with the properties shown in Example 13–4.1. If 5.5 gal of water are used with the above unit mix, how many cubic feet of concrete will result?

Answer. 4.524 ft^3

13-4.4 In a certain construction project the builder uses a concrete mix composed of dry cement, sand, and gravel in volume proportions of $1:2:3.5$, respectively, plus 6 gal of water for each sack (cubic foot) of cement. The densities of the constituents are as follows:

	Bulk density	Specific gravity
Dry cement	94 lb/ft^3	3.10
Sand	105 lb/ft^3	2.65
Gravel	95 lb/ft^3	2.60
Water	62.4 lb/ft^3	1.00
	(8.33 lb/gal)	

Air-entraining agents are added to give 4.5 v/o closed pore space. Calculate the bulk density of the wet concrete mix.

Answer. 142 lb/ft^3 with 4.5 v/o pores

13-4.5 The cement of Example 13–4.2 contains 40 w/o Ca_3SiO_5, 40 w/o Ca_2SiO_4, 15 w/o $Ca_3Al_2O_6$ and 5 w/o nonreactive material. What is the maximum temperature rise if heat loss is negligible?

Answer. 70.2°C

13-5.1 Birchwood veneer is impregnated with phenol-formaldehyde (Fig. 7–2.5) to ensure resistance to water and to increase strength in the final product. Although dry birch weighs only 0.56 g/cm^3, the true specific gravity of the cellulose-lignin combination is 1.52. (a) How many grams of phenol-formaldehyde are required to impregnate 1 in^3 of dry birchwood? (b) What is the final density?

Answer. (a) 13.4 g   (b) 1.38 g/cm^3

CONSTANTS
AND
CONVERSIONS

CONSTANTS

Atomic mass unit, amu	1.66×10^{-24} g
Avogadro's number, AN	6.0×10^{23}
Charge on the electron, q	1.6×10^{-19} coul
Natural logarithm base, e	$2.718\ldots$
Velocity of light, c	3.0×10^{10} cm/sec

CONVERSIONS

To convert from	To	Multiply by
ampere	coulomb/sec	1
angstrom	centimeter	10^{-8}
angstrom	inch	3.937×10^{-9}
angstrom	meter	10^{-10}
amu	gram	1.66×10^{-24}
Btu	calorie, gram	252
Btu	joule	1055
Btu/°F	joule/°C	1897
Btu $\cdot$ in./ft^2 $\cdot$ sec $\cdot$ °F	joule $\cdot$ cm/cm^2 $\cdot$ sec $\cdot$ °C	5.192
Btu/ft^2	joule/m^2	11,300
calorie, gram	Btu	0.00397
calorie, gram	joule	4.184
calorie/°C	Btu/°F	0.0022
centimeter	angstrom	10^8
centimeter	inch	0.3937
centimeter	micron	10^4
coulomb	ampere $\cdot$ second	1
coulomb	electronic charge	6.24×10^{18}
cubic centimeter	cubic inch	0.061
cubic foot	gallon	7.48
cubic inch	cubic centimeter	16.39
°C	(°F $-$ 32)	1.8
electron volt	joule	1.6×10^{-19}
electron charge	coulomb	1.6×10^{-19}
(°F $-$ 32)	°C	0.555
foot	centimeter	30.48

CONVERSIONS *(continued)*

To convert from	To	Multiply by
foot · pound	joule	1.356
gallon (U.S. liq.)	cubic foot	0.134
gallon (U.S. liq.)	cubic inch	231
gallon (U.S. liq.)	liter	3.79
gallon (U.S. liq.)	pound (water)	8.34
gram	amu	6.02×10^{23}
gram	pound (water)	0.0022
gram/cm^3	pound/ft^3	62.43
inch	angstrom	2.54×10^8
inch	centimeter	2.54
joule	Btu	0.000948
joule	calorie	0.239
joule	electron volt	6.24×10^{18}
joule	foot · pound	0.738
joule/m^2	Btu/ft^2	8.82×10^{-5}
joule · cm/cm^2 · sec · °C	Btu · in./ft^2 · sec · °F	0.1926
kilogram	pound	2.20
liter	gallon (U.S. liq.)	0.264
meter	angstrom	10^{10}
meter	foot	3.28
meter	inch	39.37
millimeter of Hg	atmosphere	0.001316
newton	pound force	0.225
newton/m^2	pound/in^2	1.45×10^{-4}
ohm · centimeter	ohm · inch	0.3937
ohm · inch	ohm · centimeter	2.54
pound	gram	453.6
pound/ft^3	gram/cm^3	0.016
pound/in^2	newton/m^2	6,900
watt	joule/sec	1

PROPERTIES OF
SELECTED
ENGINEERING MATERIALS

PART 1 — METALS (taken from numerous sources)

Material	Specific gravity	Thermal conductivity, $\left[\dfrac{\text{Btu} \cdot \text{in.}}{\text{ft}^2 \cdot \text{sec} \cdot {}^\circ\text{F}}\right]$ at 68°F*	Thermal expansion, in/in/°F at 68°F†	Electrical resistivity, ohm · cm at 68°F‡	Average modulus of elasticity, psi at 68°F§
Aluminum (99.9+)	2.7	0.43	12.5×10^{-6}	2.7×10^{-6}	10×10^6
Aluminum alloys	2.7(+)	0.3(±)	12×10^{-6}	$3.5 \times 10^{-6}(+)$	10×10^6
Brass (70Cu–30Zn)	8.5	0.235	11×10^{-6}	6.2×10^{-6}	16×10^6
Bronze (95Cu–5Sn)	8.8	0.16	10×10^{-6}	9.5×10^{-6}	16×10^6
Cast iron (gray)	7.15	—	5.8×10^{-6}	—	$12\text{--}25 \times 10^6$
Cast iron (white)	7.7	—	5×10^{-6}	—	30×10^6
Copper (99.9+)	8.9	0.77	9×10^{-6}	1.7×10^{-6}	16×10^6
Iron (99.9+)	7.87	0.14	6.53×10^{-6}	9.7×10^{-6}	30×10^6
Lead (99+)	11.34	0.06	16×10^{-6}	20.65×10^{-6}	2×10^6
Magnesium (99+)	1.74	0.37	14×10^{-6}	4.5×10^{-6}	6.5×10^6
Monel (70Ni–30Cu)	8.8	0.05	8×10^{-6}	48.2×10^{-6}	26×10^6
Silver	10.5	0.79	10×10^{-6}	1.6×10^{-6}	11×10^6
Steel (1020)	7.86	0.096	6.5×10^{-6}	16.9×10^{-6}	30×10^6
Steel (1040)	7.85	0.093	6.3×10^{-6}	17.1×10^{-6}	30×10^6
Steel (1080)	7.84	0.089	6.0×10^{-6}	18.0×10^{-6}	30×10^6
Steel (18Cr–8Ni stainless)	7.93	0.028	5×10^{-6}	70×10^{-6}	30×10^6

* Multiply by 5.19 to get joule · cm/cm^2 · sec · °C.
† Multiply by 1.8 to get cm/cm/°C.
‡ Divide by 2.54 to get ohm · in.
§ Multiply by 6900 to get n/m^2.

PART 2 — CERAMICS (taken from numerous sources)

Material	Specific gravity	Thermal conductivity, $\left[\dfrac{\text{Btu} \cdot \text{in.}}{\text{ft}^2 \cdot \text{sec} \cdot {}^\circ\text{F}}\right]$ at 68°F*	Thermal expansion, in/in/°F at 68°F†	Electrical resistivity, ohm · cm at 68°F‡	Average modulus of elasticity, psi at 68°F§
Al_2O_3	3.8	0.057	5×10^{-6}	—	50×10^6
Brick					
Building	2.3(±)	0.0012	5×10^{-6}	—	—
Fireclay	2.1	0.0016	2.5×10^{-6}	1.4×10^8	—
Graphite	1.5	—	3×10^{-6}	—	—
Paving	2.5	—	2×10^{-6}	—	—
Silica	1.75	0.0016	—	1.2×10^8	—
Concrete	2.4(±)	0.002	7×10^{-6}	—	2×10^6
Glass					
Plate	2.5	0.0014	5×10^{-6}	10^{14}	10×10^6
Borosilicate	2.4	0.002	1.5×10^{-6}	$>10^{17}$	10×10^6
Silica	2.2	0.0025	0.3×10^{-6}	$\sim 10^{20}$	10×10^6
Vycor	2.2	0.0025	0.35×10^{-6}	—	—
Wool	0.05	0.0005	—	—	—
Graphite (bulk)	1.9	—	3×10^{-6}	10^{-3}	1×10^6
MgO	3.6	—	5×10^{-6}	10^5 (2000°F)	30×10^6
Quartz (SiO_2)	2.65	0.025	—	10^{14}	45×10^6
SiC	3.17	0.025	2.5×10^{-6}	2.5 (2000°F)	—
TiC	4.5	0.06	4×10^{-6}	50×10^{-6}	50×10^6

* Multiply by 5.19 to get joule · cm/cm² · sec · °C.
† Multiply by 1.8 to get cm/cm/°C.
‡ Divide by 2.54 to get ohm · in.
§ Multiply by 6900 to get n/m².

PART 3 — POLYMERS (taken from numerous sources)

Material	Specific gravity	Thermal conductivity, $\left[\dfrac{Btu \cdot in.}{ft^2 \cdot sec \cdot °F}\right]$ at 68°F*	Thermal expansion, in/in/°F at 68°F†	Electrical resistivity, ohm · cm at 68°F‡	Average modulus of elasticity, psi at 68°F§
Melamine-formaldehyde	1.5	0.00057	15×10^{-6}	10^{13}	1.3×10^6
Phenol-formaldehyde	1.3	0.00032	40×10^{-6}	10^{12}	0.5×10^6
Urea-formaldehyde	1.5	0.00057	15×10^{-6}	10^{12}	1.5×10^6
Rubbers (synthetic)	1.5	0.00025	—	—	600–11,000
Rubber (vulcanized)	1.2	0.00025	45×10^{-6}	10^{14}	0.5×10^6
Polyethylene (LD)	0.92	0.00065	100×10^{-6}	10^{15}–10^{18}	14,000–50,000
Polyethylene (HD)	0.96	0.0010	70×10^{-6}	10^{16}–10^{18}	50,000–180,000
Polystyrene	1.05	0.00016	40×10^{-6}	10^{18}	0.4×10^6
Polyvinyl chloride	1.3	0.0003	75×10^{-6}	10^{14}	100,000–600,000
Polyvinylidene chloride	1.7	0.00025	105×10^{-6}	10^{13}	0.05×10^6
Polytetrafluoroethylene	2.2	0.0004	55×10^{-6}	10^{16}	50,000–100,000
Polymethyl methacrylate	1.2	0.0004	50×10^{-6}	10^{16}	0.5×10^6
Nylon	1.15	0.0005	55×10^{-6}	10^{14}	0.4×10^6

* Multiply by 5.19 to get joule · cm/cm^2 · sec · °C.
† Multiply by 1.8 to get cm/cm/°C.
‡ Divide by 2.54 to get ohm · in.
§ Multiply by 6900 to get n/m^2.

TABLE
OF
SELECTED ELEMENTS

TABLE OF SELECTED ELEMENTS

Element	Symbol	Atomic number	Atomic weight	Melting point, °C	°F	Density (solid) 20°C (68°F)	Crystal structure 20°C (68°F)	Approx. atomic radius, Å‡	Valence (most common)	Approx. ionic radius, Å§
Aluminum	Al	13	26.98	660.2	1220	2.699	fcc	1.431	3+	0.51
Antimony	Sb	51	121.75	630.5	1167	6.62	—	1.452	5+	—
Argon	Ar	18	39.95	−189.4	309	—	—	1.92	Inert	—
Arsenic	As	33	74.92	817	1503	5.72	—	1.25	3+	—
Beryllium	Be	4	9.01	1277	2332	1.85	hcp	1.14	2+	0.35
Boron	B	5	10.81	2030	3090	2.34	—	0.46	3+	~0.25
Calcium	Ca	20	40.08	838	1540	1.55	fcc	1.969	2+	0.99
Carbon	C	6	12.01	3727	6740	2.25	hex	0.71	—	—
Chlorine	Cl	17	35.45	−101	−150	—	—	0.905	1−	1.81
Chromium	Cr	24	52.00	1875	3407	7.19	bcc	1.249	3+	0.63
Cobalt	Co	27	58.93	1495	2723	8.9	hcp	1.25	2+	0.73
Copper	Cu	29	63.54	1083	1981	8.96	fcc	1.278	1+	0.96
Fluorine	F	9	19.00	−220	−364	—	—	0.6	1−	1.33
Germanium	Ge	32	72.59	937	1719	5.32	*	1.224	4+	—
Gold	Au	79	197.0	1063	1945	19.32	fcc	1.441	1+	1.37
Helium	He	2	4.003	−272.2	−458	—	—	1.76	Inert	—
Hydrogen	H	1	1.008	−259.14	−434.5	—	—	0.46	1+	Very small
Iodine	I	53	126.9	114	238	4.94	**	1.35	1−	2.20
Iron	Fe	26	55.85	1537	2798	7.87	bcc	1.241	2+	0.74
									3+	0.64

Element	Symbol	Atomic number	Atomic weight	Melting point, °C	Melting point, °F	Density (solid) 20°C (68°F)	Crystal structure 20°C (68°F)	Approx. atomic radius, Å‡	Valence (most common)	Approx. ionic radius, Å§
Lead	Pb	82	207.2	327.4	621.3	11.34	fcc	1.750	2+	1.20
Lithium	Li	3	6.94	179	354	0.534	bcc	1.519	1+	0.68
Magnesium	Mg	12	24.31	650	1202	1.74	hcp	1.61	2+	0.66
Manganese	Mn	25	54.94	1245	2273	7.4	—	1.12	2+	0.80
Mercury	Hg	80	200.6	−38.87	−38.0	—	—	1.55	2+	1.10
Neon	Ne	10	20.18	−248.6	−415.5	—	—	1.60	Inert	—
Nickel	Ni	28	58.71	1453	2647	8.90	fcc	1.246	2+	0.69
Nitrogen	N	7	14.007	−210	−346	—	—	0.71	3−	—
Oxygen	O	8	15.9994	−218.4	−361	—	—	0.60	2−	1.40
Phosphorus	P	15	30.97	44	111	1.8	—	1.1	5+	∼0.35
Potassium	K	19	39.10	64	147	0.86	bcc	2.312	1+	1.33
Silicon	Si	14	28.09	1410	2570	2.33	*	1.176	4+	0.42
Silver	Ag	47	107.87	960.8	1761	10.5	fcc	1.444	1+	1.26
Sodium	Na	11	22.99	97.8	208	0.97	bcc	1.857	1+	0.97
Sulfur	S	16	32.06	119.0	246	2.07	—	1.06	2−	1.84
Tin	Sn	50	118.69	231.9	449.4	7.3	—	1.509	4+	0.71
Titanium	Ti	22	47.90	1668	3035	4.51	hcp	1.46	4+	0.68
Tungsten	W	74	183.9	3410	6170	19.3	bcc	1.367	4+	0.70
Uranium	U	92	238	1132	2070	18.7	—	1.38	4+	0.97
Zinc	Zn	30	65.37	419.4	787	7.135	hcp	1.39	2+	0.74

* Diamond cubic. See Fig. 12–2.2(a).
** Noncubic. See Fig. 7–3.1.
‡ One-half the closest approach of two atoms in the elemental solid. Also, for hcp, see footnote to Table 3–1.1.
§ Radii for CN = 6. (Also see Table 10–1.1.) Patterned after Ahrens.

GLOSSARY OF SELECTED TERMS AS APPLIED TO MATERIALS

Abrasives Mechanically hard materials. The two most widely used abrasives are Al_2O_3 (emery) and SiC.

Absolute temperature (°K) Zero at $-273°C$ ($-460°F$); °K = °C + 273.

Absorptivity (β) Measure of light absorption; units of cm^{-1}.

$A_mB_nX_p$ *compounds* Ternary compounds with integer atom ratios. In this text, X is generally limited to oxygen.

Acceptor Energy level slightly above the valence band, which accepts electrons (and therefore produces an electron hole in the valence band).

Acceptor saturation Filling of acceptor sites. As a consequence, additional thermal activation does not increase the number of extrinsic carriers.

Additional polymerization Polymerization by a double bond opening up into two single bonds as the monomer joins a reactive site of a growing chain (chain-reaction polymerization).

Additives Nonpolymeric materials incorporated into plastics for purposes of filling, stabilizing, plasticizing, coloring, etc.

Age-hardening Hardening with time by incipient precipitation.

Aging Process of age-hardening.

AISI-SAE steels Standardized identification code for plain-carbon and low-alloy steels which is based on composition. Last two numbers indicate the carbon content.

Alloy A metal containing two or more components.

Amorphous solids Solids which are noncrystalline and without long-range order.

Annealing Heating and slow cooling to produce softening, toughening, or stress relief.

Annealing (*full*) Heating steel 50°F into austenite range and slow cooling to soften the steel.

Annealing point (*glass*) Temperature at which the viscosity, η, of glass is 10^{13} poises.

Annealing (*process*) Heating steel to relieve stresses and strain-hardening without forming austenite.

Atactic polymer A polymer which does not have long-range repetition of its mers (as contrasted with *isotactic*).

Atomic mass unit (amu) Units for expressing the relative mass of individual atoms and molecules; 6×10^{23} amu = 1 g.

Atomic packing factor (APF) Fraction of space occupied by "spherical" atoms.

Atomic radius One-half the distance of closest approach of two atoms of a metal.

Austenite (γ) Iron-rich fcc phase.

Avogadro's number (AN) Number of atoms per gram-atomic weight (or number of molecules per gram-molecular weight); 6×10^{23}.

AX *compounds* Binary compounds with a 1 : 1 ratio of the two elements; commonly ionic.

$A_m X_p$ *compounds* Binary compounds with an integer ratio of atoms, other than 1 : 1.

Base The center zone of a transistor.

Basic oxygen furnace (BOF) Refining furnace for steel in which impurities are oxidized by oxygen injection.

$BaTiO_3$-*type structure* Ternary compound with 8-fold and 6-fold coordinations for the two cations.

Bifunctional molecule Molecule with two reaction sites for joining with adjacent molecules.

Birefringence ($n_2 - n_1$) Difference between index of refraction of light in the two directions of vibration.

Blast furnace Reducing furnace which removes oxygen from ore.

Block copolymer Copolymer with a clustering of similar mers along its chain.

Blow molding Processing by expanding a parison into a mold by air pressure. (Also see *Glass-bottle making*.)

Body centered cubic (bcc) The lattice of a cubic crystal which has equivalent sites at the corners *and* at the center of the cube.

Boule Large single crystal grown by solidification onto a seed crystal.

Branching Bifurcation in chain polymers.

Brass An alloy of copper and zinc.

Brazing Joining metals with a filler metal at a temperature above 800°F (425°C) but below the melting point of the metals.

Breaking strength (BS) Stress at fracture. This value is most useful when expressed in terms of the actual area.

Brinell hardness number (BHN) Hardness index derived by standardized procedures. The hardness is calibrated by measuring the diameter of the indentation.

Brittle Opposite of tough. A brittle material fractures with very little consumption of energy.

Bronze An alloy of copper and tin (if not specified otherwise). Other bronzes include Al bronze, a Cu-Al alloy; Be bronze, a Cu-Be alloy; and Mn bronze, a Cu-Mn alloy.

Bubble processing Process for biaxial straining of plastic film by air pressure inside a cylindrical extrusion.

CaF_2-type structure Binary compound with simple cubic anion lattice and cations in half the 8-fold sites. (Alternatively, fcc cation lattice with anions in all the 4-fold sites.)

Calcine Decomposition by gas evolution.

Carbide Any binary metal-carbon compound. However, the word usually refers to Fe_3C or a related phase with other metals replacing iron.

Carburization Process of hardening steel by diffusion of carbon into the surface.

Case-hardening See *Carburization.*

Casting The process of pouring molten metal or a suspension into a mold; or the metal object produced by this process.

Cast iron Fe-C alloy, sufficiently rich in carbon to produce a eutectic during solidification. In practice this generally means more carbon than can be dissolved in austenite ($>2\%$).

Cast iron (*gray*) Cast iron with graphite flakes; hence its fractured surface is gray in color.

Cast iron (*malleable*) Cast iron in which the Fe_3C dissociates after solidification. Graphite flakes are avoided and some ductility is exhibited.

Cast iron (*nodular*) Cast iron in which graphite spherulites form during solidification. Sometimes called ductile iron because of its properties.

Cast iron (*white*) Cast iron which was chilled to avoid graphitization.

Cellulose Natural polymer of wood and cotton. See Fig. 8–1.1 with R as —OH.

Cemented carbides Refractory carbide powders, for example, WC or TiC, bonded by a metal, usually nickel or cobalt.

Cementite The phase Fe_3C.

Ceramic materials Materials consisting of compounds of metallic and nonmetallic elements. (See Fig. 1–2.2.)

Chain-reaction polymerization See *addition polymerization.*

Chain polymers Macromolecules of bifunctional mers which give a linear structure.

Chain silicates Polymerized silicates with two oxygens of each tetrahedron, jointly shared with two neighboring tetrahedra to provide a chainlike structure.

Charge carriers Electrons in the conduction band provide *n*-type (negative) carriers. Electron holes in the valence band provide *p*-type (positive) carriers.

Charge density ($\mathscr{D}$) Coulombs per unit area.

Charpy test The most common procedure for the testing of toughness (or impact "strength"). The test specimen contains one of two standardized notches: (a) V-notch, or (2) keyhole notch. (See Fig. 2–5.1.)

Clay Fine soil particles (<0.1 mm). In ceramics, clays are specifically sheetlike alumino-silicates.

Clustering Grouping of atoms in a solid solution prior to precipitation.

Coatings Adherent surface layer formed on a material.

Coatings (*polymer*) Surface applications. Hardening is accomplished by cooling (thermoplastic coatings), evaporation (diluent coatings) and/or polymerization (reactive coatings).

Cold working (*CW*) Deformation below the recrystallization temperature so that the strain-hardening remains; CW $= (A_0 - A_f)/A_0$.

Collector The zone of a transistor which receives the charge carriers across the base from the emitter.

Colorants Dyes (soluble) or pigments (particles) used as additives for coloring purposes.

Components Basic chemical substances required to create a chemical compound or solution.

Composite Material containing two or more integrated components.

Compound Material containing two or more types of elements, in more or less definite ratios.

Concrete Composite of an aggregate and a hydraulic cement.

Condensation polymerization See *Step-reaction polymerization*.

Conduction band Energy band of conduction electrons. Electrons must be in this band to be carriers.

Conduction electron (*n*) Electron raised above the energy gap to serve as negative charge carrier.

Configuration Arrangement of mers along a polymer chain.

Conformation Twisting and/or kinking of a polymer chain.

Cooling rate Decrease in temperature per second; specifically, the rate of change at the transformation temperature.

Coordination number (*CN*) Number of closest ionic or atomic neighbors.

Copolymer Polymers with more than one type of mer.

Coulomb (*Q*) Unit of electric charge generated by one ampere of current lasting for one second. Each electron has a charge of 1.6×10^{-19} coul (or 6.25×10^{18} electrons per coulomb).

Covalent bond Interatomic bond created when two adjacent atoms share a pair of electrons.

Creep A slow deformation by stresses below the normal yield strength (commonly occurring at elevated temperatures).

Creep rate Creep strain per unit time.

Cr_2O_3-type structure Hexagonal array of oxygen ions with cations at two-thirds of the 6-fold sites.

Crystal A structurally uniform solid in three dimensions, with a long-range repetitive pattern of atoms.

Crystal lattice The spatial arrangement of equivalent sites within a crystal.

Crystallinity Volume fraction of a solid which has a crystalline (as contrasted to an amorphous) structure.

Crystallization shrinkage Decrease in volume during crystallization, attributable to the more efficient packing of atoms in a crystal array.

CsCl-type structure Simple cubic lattice, with one ion at the cube corner and the second ion at the cube center, each in 8-fold coordination.

Dashpot Model for viscous flow.

Defect structures Compounds with noninteger ratios of atoms or ions. These compounds contain either vacancies or interstitials within the structure. Defects occur most commonly when multivalence ions such as Fe^{2+} or Fe^{3+} are present. They may also have the effect of balancing a charge in solid solution, for example, K^+Al^{3+} for Si^{4+}.

Deflocculation Dispersion of very fine particles in a suspension.

Deformation crystallization Crystallization occurring as polymers are unkinked into parallel linear orientations.

Degradation Deterioration of polymers through breaking bonds.

Density (ρ) Mass/unit volume; for example, g/cm^3 or lb/ft^3.

Density, apparent Mass/(true volume + closed pores).

Density, bulk Mass/(true volume + closed pores + open pores).

Density, true Mass/true volume.

Depolymerization Reversal of polymerization; this produces smaller molecules.

Devitrification Crystallization from a glass.

Die-casting alloys Alloys with melting points sufficiently low so they can be cast into reusable metal dies; commonly zinc-base or aluminum-base alloys.

Dielectric An insulator. A material which can be placed between two electrodes without conduction.

Dielectric constant (κ) See *Relative dielectric constant*.

Diffusion The movement of atoms or molecules through a material.

Diffusion coefficient (D) Diffusion flux per unit concentration gradient.

Dislocation Linear defect within a crystal. Particularly significant in plastic deformation.

Displacement distance (b) "Error of closure" around a dislocation. Repeating step for dislocation formation and movement.

Dissociation (by dehydration) Calcination which releases water.

Distribution coefficient (k) Ratio of solute composition in the equilibrium solid to solute composition in the equilibrium liquid.

Donor level Energy level slightly below the conduction band which donates electrons to the conduction band.

Donor exhaustion Depletion of donor electrons. Because of it, additional thermal activation does not increase the number of extrinsic carriers.

Dopant Impurity solute element which provides donors or acceptors.

Double-end pressing Pressing of powders between pistons, in which both the upper and lower pistons move and the die remains stationary.

Drawing Mechanical forming by tension through a die, e.g., wire drawing and sheet drawing; usually carried out at temperatures below the recrystallization temperature.

Drift velocity ($\bar{v}$) Net velocity of electrons in an electric field.

Drying Water-removal step in making hydroplastic and slip-cast ceramic products.

Ductile fracture Fracture accompanied by plastic deformation, and therefore by energy absorption.

Ductility Permanent deformation before fracture; measured as elongation or reduction in area.

Dye Colorant which is dissolved in the polymer structure.

Edge dislocation Linear defect at the end of an extra (half) plane of atoms.

Elastic deformation Reversible deformation without permanent atomic (or molecular) displacements.

Elastomer Polymer with a large elastic strain. This strain arises from the unkinking of the polymer chain.

Electric dipole Polarity with a positively charged end and a negatively charged end. The dipole moment μ is the product of the charge Q and the distance d between the centers of opposing charges.

Electric field ($\mathscr{E}$) Voltage gradient, volts/cm.

Electrical conductivity (σ) Coefficient between charge flux and electric field. Reciprocal of electrical resistivity.

Electrical resistivity (ρ) Resistance of a material in which $A = l$. (See Eq. 2–1.1.) Reciprocal of electric conductivity.

Electron charge (q) The charge of 1.6×10^{-19} coul (or 1.6×10^{-19} amp · sec) carried by each electron.

Electron hole (p) Electron vacancy in the valence band which serves as a positive charge carrier.

Electron-hole pairs A conduction electron in the conduction band and an accompanying electron hole in the valence band which result when an electron jumps the gap in an intrinsic semiconductor.

Electronic conduction Conduction by electron transfer (as opposed to conduction by ion transfer).

Elongation Axial strain accompanying fracture. (A gage length must be stated.)

Emitter The zone of a transistor that provides the charge carriers to cross the base to the collector.

End-quench test Standardized test by quenching from one end only for determining hardenability.

Endurance limit The maximum stress allowable for unlimited stress cycling.

Energy gap (E_g) Unoccupied energies between the valence band and the conduction band.

Equiaxial Comparable dimensions in the three principal directions.

Equilibrium diagram See *Phase diagram.*

Eutectic composition Composition of minimum melting temperature.

Eutectic reaction $L_2 \underset{\text{heating}}{\overset{\text{cooling}}{\rightleftharpoons}} S_1 + S_3.$

Eutectic temperature Equilibrium temperature of the eutectic reaction.

Eutectoid composition Composition which produces the eutectoid reaction directly.

Eutectoid ferrite Ferrite formed from austenite by the eutectoid reaction (as contrasted to proeutectoid ferrite).

Eutectoid reaction $S_2 \underset{\text{heating}}{\overset{\text{cooling}}{\rightleftharpoons}} S_1 + S_3.$

Eutectoid temperature Equilibrium temperature of eutectoid reaction.

Expansion coefficient (α) Fractional change in dimension accompanying change in temperature. Both volume expansion coefficients α_v and linear expansion coefficients α_l are used ($\alpha_v \cong 3\alpha_l$).

Extraction Removal of metal by reduction from ores.

Extrinsic semiconductor A material which is semiconducting by virtue of its impurities, which either donate electrons to the conduction band (*n*-type) or accept electrons from the valence band (*p*-type).

Extrusion Mechanical forming by compression through open-ended dies (cf. Drawing).

Face-centered cubic (*fcc*) The lattice of a cubic crystal which has equivalent sites at the corners *and* at the center of the faces of the cube.

Faraday ($\mathscr{F}$) A unit of charge per 6×10^{23} electrons (= 96,500 coul).

Fatigue Fracture arising from time in service.

Ferrite (α) Iron-rich bcc phase.

Fibers Product drawn through a bushing with a diameter less than 0.01 inch.

Fillers Additives used for the purpose of strengthening and/or extending the basic polymer.

Films Two-dimensional products; by definition, less than 0.01-in. thick.

Firing The sintering process used to agglomerate ceramic powders.

Flame-hardening Hardening by means of surface heating by flames (followed by quenching).

Flame retardants Additives which reduce combustion, commonly by providing an oxygen-smothering gas.

Flocculation Clustering of fine particles in a suspension.

Forging Mechanical forming by pressure, usually by hot working.

Fracture Failure by propagation of cracks.

Fusion (heat of) Thermal energy required for the melting of a crystalline solid.

Gage length Initial dimensions, for determining elongation. (See Fig. 2–3.2.)

Gates Entrances through which castings are fed.

Glass An amorphous solid below its transition temperature. A glass lacks long-range crystalline order, but normally has short-range order.

Glass-ceramics Devitrified glass.

Glass-reinforced plastics (GRP) Composite of glass fibers in plastic.

Glass transition temperature (T_g) Transition temperature between a super-cooled liquid and its glassy solid.

Grain boundary The zone of crystalline mismatch between adjacent grains.

Grain boundary area (S_v) Area/unit volume; for example, in^2/in^3 or cm^2/cm^3.

Grain growth Increase in average grain size by atoms diffusing across grain (or phase) boundaries.

Grain size number (GS#) An index of grains per standard area on a 2-dimensional section through a solid. (See Fig. 3–5.5.)

Grains Individual crystals within a material.

"Hard-ball" model The concept of a rigid spherical atom. This is a very useful concept, but requires modification if one wants to explain nuclear reactions.

Hardenability The ability to develop maximum hardness by avoiding the ($\gamma \rightarrow \alpha$ + carbide) reaction.

Hardenability curve Hardness profile of end-quench test bar.

Hardness Resistance to penetration. (See *Brinell hardness number* and *Rockwell hardness number*.)

Heat of hydration Energy released as heat during the hydration reaction, e.g., of portland cement.

Hexagonal close-packed (hcp) The lattice of a hexagonal crystal with equivalent sites at the corners *and* in offset positions at midheight. (See Fig. 3–3.2a.)

High-speed steels Steels with their carbides stabilized against overtempering.

Hole See *Electron hole.*

Hot working Deformation which is performed above the recrystallization temperature, so that annealing occurs concurrently.

Humidity drying Initial heating of a material in a humid atmosphere so that drying occurs with a less steep moisture gradient.

Hydroplastic forming Shaping of a fine-particle product which is plasticized with water.

Impact strength See *Toughness.*

Induction hardening Hardening by high-frequency induced currents which heat the surfaces only.

Ingot A piece of metal which is to be subsequently rolled or forged.

Injection molding Process of molding a material in a closed die. For thermoplasts, the die is appropriately cooled. For thermosets, the die is maintained at the curing temperature for the plastic.

Insulator Nonconductor of (a) electrical or (b) thermal energy; in either case, the insulator has significant electronic resistivity.

Internal structure Arrangements of atoms, molecules, crystals, and grains within a material.

Intrinsic semiconductor A material which is a semiconductor without the addition of impurities. For conduction, electrons must jump the energy gap between the valence and conduction bands.

Ion exchange Exchange of ions in a solid solution.

Ion stuffing Ion exchange which produces compressive forces because the new ions are larger than the original ion sites.

Ion vacancy (□) Unoccupied ion site within a crystal structure. The charge of the missing ion must be appropriately compensated.

Ionic compounds Compounds of two or more varieties of ions bonded by the attraction of unlike charges.

Isomer Molecules of the same composition but different structures.

Isostatic molding Pressure fabrication in a rubber mold by hydrostatic pressure.

Isotactic polymer Identically repetitive mer configurations (as contrasted to atactic polymer).

Kaolinite The most common clay, $Al_2Si_2O_5(OH)_4$.

Lattice points Equivalent sites within a crystal. In metals, we commonly assign atoms to these points.

Piezoelectrics Dielectric materials with structures that are asymmetric, so that their centers of positive and negative charges are not coincident. As a result the polarity is sensitive to pressures which change the dipole distance, and the polarization.

Pigment Colorant from insoluble particles.

Plastic Capable of being permanently deformed.

Plastics Materials consisting predominantly of nonmetallic elements or compounds. (See *Polymers*.) Moldable organic resins.

Plastic deformation Permanent deformation arising from the displacement of atoms (or molecules) to new lattice sites.

Plasticizer Micromolecules added among macromolecules to induce deformation and flexibility.

Plywood Modified wood sheet with alternate layers rotated 90° in high-strength directions.

Poisson's ratio (*v*) Ratio (negative) of lateral to axial strain.

Polar group A local electric dipole within the polymer molecule.

Polarization ($\mathscr{P}$) The dipole moment μ ($= Qd$) per unit volume.

Polyblend A mixture of two or more polymers blended into a plastic.

Polycrystalline Materials with more than one crystal; therefore with grain boundaries.

Polyester Condensation, or step-reaction, polymer which contains

linkages.

Polyfunctional Molecule with three or more sites at which there can be joining reactions with adjacent molecules.

Polymers Nonmetallic materials consisting of (large) macromolecules composed of many repeating units; the technical term for plastics.

Porcelain Glass-bonded ceramic.

Porosity, open Open-pore volume/total volume.

Porosity, total Open- + closed-pore volume/total volume.

Portland cement A hydraulic calcium silicate cement.

Precipitation hardening Hardening by the formation of clusters prior to precipitation.

Pressure fabrication Compacting of particles into a product by pressure.

Proeutectoid ferrite Ferrite which separates from austenite above the eutectoid temperature. The metal must contain less carbon than the eutectoid composition.

Propagation Polymer growth step through the reaction: reactive site + monomer → new reactive site. (See Eqs. 8–2.2 and 8–2.3.)

Properties Quantitative attributes of materials, e.g., density, strength, conductivity.

Quench Cooling accelerated by immersion in agitated water or oil.

Radius ratio (r/R) Ratio of ionic radii of coordinated cations, r, and anions, R.

Rayon A cellulose-base polymer in which about two-thirds of the —OH radicals have been replaced by

$$-\text{OC}-\text{CH}_3,$$
$$\underset{\text{O}}{\overset{\|}{}}$$

that is, acetate radicals. (See Example 8–1.1.)

Reactive site (●) Open end of a free radical, in which there are not enough bonds to meet the requirement of Table 7–1.1. Also see Eq. (8–2.2).

Recombination Annihilation of electron-hole pairs.

Recrystallization The formation of new annealed grains from former strained grains.

Recrystallization temperature Temperature at which recrystallization is spontaneous, usually about one-third to one-half the absolute melting temperature.

Rectifier Electric "valve" which permits forward current and prevents reverse current.

Reduction Removal of oxygen from an oxide. The lowering of the valence level of an element.

Reduction of area (*Red. of A*) Measure of plastic deformation at the point of fracture. (See Eq. 2–3.2.)

Refining Removal of impurities from a liquid.

Refraction (*n*) Bending of a light ray at a phase boundary. The index of refraction, n, is inversely proportional to the velocity of light through the material.

Refractories Materials capable of withstanding extremely high temperatures.

Relative dielectric constant (κ) Ratio of charge density arising from an electric field (1) with, and (2) without, a dielectric material present.

Relaxation time (λ) Time required to decay an exponentially dependent value to 37%, that is, $1/e$, of the original value.

Residual stress Stresses induced as a result of differences in temperature or volume.

Resistivity (ρ) Reciprocal of conductivity (usually expressed in ohm · cm).

Resistivity, solid solution (ρ_x) Resistivity arising from additions of solute.

Resistivity, strain-hardening (ρ_s) Resistivity arising from cold work and the presence of dislocations.

Resistivity, thermal (ρ_T) Electrical resistivity arising from thermal agitation.

Riser Liquid metal reservoir for a casting.

Rockwell hardness number (R) Hardness indexes derived by standardized procedures. The hardness is calibrated against the depth of the indentation made by an indenter. There are a number of Rockwell scales (see Table 2–4.1).

Rolling Deformation of metals between rolls. Used for both hot and cold deformation.

Rubber A polymer with high elastic strain.

Rubbery plateau For a plastic, the range of temperature between the glass temperature and melting temperature, which has a viscoelastic modulus that is relatively constant.

Scission The breaking of intramolecular bonds by radiation.

Semiconducting compounds Compounds of two or more elements with an average of four shared electrons per atom.

Semiconductors Materials with conductivities intermediate between those of conductors and insulators, that is, $\sigma \approx 10^{-2 \pm 4}$ ohm$^{-1} \cdot$ cm^{-1}.

Setting (*cement*) Hardening by reaction (*not* by drying).

Shear modulus (G) Shear stress τ per unit shear strain γ.

Shear strain (γ) Tangent of shear angle α developed from shear stress.

Shear stress (τ) Shear force per unit area.

Sheet Two-dimensional products. (When applied to plastics, >0.01 in. thick.)

Sheet silicate Polymerized silicate with three oxygens of each tetrahedron shared to provide a two-dimensional, sheetlike structure, e.g., mica.

Shot peening Mechanical treatment of a surface by bombarding it with hardened steel shot. The surface undergoes plastic deformation and receives a residual compressive stress.

Silicones "Silicates" with organic side radicals; thus silicon-based polymeric molecules.

Single-phase materials Materials containing only one basic structure.

Sintering (*solid*) Agglomeration of particles by diffusion through the solid.

Sintering (*vitreous*) Agglomeration of particles by viscous flow of a glass phase.

Slip casting Forming process in which a suspension of particles is dewatered by a porous mold.

Slip, ceramic Thick suspension of particles.

Slip, deformation Relative displacement along a structural direction.

Soda-lime glass The most widely produced glass; it serves as a basis for window and container glass. Contains $Na_2O/CaO/SiO_2$ in approximately a 1/1/6 ratio.

Solder Metals which melt below 800°F which are used for joining. Commonly, Pb-Sn alloys, but may also be other materials, even glass.

Solid solution, interstitial Crystals which contain a second component in their interstices. The basic structure is unaltered.

Solid solution, substitutional Crystals with a second component substituted for solvent atoms in the basic structure.

Solidification shrinkage Change in volume during freezing, encountered in metal castings.

Solubility limit Maximum solute addition without supersaturation.

Solution hardening Increased strength arising from the creation of solid solutions (or from pinning of dislocations by solute atoms).

Solution treatment Heating to induce solid solutions.

Spheroidite Microstructure of coarse, equiaxial, approximately spherical carbide particles within a ferrite matrix.

Spheroidization Process of making spheroidite, generally by extensive over-tempering.

Spinning Mechanical deformation of sheet metal over a mandrel.

Spring wood The spring growth of wood in a tree, which is less dense and contains larger cells than the summer growth. Alternation of these two growths create the grain in wood.

Stabilizer Additive put into plastics for the purpose of retarding chemical reactions while the article is in service.

Stampings Cold-shaping processes between mated molds.

Standard deviation (s) Measure of variation of data.

Steel Iron-base alloys, commonly containing carbon. In practice the carbon can all be dissolved by heat treatment; hence < 2.0 w/o C.

Steels, low-alloy Steels containing up to 5% alloying elements other than carbon. Phase equilibria are related to the Fe-C diagram.

Steels, plain carbon Basically Fe-C alloys with minimal alloy content.

Steels, stainless High-alloy steels (usually containing Cr or Cr + Ni) designed for resistance to corrosion and/or oxidation.

Steels, tool Steels with high tempering temperatures, usually containing carbide stabilizers such as Cr, Mn, Mo, V, W.

Step-reaction polymerization The polymerization reaction produces a by-product.

Stereoisomers Configuration isomers of polymers.

Sterling silver Alloy of 92.5 w/o Ag and 7.5 w/o Cu. (This corresponds to nearly the maximum solubility of copper in silver.)

Strain (ε) Deformation from an applied stress.

Strain hardening Increased hardness (and strength) arising from plastic deformation.

Strain point, glass Temperature at which the viscosity, η, of a glass is $10^{14.5}$ poises.

Stress (σ) Force per unit area.

Stress relaxation Decay of stress at constant strain by molecular re-arrangement.

Stress relief Removal of residual stresses by heating.

Super-alloys Heat-resistant, high-temperature alloys, usually based on cobalt or nickel, or the refractory metals such as Cb, Mo, Ta, W, and Zr.

Supersaturation Excess solute beyond the solubility limit; most commonly achieved by supercooling.

Surface resistivity (ρ_s) Resistivity across a surface, as contrasted to resistivity through the bulk.

Swelling Absorption of a solute with an increase in volume.

Symmetry Structural correspondence of size, shape, and relative position.

Syndiotactic Multiple repetition configuration of chain polymers. (See Fig. 7–2.1.)

Tempered glass Glass with surface compressive stresses induced by heat treatment.

Tempered martensite A microstructure of ferrite and carbide obtained by heating martensite.

Tempering A toughening process in which martensite is heated to initiate a ferrite-plus-carbide microstructure.

Tensile strength (*TS*) Maximum load per unit of *original* area. This is the ultimate strength used for design purposes.

Tentering Biaxial straining process of plastic sheet or film by conical rolls.

Termination Finalizing step of addition polymerization. A common reaction involves the joining of the reactive sites at the growing ends of two propagating molecules.

Tetragonal crystal Noncubic crystal with $a = b \neq c$, and all axial angles equal to 90°.

Thermal agitation Thermally induced movements of atoms and molecules.

Thermal conductivity (k) Coefficient between thermal flux and thermal gradient.

Thermal expansion Expansion caused by increased atomic vibrations due to increased thermal energy.

Thermistors Semiconductor devices with a high resistance dependence on temperature. They may be calibrated as thermometers.

Thermoplasts Plastics which soften and are moldable due to the effect of heat. They are hardened by cooling, but soften again during subsequent heating cycles.

Thermosets Plastics which polymerize further on heating; therefore heat causes them to take on an additional set. They do not soften with subsequent heating.

Toughness A measure of the energy required for mechanical failure.

Transducer A material or device which converts energy from one form to another, specifically electrical energy to or from mechanical energy.

Transistor Semiconductor device for amplification of current.

Transition temperature Temperature at which ductile fracture changes to brittle fracture.

Translucency Transmission of light with internal scattering.

Transmittance (*T*) Ratio of incoming light to transmitted light.

Transparency Transmission of light without internal scattering.

Triaxial composition Three-component material; specifically a porcelain made of clay, feldspar, and silica.

Triaxial diagram Graphical presentation of a three-component material. Each apex of a triangle represents one of the components.

True stress (σ_{tr}) Stress based on actual rather than original area.

Types (*ASTM*) *of cement* Standardized categorization of portland cement according to setting characteristics. Type I is general purpose; Type III is high, early strength; Type IV is low heat of hydration.

Unit cell A small (commonly the smallest) repetitive volume that comprises the complete lattice pattern of a crystal.

Unit mix (*concrete*) The sand : gravel : water ratio based on 1 sack of portland cement (94 lb = 1 ft³).

Vacuum forming Process of shaping sheet into an unmated mold by suction.

Valence band Filled energy band below the energy gap. Conduction in this band requires holes.

Valence electron Electron from the outer shell of an electron.

Vinyl compounds Compounds of the general form

which may be polymerized into chain macromolecules. (See Table 7–1.3 for various possibilities for R.)

Viscoelastic modulus (M_{ve}) Ratio of shear stress to the sum of elastic deformation, γ_e, and viscous flow, γ_f.

Viscosity (η) The ratio of shear stress to velocity gradient.

Viscous flow (γ_f) Nonreversible flow due to a shear stress.

Vitreous Glassy or glasslike.

Volume resistivity (ρ_v) See *Resistivity*.

Vulcanization Treatment of rubber with sulfur to cross-link the molecular chains.

Warpage, wood Irregular shrinkage arising from anisotropic drying coefficients.

Welding A joining of metals by fusion.

White metals Nonferrous metals other than copper-base alloys. Usually low-melting alloys containing Bi, Cd, Pb, Sn, and/or Zn.

Wiedemann-Franz ratio (k/σ) Ratio of thermal to electrical conductivity.

Working range (*glass*) Temperature range lying between viscosities of $10^{3.5}$ to 10^7.

Yield strength (*YS*) Stress of first significant plastic deformation.

Young's modulus (*Y*) Axial stress per unit axial strain.

ZnS-type structure Fcc arrangement of one element of the AX compound, with the other at half the 4-fold sites.

Zone refining Purification by solidification segregation.

INDEX

Cubic, face-centered, 31, *324*
Cubic closed-packed, 35
Cubic metals, 30
Cutting tools, 120

Darken, L. S., 49
Dashpot, 190, *322*
Deep drawing, 118
Defect compounds, conductivity of, 271
Defect structure, 218, *322*
Deflocculation, 252, *322*
Deformation, elastic, 44, *323*
Degradation, 173, 195, *322*
Degree of polymerization, 146
Density, 4, 33, *322*
    apparent, 241, *322*
    bulk, 241, *322*
    true, 241, *322*
Density of elements, 316, 317
Dental materials, 304
Dentin, 304
Depolymerization, 195, *322*
Deterioration by oxidation, 173
Deterioration of polymers, 195
Devices, semiconducting, 282
Devitrification, 264, *322*
Diagram, equilibrium, 84, *324*
Diamond, 227, *324*
Die, 117
Die-casting alloys, 76, 104, *322*
Dielectric, 23, *322*
Dielectric constant, 199, 234, *322*
    relative, 22, *330*
Dielectric strength, 234
Dielectrics, ceramic, 234
    properties of, 234
Diffusion, 63, *322*
Diffusion coefficient, 64, *322*
Dipole, electric, 200, 235, *323*
Dipole moment, 236
Dislocation, edge, 48, 49, *322, 323*
Displacement distance, 49, *323*
Dissociation, 248, *323*
Distances, interatomic, 29
Distribution coefficient, 285, *323*
Donor, 280
Donor exhaustion, 281, *323*
Donor level, 280, *323*
Dopants, 285, *323*
Double-end pressing, 250, 251, *323*
Drawing, 117, *323*
    deep, 118

wire, 117
Drift velocity, 21, 271, *323*
Drying, 253, *323*
    humidity, 255, *326*
Drying shrinkage, 254
Ductile fracture, 55, *323*
Ductility, 13, *323*
Dye, 174, 204, *323*

e, 311
E-glass, 225, 234
East, W. H., 247
Edge dislocation, 48, 49, *323*
Elastic deformation of metals, 44, *323*
Elastic modulus, 44
    vs. expansion, 61
    vs. melting temperature, 44, 61
    vs. temperature, 45
Elastic strain, 11
Elastic stress, 11
Elasticity, modulus of, 11, 313–315, 327
Elastomer, 190, *323*
Electric dipole, 235, *323*
Electric field, 22, *323*
Electric motor, 1
Electrical behavior, 58
Electrical ceramics, 232
Electrical conductivity, 10, 21, *323*
Electrical porcelain, 234, 257, 258, 259, *329*
Electrical properties, 9, 21
    of polymers, 199, 315
Electrical resistivity, 21, 313–315, *323*
Electroluminescence, 278
Electrolytic extraction, 214
Electron charge, 3, 311, *323*
Electron hole, 272, *323*
Electron-hole pair, 227, *324*
Electron mobility, 272
Electronic conduction, 218, *324*
Electrons, 3, 271
Elements, atomic number, 316
    atomic radius, 316
    atomic weight, 316
    crystal structure, 316
    density, 316
    ionic radius, 316
    melting point, 316
    nonmetallic, 5, *328*
    properties of, 316
Elements of Ceramics, 252, 254, 262
Elements of Physical Metallurgy, 45, 130, 137

Elongation, 13, *324*
Emery, 227
Emitter, 284, *324*
Emitter junction, 284
Enamel, vitreous, 296
End-quench test, 130, *324*
Endurance limit, 56, *324*
Energy, surface, 258
Energy gap, 273, *324*
Energy of formation, 111, 112
Engineering, materials of, 1
Equiaxed, 33, *324*
Equilibrium diagram, 84, *324*
Erickson, M. A., 56, 57
Ethane, 147
Ethylene, 147
Eutectic composition, 79, *324*
Eutectic microstructure, 79
Eutectic reaction, 79, *324*
Eutectic temperature, 78, *324*
Eutectoid composition, 86, 88, *324*
Eutectoid ferrite, 89, *324*
Eutectoid reaction, 87, *324*
Eutectoid shift, 94, 95
Eutectoid steels, microstructure of, 91
Eutectoid temperature, 95, *324*
Exhaustion, donor, 281, *323*
Expansion, thermal, 18, 61, 313–315, *333*
Expansion vs. temperature, 62
Expansion coefficient, linear, 18, 19, 313–315, *324*
    volume, 18
    vs. melting temperature, 61
    vs. modulus, 61
Extraction, 111, *324*
    electrolytic, 114
Extrinsic semiconductors, 280, *324*
Extrusion, 117, 177, *324*

Fabrication, pressure, 250, *329*
Fabrication processes, 135
Face-centered cubic, 31, *324*
Face diagonal, 50
Faraday, 116, *324*
Farrel Company, 297
Fatigue, 56, *324*
    static, 56
FCC, 31
FCC metals, 30
Fe-C phase diagram, 85, 86
Fe-Fe$_3$C phase diagram, 85
Fe-O phase diagram, 219

Natural rubber, 149
Nature of Metals, 39
Naval brass, 98
Network, silicate, 223
  tetrahedral, 223
Network modifiers, 224
Network polymers, 153, 176, *328*
Network silicates, 223, *328*
Neutron radiation, 197
Ni-Cu phase diagram, 107
Nickel silver, 99
$NiFe_2O_4$-type structure, 217, *328*
Nitriding, 298, *328*
Nodular cast iron, 96, *320*
Nonbridging oxygens, 225
Noncrystalline materials, viscosity of, 261
Noncubic metals, 34
Nonmetallic elements, 5
Nonmetals, 4, 5, *328*
Nonstoichiometric structures, 218
Normalizing, 122, 123, *328*
NORTON, F. H., 252, 254, 262
Nuclei, 38

O-Fe phase diagram, 219
Offset, 12
Ohms/square, 20
Open porosity, 241, *329*
Optical characteristics of plastics, 203
Order, long-range, 60
Orientation, 179, *328*
  molecular, 179
Orthorhombic, 157
OSBORN, E. F., 230, 233, 257
OSBORN, H. B. JR., 297
Overaging, 126, *328*
Overtempering, *328*
OWENS-ILLINOIS GLASS CO., 264
Oxidation, *328*
  deterioration by, 173
Oxygens, bridging, 220
  nonbridging, 225

p-type semiconductor, 280
Packing factor, atomic, 30, *319*
Paint, 296
Pair, electron-hole, 277, *324*
Parison, 178
Pb-Sn phase diagram, 79
PCE, 228
Pearlite, 87, *328*

formation of, 87
Peening, shot, 296, *331*
Periodic table, 4, *328*
Permittivity, 24
Phase, 43, 75, *328*
Phase amounts, 82, *328*
  vs. properties, 90
Phase boundary, 75
Phase composition, 78, *328*
Phase diagram, 79, *328*
  Ag-Cu, 78
  Al-Cu, 122
  Al-Mg, 107
  Al_2O_3-SiO_2, 230
  Al-Si, 286
  Cu-Ni, 107
  Cu-Zn, 81
  Fe-C, 85, 86
  Fe-Fe_3C, 85
  Fe-O, 219
  FeO-MgO, 218
  FeO-SiO_2, 231
  K_2O-Al_2O_3-SiO_2, 257, 260
  MgO-Al_2O_3-SiO_2, 233, 238
  Pb-Sn, 79
Phase Equilibria Among Oxides in Steel-making, 230, 233, 257
Phase shape vs. properties, 91
Phase size vs. properties, 91
Phenol, 147
Phenol-formaldehyde, 154, 171, *328*
Photoconduction, 276, *328*
Photoelasticity, 205, *328*
Photoluminescence, 278
Photon, 277, 278
Physical Metallurgy for Engineers, 97, 98, 99, 101, 102, 113, 117
Piezoelectrics, 234, *329*
Pigment, 174, 204, *329*
Plain-carbon steel, 94, *332*
Planar transistor, 288
Planck's constant, 198
Planes, 47, 48
Plastic deformation of metals, 47
Plastic production, 166
Plastic strain, 12
Plastic stress, 12
Plasticizers, 172, *329*
Plastics, 5, 145, *329*
  characteristics of, 184
  dielectric constants of, 205
  glass-reinforced, 294, *325*

making and shaping of, 165
mechanical behavior of, 187
optical characteristics of, 204
reinforced, 172
shaping of, 175
Plate glass, 225
Plywood, 165, 307, 309
P/M preforms, 139
p-n junction, 283, *328*
Poisson's ratio, 45, *329*
Polar group, 199, *329*
Polar site, 199
Polarization, 236, *329*
  saturation, 237
Polyblend, 194, *329*
Polycrystalline, 38, *329*
Polyester, 169, *329*
Polyethylene, 156
  high-density, 184
  low-density, 184
  thermal properties of, 185
Polyethlyene unit cell, 156
Polyfunctional, 153, *329*
Polyisoprene, 155
Polymerization, addition, 167, *318*
  chain-reaction, 167, *320*
  condensation, 168, *321*
  degree of, 146
  step-reaction, 168, *332*
Polymers, 5, 145, *329*
  amorphous, 157
  chain, 150, *320*
  cross-linked, 189
  crystallinity of, 155
  deterioration of, 195
  electrical properties of, 199, 315
  linear, 153, 175, *327*
  mechanical failure of, 192
  network, 153, 176, *328*
  properties of, 315
  softening of, 185
Polymethyl methacrylate, 185
Polypropylene, 151
Polystyrene, 148
Polyvinyl chloride, 146, 147, 161
Porcelain, electrical, 234, 257, 258, 259, *329*
  microstructure of, 258
  triaxial, 257
Pore water, 253
Porosity, open, 241, *329*
  true, 241
Portland cement, 300, *329*